T0295741

Introduction to the Theory of Optimization in Euclidean Space

Series in Operations Research

Series Editors:
Malgorzata Sterna, Marco Laumanns

About the Series

The CRC Press Series in Operations Research encompasses books that contribute to the methodology of Operations Research and applying advanced analytical methods to help make better decisions.

The scope of the series is wide, including innovative applications of Operations Research which describe novel ways to solve real-world problems, with examples drawn from industrial, computing, engineering, and business applications. The series explores the latest developments in Theory and Methodology, and presents original research results contributing to the methodology of Operations Research, and to its theoretical foundations. Featuring a broad range of reference works, textbooks and handbooks, the books in this Series will appeal not only to researchers, practitioners and students in the mathematical community, but also to engineers, physicists, and computer scientists. The inclusion of real examples and applications is highly encouraged in all of our books.

Rational Queueing
Refael Hassin

Introduction to the Theory of Optimization in Euclidean Space
Samia Challal

For more information about this series please visit: https://www.crcpress.com/Chapman--HallCRC-Series-in-Operations-Research/book-series/CRCOPSRES

Introduction to the Theory of Optimization in Euclidean Space

Samia Challal

Glendon College-York University

Toronto, Canada

CRC Press
Taylor & Francis Group
Boca Raton London New York

CRC Press is an imprint of the
Taylor & Francis Group, an **informa** business

A CHAPMAN & HALL BOOK

CRC Press
Taylor & Francis Group
6000 Broken Sound Parkway NW, Suite 300
Boca Raton, FL 33487-2742

Visit the Taylor & Francis Web site at
http://www.taylorandfrancis.com

and the CRC Press Web site at
http://www.crcpress.com

To my parents

Contents

viii   *Contents*

Preface

The book is intended to provide students with a useful background in optimization in Euclidean space. Its primary goal is to demystify the theoretical aspect of the subject.

In presenting the material, we refer first to the intuitive idea in one dimension, then make the jump to n dimension as naturally as possible. This approach allows the reader to focus on understanding the idea, skip the proofs for later and learn to apply the theorems through examples and solving problems. A detailed solution follows each problem constituting an image and a deepening of the theory. These solved problems provide a repetition of the basic principles, an update on some difficult concepts and a further development of some ideas.

Students are taken progressively through the development of the proofs where they have the occasion to practice tools of differentiation (Chain rule, Taylor formula) for functions of several variables in abstract situation. They learn to apply important results established in advanced Algebra and Analysis courses, like, Farkas-Minkowski Lemma, the implicit function theorem and the extreme value theorem.

The book starts, in Chapter 1, with a short introduction to mathematical modeling leading to formulation of optimization problems. Each formulation involves a function and a set of points. Thus, basic properties of open, closed, convex subsets of $\mathbb{R}^n$ are discussed. Then, usual topics of differential calculus for functions of several variables are reminded.

In the following chapters, the study is devoted to the optimisation of a function of several variables f over a subset S of $\mathbb{R}^n$. Depending on the particularity of this set, three situations are identified. In Chapter 2, the set S has a nonempty interior; in Chapter 3, S is described by an equation $g(x) = 0$ and in Chapter 4 by an inequality $g(x) \leqslant 0$ where g is a function of several variables. In each case, we try to answer the following questions:

- If the extreme point exists, then where is it located in S? Here, we look for necessary conditions to have candidate points for optimality. We make the distinction between local and global points.

- Among the local candidate points, which of them are local maximum or local minimum points? Here, we establish sufficient conditions to identify a local candidate point as an extreme point.

- Now, among the local extreme points found, which ones are global extreme points? Here, the convexity/concavity property intervenes for a positive answer.

Finally, we explore how the extreme value of the objective function f is affected when some parameters involved in the definition of the functions f or g change slightly.

Acknowledgments

I am very grateful to my colleagues David Spring, Mario Roy and Alexander Nenashev for introducing the course on optimization, for the first time, to our math program and giving me the opportunity to teach it. I, especially, thank Professor Vincent Hildebrand, Chair of the Economics Department for the useful discussions during the planning of the course content to support students majoring in Economics.

My thanks are also due to Sarfraz Khan and Callum Fraser from Taylor and Francis Group, to the reviewers for their invaluable help, and to Shashi Kumar for the expert technical support.

I have relied on the various authors cited in the bibliography, and I am grateful to all of them. Many exercises are drawn or adapted from the cited references for their aptitude to reinforce the understanding of the material.

Symbol Description

$\forall$ For all, or for each

$\exists$ There exists

$\exists!$ There exists a unique

$\emptyset$ The empty set

s.t Subject to

$\overset{\circ}{S}$ Interior of the set S

∂S Closure of the set S

$\overline{S}$ Boundary of the set S

C^S The complement of S.

i, j, k $i = (1,0,0)$, $j = (0,1,0)$, $k = (0,0,1)$ standard basis of $\mathbb{R}^3$

$B_r(x_0)$ Ball centered at x_0 with radius r

$\mathbb{B}_r(x_0)$ Bordered Hessian of order r at x_0.

$\langle . , . \rangle$ or $[\ .\ ,\ .\ ]$ brackets for vectors

∇f gradient of f

x^* $= \begin{bmatrix} x_1^* \\ \vdots \\ x_n^* \end{bmatrix}$ column vector identified sometimes to the point $(x_1^*, \ldots, x_n^*)$

$\|x\|$ $= \sqrt{x_1^2 + x_2^2 + \ldots + x_n^2}$ norm of the vector x

M_{mn} set of matrices of m rows and n columns

A $= (a_{ij})\ {\substack{i-1,\ldots,m, \\ j=1,\ldots,n}}$ is an $m \times n$ matrix

$\|A\|$ $= \left(\sum_{i,j=1}^{n} a_{ij}^2 \right)^{1/2}$ norm of the matrix $A = (a_{ij})_{i,j=1,\ldots,n}$

$rankA$ rank of the matrix A

$detA$ determinant of the matrix A

$KerA$ $= \{x : Ax = 0\}$ Kernel of the matrix A

${}^t h$ $= \begin{bmatrix} h_1 & \ldots & h_n \end{bmatrix}$ transpose of $h = \begin{bmatrix} h_1 \\ \vdots \\ h_n \end{bmatrix}$

${}^t h.x^*$ $= \sum_{k=1}^{n} h_k.x_k$ dot product of the vectors h and x^*

$C^1(D)$ set of continuously differentiable functions on D

$C^k(D)$ set of continuously differentiable functions on D up to the order k

$C^\infty(D)$ set of continuously differentiable functions on D for any order k

$H_f(x)$ $= (f_{x_i x_j})_{n \times n}$ Hessian of f

$D_k(x)$ $= \begin{vmatrix} f_{x_1 x_1} & f_{x_1 x_2} & \cdots & f_{x_1 x_k} \\ f_{x_2 x_1} & f_{x_2 x_2} & \cdots & f_{x_2 x_k} \\ \vdots & \vdots & \vdots & \vdots \\ f_{x_k x_1} & f_{x_k x_2} & \cdots & f_{x_k x_k} \end{vmatrix}$ leading minor of order k of the Hessian H_f

Author

Samia Challal is an assistant professor of Mathematics at Glendon College, the bilingual campus of York University. Her research interests include homogenization, optimization, free boundary problems, partial differential equations and problems arising from mechanics.

Chapter 1

Introduction

Optimization problems arise in different domains. In Section 1.1 of this chapter, we introduce some applications and learn how to model a situation as an optimization problem.

The points where an optimal quantity is attained are looked for in subsets that can be one dimensional, multi-dimensional, open, closed, bounded or unbounded, ... etc. We devote Section 1.2 to study some topological properties of such subsets of $\mathbb{R}^n$.

Finally, since, the phenomena analyzed are often complex, because of the many parameters that are involved, this requires an introduction to functions of several variables that we study in Section 1.3.

1.1   Formulation of Some Optimization Problems

The purpose of this short section is to show, through some examples, the main elements involved in an optimization problem.

<u>Example 1</u>. *Different ways in modeling a problem.*

To minimize the material in manufacturing a closed can with volume capacity of V *units*, we need to choose a suitable radius for the container.

i)   Show how to make this choice without finding the exact radius.

ii) How to choose the radius if the volume V may vary from one liter to two liters?

Solution: Denote by h and r the height and the radius of the can respectively. Then, the area and the volume of the can are given by

$$\text{area} = A = 2\pi r^2 + 2\pi rh, \qquad\qquad \text{volume} = V = \pi r^2 h.$$

i) * The area can be expressed as a function of r and the problem is reduced to find $r \in (0, +\infty)$ for which A is minimum:

$$\begin{cases} \text{minimize } A = A(r) = 2\pi r^2 + \dfrac{2V}{r} & \text{over the set } S \\[2mm] S = (0, +\infty) = \{r \in \mathbb{R} \ / \ r > 0\}. \end{cases}$$

Note that the set S, as shown in Figure 1.1, is an open unbounded interval of $\mathbb{R}$.

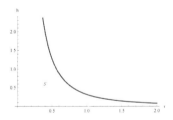

FIGURE 1.1: $S = (0, +\infty) \subset \mathbb{R}$

** We can also express the problem as follows:

$$\begin{cases} \text{minimize } A(r, h) = 2\pi r^2 + 2\pi rh & \text{over the set } S \\[2mm] S = \{(r, h) \in \mathbb{R}^+ \times \mathbb{R}^+ \ / \ \pi r^2 h = V\}. \end{cases}$$

Here, the set S is a curve in $\mathbb{R}^2$ and is illustrated by Figure 1.2 below:

FIGURE 1.2: S is a curve $h = \pi^{-1}/r^2$ in the plane (V=1 liter)

ii) In the case, we allow more possibilities for the volume, for example $1 \leqslant V \leqslant 2$, then we can formulate the problem as a two dimensional problem

$$\begin{cases} \text{minimize } A(r,h) = 2\pi r^2 + 2\pi r h \qquad \text{over the set } S \\[2mm] S = \{(r,h) \in \mathbb{R}^+ \times \mathbb{R}^+ \quad / \quad \dfrac{1}{\pi r^2} \leqslant h \leqslant \dfrac{2}{\pi r^2}\}. \end{cases}$$

The set S is the plane region, in the first quadrant, between the curves $h = \dfrac{1}{\pi r^2}$ and $h = \dfrac{2}{\pi r^2}$ (see Figure 1.3).

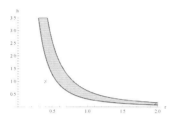

FIGURE 1.3: S is a plane region between two curves

A three dimensional formulation of the same problem is

$$\begin{cases} \text{minimize } A(r,h,V) = 2\pi r^2 + \dfrac{2V}{r} \qquad \text{over the set } S \\[2mm] S = \{(r,h,V) \in \mathbb{R}^+ \times \mathbb{R}^+ \times \mathbb{R}^+ \quad / \quad \pi r^2 h = V, \quad 1 \leqslant V \leqslant 2\} \end{cases}$$

where, the set $S \subset \mathbb{R}^3$ is the part of the surface $V = \pi r^2 h$ located between the planes $V = 1$ and $V = 2$ in the first octant; see Figure 1.4.

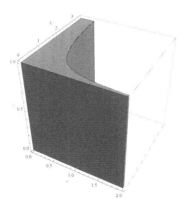

FIGURE 1.4: S is a surface in the space

Example 2. *Too many variables and linear inequalities.*

Diet Problem. * One can buy four types of aliments where the nutritional content per unit weight of each food and its price are shown in Table 1.1 [5]. The diet problem consists of obtaining, at the minimum cost, at least twelve calories and seven vitamins.

	type1	type2	type3	type4
calories	2	1	0	1
vitamins	3	4	3	5
price	2	2	1	8

TABLE 1.1: A diet problem with four variables

Solution: Let u_i be the weight of the food of type i. The total price of the four aliments consumed is given by the relation

$$2u_1 + 2u_2 + u_3 + 8u_4 = f(u_1, u_2, u_3, u_4).$$

To ensure that at least twelve calories and seven vitamins are included, we can express these conditions by writing

$$2u_1 + u_2 + u_4 \geqslant 12 \qquad \text{and} \qquad 3u_1 + 4u_2 + 3u_3 + 5u_4 \geqslant 7.$$

Hence, the problem would be

$$\begin{cases} \text{minimize } f(u_1, u_2, u_3, u_4) \quad \text{over the set } S = \Big\{(u_1, u_2, u_3, u_4) \in \mathbb{R}^4 : \\ 2u_1 + u_2 + u_4 \geqslant 12, \quad 3u_1 + 4u_2 + 3u_3 + 5u_4 \geqslant 7\Big\}. \end{cases}$$

** The above problem is rendered more complex if more factors (fat, proteins) and types of food (steak, potatoes, fish, ...) were to be considered. For example, from Table 1.2, we deduce that the total price of the seven

	type1	type2	type3	type4	type5	type6	type7
protein	3	1	2	7	8	5	10
fat	0	1	0	8	15	10	6
calories	2	1	0	1	5	7	9
vitamins	3	4	3	5	1	2	5
price	2	2	1	8	12	10	8

TABLE 1.2: A diet problem with seven variables

aliments consumed is

$$2u_1 + 2u_2 + u_3 + 8u_4 + 12u_5 + 10u_6 + 8u_7 = p(u_1, u_2, u_3, u_4, u_5, u_6, u_7).$$

To ensure that at least twelve calories, seven vitamins, twenty proteins are included, and less than fifteen fats are consumed, the problem would be formulated as

$$\left\{ \begin{array}{l} \text{minimize } p(u_1, u_2, u_3, u_4, u_5, u_6, u_7) \quad \text{over the set} \\[2mm] S = \Big\{ (u_1, u_2, u_3, u_4, u_5, u_6, u_7) \in \mathbb{R}^7 : \\[2mm] \quad 3u_1 + u_2 + 2u_3 + 7u_4 + 8u_5 + 5u_6 + 10u_7 \geqslant 20 \\[2mm] \quad u_2 + 8u_4 + 15u_5 + 10u_6 + 6u_7 \leqslant 15 \\[2mm] \quad 2u_1 + u_2 + u_4 + 5u_5 + 7u_6 + 9u_7 \geqslant 12 \\[2mm] \quad 3u_1 + 4u_2 + 3u_3 + 5u_4 + u_5 + 2u_6 + 5u_7 \geqslant 7. \Big\} \end{array} \right.$$

Example 3. *Too many variables and nonlinearities.*

* A company uses x units of capital and y units of labor to produce $x\,y$ units of a manufactured good. Capital can be purchased at 3\$/ unit and labor can be purchased at 2\$/ unit. A total of 6\$ is available to purchase capital and labor. How can the firm maximize the quantity of the good that can be manufactured?

Solution: We need to maximize the quantity $x\,y$ on the set of points (see Figure 1.5)

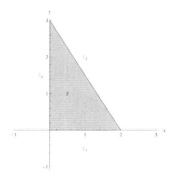

FIGURE 1.5: S is a triangular region in the plane

$$S = \{(x, y) \in \mathbb{R}^2 : \quad 3x + 2y \leqslant 6, \quad x \geqslant 0, \quad y \geqslant 0\}.$$

The set S is the triangular plane region bounded by the sides L_1, L_2 and L_3, defined by: $L_1 = \{(x,0),\ 0 \leqslant x \leqslant 2\}$,

$$L_2 = \{(0,y),\ 0 \leqslant y \leqslant 3\}, \qquad L_3 = \{(x,(6-3x)/2),\ 0 \leqslant x \leqslant 2\}.$$

Here, the objective function $f(x,y) = xy$ is nonlinear and the set S is described by linear inequalities.

** Such a model may work for a certain production process. However, it may not reflect the situation as other factors involved in the production process cannot be ignored. Therefore, new models have to be considered. For Example [7]:

- The Canadian manufacturing industries for 1927 is estimated by:

$$P(l,k) = 33l^{0.46}k^{0.52}$$

where P is product, l is labor and k is capital.

- The production P for the dairy farming in Iowa (1939) is estimated by:

$$P(A,B,C,D,E,F) = A^{0.27}B^{0.01}C^{0.01}D^{0.23}E^{0.09}F^{0.27}$$

where A is land, B is labor, C is improvements, D is liquid assets, E is working assets and F is cash operating expenses.

Each of these nonlinear production function P is optimized on a suitable set S that describes well the elements involved.

As seen above, the main purpose, of this study, is to find a solution to the following optimization problems

$$\text{find } u \in S \qquad \text{such that} \qquad f(u) = \min_{S} f(v)$$

or

$$\text{find } u \in S \qquad \text{such that} \qquad f(u) = \max_{S} f(v)$$

where $f : S \subset \mathbb{R}^n \longrightarrow \mathbb{R}$ is a given function and S a given subset of $\mathbb{R}^n$.

It is obvious that establishing existence and uniqueness results of the extreme points, depends on properties satisfied by the set S and the function f. So, we need to know some categories of subsets in $\mathbb{R}^n$ as well as some calculus on multi-variable functions. But, first look at the following remark:

Remark 1.1.1 *The extreme point may not exist on the set S. In our study, we will explore the situations where $\min_{S} f$ and $\max_{S} f$ are attained in S.*

For example

$$\min_{(0,1)} f(x) = x^2 \qquad\qquad \text{does not exist.}$$

Indeed, suppose there exists $x_0 \in (0,1)$ such that $f(x_0) = \min_{(0,1)} f(x)$. Then,

$$0 < \frac{x_0}{2} < x_0 \qquad \Longrightarrow \qquad \frac{x_0}{2} \in (0,1)$$

f is a strictly increasing function on $(0,1)$ $\Longrightarrow$ $f(\frac{x_0}{2}) < f(x_0)$,

which contradicts the fact that x_0 is a minimum point of f on $(0,1)$. However, we remark that

$$f(x) > 0 \qquad \forall x \in (0,1).$$

To include these limit cases, usually, instead of looking for a minimum or a maximum, we look for

$$\inf_{S} f(x) = \inf\{f(x) : x \in S\} \qquad \text{and} \qquad \sup_{S} f(x) = \sup\{f(x) : x \in S\}$$

where $\inf E$ and $\sup E$ of a nonempty subset E of $\mathbb{R}$ are defined by [2]

$$\sup E = \text{ the least number greater than or equal to all numbers in } E$$

$$\inf E = \text{ the greatest number less than or equal to all numbers in } E.$$

If E is not bounded below, we write $\inf E = -\infty$. If E is not bounded above, we write $\sup E = +\infty$. By convention, we write $\sup \emptyset = -\infty$ and $\inf \emptyset = +\infty$.

For the previous example, we have

$$\inf_{(0,1)} x^2 = 0, \qquad \text{and} \qquad \sup_{(0,1)} x^2 = 1.$$

1.2   Particular Subsets of $\mathbb{R}^n$

We list here the main categories of sets that we will encounter and give the main tools that allow their identification easily. Even though the purpose is not a topological study of these sets, it is important to be aware of the precise definitions and how to apply them accurately [18], [13].

Open and Closed Sets

In one dimension, the distance between two real numbers x and y is measured by the absolute value function and is given by

$$d(x, y) = |x - y|.$$

d satisfies, for any x, y, z, the properties

$$d(x, y) \geq 0 \qquad\qquad d(x, y) = 0 \iff x = y$$
$$d(y, x) = d(x, y) \qquad\qquad \text{symmetry}$$
$$d(x, z) \leq d(x, y) + d(y, z) \qquad\qquad \text{triangle inequality.}$$

These three properties induce on $\mathbb{R}$ a metric topology where a set $\mathcal{O}$ is said to be **open** if and only if, at each point $x_0 \in \mathcal{O}$, we can insert a **small interval** centered at x_0 that remains included in $\mathcal{O}$, that is,

$$\mathcal{O} \quad \text{is open} \quad \iff \quad \forall x_0 \in \mathcal{O} \quad \exists \epsilon > 0 \quad \text{such that} \quad (x_0 - \epsilon, x_0 + \epsilon) \subset \mathcal{O}.$$

In higher dimension, these tools are generalized as follows:

The distance between two points $x = (x_1, \cdots, x_n)$ and $y = (y_1, \cdots, y_n)$ is measured by the quantity

$$d(x, y) = \|x - y\| = \sqrt{(x_1 - y_1)^2 + \ldots + (x_n - y_n)^2}.$$

d is called the Euclidean distance and satisfies the three properties above. A set $\mathcal{O} \subset \mathbb{R}^n$ is said to be open if and only if, at each point $x_0 \in \mathcal{O}$, we can insert a small **ball**

$$B_\epsilon(x_0) = \{x \in \mathbb{R}^n \ : \ \|x - x_0\| < \epsilon\}$$

centered at x_0 with ϵ that remains included in $\mathcal{O}$, that is,

$$\mathcal{O} \quad \text{is open} \quad \Longleftrightarrow \quad \forall x_0 \in \mathcal{O} \quad \exists \epsilon > 0 \quad \text{such that} \quad B_\epsilon(x_0) \subset \mathcal{O}.$$

The point x_0 is said to be **an interior point** to $\mathcal{O}$.

<u>**Example 1**</u>. As n varies, the ball takes different shapes; see Figure 1.6.

$$n = 1 \quad a \in \mathbb{R} \quad\quad B_r(a) = (a - r, a + r) \quad : \quad \text{an open interval}$$

$$n = 2 \quad a = (a_1, a_2) \quad B_r(a) = \{(x_1, x_2) : (x_1 - a_1)^2 + (x_2 - a_2)^2 < r^2\} :$$
$$\text{an open disk}$$

$$n = 3 \quad\quad a = (a_1, a_2, a_3)$$
$$B_r(a) = \{(x_1, x_2, x_3) : (x_1 - a_1)^2 + (x_2 - a_2)^2 + (x_3 - a_3)^2 < r^2\} :$$
$$\text{set of points delimited by the sphere centered at } a \text{ with radius } r$$

$$n > 3 \quad\quad a = (a_1, \dots, a_3) \quad\quad B_r(a) \text{ is the set of points delimited by}$$
$$\text{the hyper sphere of points } x \text{ satisfying } d(a, x) = r.$$

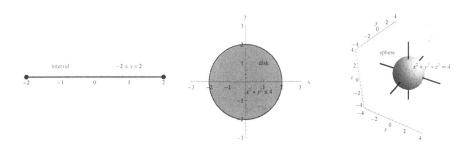

FIGURE 1.6: Shapes of balls in $\mathbb{R}$, $\mathbb{R}^2$ and $\mathbb{R}^3$

Using the distance d, we define

Definition 1.2.1 *Let S be a subset of $\mathbb{R}^n$.*

*– $\overset{\circ}{S}$ is the **interior** of S, the set of all interior points of S.*

*– S is a **neighborhood** of a if a is an interior point of S.*

– S is a **closed** *set* $\iff$ C^S *is open.*

– ∂S *is the **boundary** of S, the set of boundary points of S, where*

$$x_0 \in \partial S \iff \forall r > 0, \ B_r(x_0) \cap \left(S\right) \neq \emptyset \quad and \quad B_r(x_0) \cap \left(C^S\right) \neq \emptyset.$$

– $\overline{S} = S \cup \partial S$ *is the **closure** of S.*

– S *is **bounded*** $\iff$ $\exists \, M > 0$ *such that* $\|x\| \leqslant M$ $\forall x \in S$.

– S *is **unbounded** if it is not bounded.*

Example 2. For the sets, $\quad S_1 = [-2, 2] \subset \mathbb{R}$

$S_2 = \{(x,y): \ x^2 + y^2 \leqslant 4\} \subset \mathbb{R}^2, \qquad S_3 = \{(x,y,z): \ x^2 + y^2 + z^2 < 4\} \subset \mathbb{R}^3,$

we have

S	$\overset{\circ}{S}$	∂S	$\overline{S}$
S_1	$(-2, 2)$	$\{-2, 2\}$	S_1
S_2	$B_2(0)$	$C_2(0):$ circle	S_2
S_3	$S_3 = B_2(0)$	$S_2(0):$ sphere	$S_3 \cup S_2(0)$

where

$$C_2(0) = \{(x,y): \ x^2 + y^2 = 4\}, \qquad S_2(0) = \{(x,y,z): \ x^2 + y^2 + z^2 = 4\}.$$

We have the following properties:

Remark 1.2.1 – $\mathbb{R}^n$ *and $\emptyset$ are open and closed sets*

– *The union (resp. intersection) of arbitrary open (resp. closed) sets is open (resp. closed).*

– *The finite intersection (resp. union) of open (resp. closed) sets is open (resp. closed).*

– S *is open* $\iff$ $S = \overset{\circ}{S}$.

– S *is closed* $\iff$ $S = \overline{S}$.

– If f is continuous on an open subset $\Omega \subset \mathbb{R}^n$ (see Section 1.3), then

$$f^{-1}\Big((-\infty, a]\Big) = [f \leqslant a], \quad [f \geqslant a], \quad [f = a] \text{ are closed sets in } \mathbb{R}^n$$

$$f^{-1}\Big((-\infty, a)\Big) = [f < a], \qquad [f > a] \qquad \text{are open sets in } \mathbb{R}^n.$$

Example 3. Sketch the set S in the xy-plane and determine whether it is open, closed, bounded or unbounded. Give $\overset{\circ}{S}$, ∂S and $\overline{S}$.

$$S = \{(x, y): \quad x \geqslant 0, \quad y \geqslant 0, \quad xy \geqslant 1\}$$

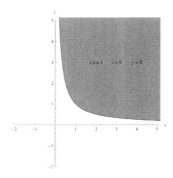

FIGURE 1.7: An unbounded closed subset of $\mathbb{R}^2$

∗ Note that the set S, sketched in Figure 1.7, doesn't contain the points on the x and y axis. So

$$S = \{(x, y): x > 0, \quad y > 0, \quad xy \geqslant 1\}$$

and can be described using the continuous function $f : (x, y) \longmapsto xy$ on the open set $\Omega = \{(x, y): x > 0, \ y > 0\}$ as

$$S = \{(x, y) \in \Omega : f(x, y) \geqslant 1\} = f^{-1}\Big([1, +\infty)\Big).$$

Therefore, S is a closed subset of $\mathbb{R}^2$. Thus $\overline{S} = S$.

∗∗ The set is unbounded since it contains the points $(x(t), y(t)) = (t, t)$ for $t \geqslant 1$ $(xy = t.t = t^2 \geqslant 1)$ and

$$\|(x(t), y(t))\| = \|(t, t)\| = \sqrt{t^2 + t^2} = \sqrt{2}t \longrightarrow +\infty \qquad \text{as} \qquad t \longrightarrow +\infty.$$

$* * *$ We have

$$\overset{\circ}{S} = \{(x,y): \; x > 0, \; y > 0, \; xy > 1\}$$

the region in the 1st quadrant above the hyperbola $y = \dfrac{1}{x}$

$$\partial S = \{(x,y): \quad x > 0, \quad y > 0, \quad xy = 1\}$$

the arc of the hyperbola in the 1st quadrant.

Example 4. A person can afford any commodities $x \geqslant 0$ and $y \geqslant 0$ that satisfies the budget inequality $x + 3y \leqslant 7$.

Sketch the set S described by these inequalities in the xy-plane and determine whether it is open, closed, bounded or unbounded. Give $\overset{\circ}{S}$, ∂S and $\overline{S}$.

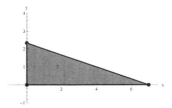

FIGURE 1.8: Closed set as intersection of three closed sets of $\mathbb{R}^2$

$*$ Figure 1.8 shows that S is the triangular region formed by all the points in the first quadrant below the line $x + 3y = 7$:

$$S = \{(x,y): \quad x + 3y \leqslant 7, \quad x \geqslant 0, \quad y \geqslant 0\}$$

and can be described using the continuous functions

$$f_1 : (x,y) \longmapsto x + 3y, \qquad f_2 : (x,y) \longmapsto x, \qquad f_3 : (x,y) \longmapsto y$$

on $\mathbb{R}^2$ as

$$S = \{(x,y) \in \mathbb{R}^2 : f_1(x,y) \leqslant 7, \quad f_2(x,y) \geqslant 0, \quad f_3(x,y) \geqslant 0\}$$

$$= f_1^{-1}\big((-\infty, 7]\big) \bigcap f_2^{-1}\big([0, +\infty)\big) \bigcap f_3^{-1}\big([0, +\infty)\big).$$

Therefore, S is a closed subset of $\mathbb{R}^2$ as the intersection of three closed subsets of $\mathbb{R}^2$. Thus $\overline{S} = S$.

** The set S is bounded since

$$x + 3y \leqslant 7, \qquad x \geqslant 0, \qquad y \geqslant 0 \qquad \implies \qquad 0 \leqslant x \leqslant 7, \qquad 0 \leqslant y \leqslant \frac{7}{3}$$

from which we deduce

$$\|(x,y)\| = \sqrt{x^2 + y^2} \leqslant \sqrt{7^2 + \left(\frac{7}{3}\right)^2} = \frac{7}{3}\sqrt{10} \qquad \forall (x,y) \in S.$$

*** We have

$\overset{\circ}{S} = \{(x,y) : x > 0, \ y > 0, \ x + 3y < 7\}$ the region S excluding its three sides

$\partial S = \quad$ the three sides of the triangular region.

Convex sets

The category of **convex** sets, deals with sets $S \subset \mathbb{R}^n$ where any two points $x, y \in S$ can be joined by a line segment that remains entirely into the set. Such sets are without holes and do not bend inwards. Thus

$$S \quad \text{is convex} \qquad \Longleftrightarrow \qquad (1-t)x + ty \in S \qquad \forall x, y \in S \qquad \forall t \in [0,1].$$

We have the following properties:

Remark 1.2.2 – $\mathbb{R}^n$ *and* $\emptyset$ *are convex sets*

– *A finite intersection of convex sets is a convex set.*

Example 5. "*Well known convex sets*" (see Figure 1.9)

* A line segment joining two points x and y is convex. It is described by

$$[x,y] = \{z \in \mathbb{R}^n : \quad \exists t \in [0,1] \quad \text{such that} \quad z = x + t(y-x) = (1-t)x + ty\}.$$

** A line passing through two points x_0 and x_1 is convex. It is described by

$$\mathcal{L} = \{x \in \mathbb{R}^n : \quad \exists t \in \mathbb{R} \quad \text{such that} \quad x = x_0 + t(x_1 - x_0)\}.$$

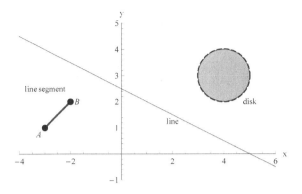

FIGURE 1.9: Convex sets in $\mathbb{R}^2$

$*\,*\,*$ A ball $B_r(x_0) = \{x \in \mathbb{R}^n : \quad \|x - x_0\| < r\}$ is convex.

Indeed, let a and b in $B_r(x_0)$ and $t \in [0,1]$, we have

$$\|[(1-t)a + tb] - x_0\| = \|(1-t)(a - x_0) + t(b - x_0)\|$$

$$\leqslant \|(1-t)(a - x_0)\| + \|t(b - x_0)\| = |1-t|\|a - x_0\| + |t|\|b - x_0\|$$

$$< |1-t|r + |t|r = r \quad \text{since} \quad \|a - x_0\| < 1 \ \text{ and } \ \|b - x_0\| < 1.$$

Hence $(1-t)a + tb \in B_r(x_0)$ for any $t \in [0,1]$; that is, $[a,b] \subset B_r(x_0)$.

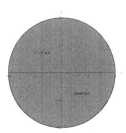

FIGURE 1.10: A closed ball is convex

$*\,*\,*\,*$ A closed ball $\overline{B_r(x_0)} = \{x \in \mathbb{R}^n : \quad \|x - x_0\| \leqslant r\}$ is convex.

For example, in the plane, the set in Figure 1.10, defined by

$$\{(x,y) : \quad x^2 + y^2 \leqslant 4\} = \overline{B_2((0,0))} \qquad \text{is convex.}$$

The set is the closed disk with center $(0,0)$ and radius 2. It is closed since it includes its boundary points located on the circle with center $(0,0)$ and radius 2. This set is bounded since $\|(x,y)\| \leqslant 2 \quad \forall (x,y) \in \overline{B_2((0,0))}$.

Example 6. *"Convex sets described by linear expressions"*

$*$ For $a = (a_1, \ldots, a_n) \in \mathbb{R}^n$, $b \in \mathbb{R}$, the set of points

$$x = (x_1, \ldots, x_n) \in \mathbb{R}^n : \quad a_1 x_1 + a_2 x_2 + \ldots + a_n x_n = a.x = b$$

is convex and called hyperplane.

Indeed, consider x^1, x^2 in the hyperplane and $t \in [0,1]$, then

$$a.[(1-t)x^1 + tx^2] = (1-t)a.x^1 + ta.x^2 = (1-t)b + tb = b$$

thus $(1-t)x^1 + tx^2$ belongs to the hyperplane.

As illustrated in Figure 1.11, the graph of an hyperplane is reduced to the point $x_1 = b/a_1$ when $n = 1$, to the line $a_1 x_1 + a_2 x_2 = b$ in the plane when $n = 2$, and to the plane $a_1 x_1 + a_2 x_2 + a_3 x_3 = b$ in the space when $n = 3$.

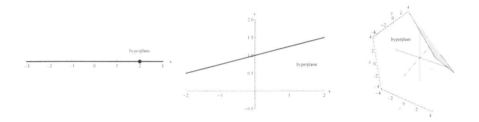

FIGURE 1.11: Hyperplane in $\mathbb{R}$, $\mathbb{R}^2$ and $\mathbb{R}^3$

$**$ The set of points in $x = (x_1, \ldots, x_n) \in \mathbb{R}^n$ defined by a linear inequality

$$a_1 x_1 + a_2 x_2 + \ldots + a_n x_n = a.x \leqslant b \qquad (\text{resp.} \geqslant, <, >) \qquad \text{is convex.}$$

Indeed, as above, consider x^1, x^2 in the region $[a.x \leqslant b]$ and $t \in [0,1]$, then

$$a.x^1 \leqslant b \quad \implies \quad (1-t)a.x^1 \leqslant (1-t)b \qquad \text{since} \quad (1-t) \geqslant 0$$

$$a.x^2 \leqslant b \quad \implies \quad ta.x^2 \leqslant tb \qquad \text{since} \quad t \geqslant 0$$

Adding the two inequalities, we get

$$a.[(1-t)x^1 + tx^2] = (1-t)a.x^1 + ta.x^2 \leqslant (1-t)b + tb = b$$

thus $(1-t)x^1 + tx^2$ belongs to the region $[a.x \leqslant b]$.

The set $a.x \leqslant b$ describes the region of points located below the hyperplane $a.x = b$.

$* * *$ A set of points in $\mathbb{R}^n$ described by linear equalities and inequalities is convex as it can be seen as the intersection of convex sets described by equalities and inequalities.

For example, in Figure 1.12, the set

$$S = \{(x,y): \ 2x+3y \leqslant 19, \ -3x+2y \leqslant 4, \ x+y \leqslant 8, \ 0 \leqslant x \leqslant 6, \ x+6y \geqslant 0\}$$

can be described as $S = S_1 \cap S_2 \cap S_3 \cap S_4 \cap S_5 \cap S_6$ where

$$S_1 = \{(x,y) \in \mathbb{R}^2: \ x+6y \geqslant 0\} \quad S_2 = \{(x,y) \in \mathbb{R}^2: \ x \leqslant 6\}$$

$$S_3 = \{(x,y) \in \mathbb{R}^2: \ x+y \leqslant 8\} \quad S_4 = \{(x,y) \in \mathbb{R}^2: \ 2x+3y \leqslant 19\}$$

$$S_5 = \{(x,y) \in \mathbb{R}^2: \ -3x+2y \leqslant 4\} \quad S_6 = \{(x,y) \in \mathbb{R}^2: \ x \geqslant 0\}.$$

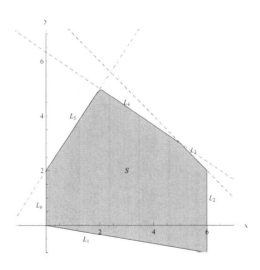

FIGURE 1.12: A convex set described by linear inequalities

S is the region of the plan xy, bounded by the lines

$$L_1 : \ x + 6y = 0 \qquad L_2 : \ x = 6, \qquad L_3 : \ x + y = 8,$$

$$L_4 : \ 2x + 3y = 19, \qquad L_5 : \ -3x + 2y = 4 \qquad L_6 : x = 0.$$

Often, such sets are described using matrices and vectors;

$$S = \left\{ \begin{bmatrix} x \\ y \end{bmatrix} \in \mathbb{R}^2 : \begin{bmatrix} 2 & 3 \\ -3 & 2 \\ 1 & 1 \\ 1 & 0 \\ -1 & -6 \\ -1 & 0 \end{bmatrix} \begin{bmatrix} x \\ y \end{bmatrix} \leqslant \begin{bmatrix} 19 \\ 4 \\ 8 \\ 6 \\ 0 \\ 0 \end{bmatrix} \right\}.$$

Example 7. "*Well-known non convex sets*"

$*$ The hyper-sphere (see Figure 1.13 for an illustration in the plane)

$$\partial B_r(x^*) = \{x \in \mathbb{R}^n : \quad \|x - x_0\| = r\} \qquad \text{is not convex.}$$

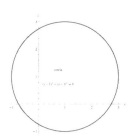

FIGURE 1.13: Circle $\partial B_2((1,1))$ is not convex

Indeed, we have

$$(x_1^*, \ldots, x_n^* \pm r) \in \partial B_r(x^*) \quad \text{since} \quad \|(0, \ldots, \pm r)\| = r$$

$$\left\| \frac{1}{2}(x_1^*, \ldots, x_n^* + r) + (1 - \frac{1}{2})(x_1^*, \ldots, x_n^* - r) - x^* \right\|$$

$$= \left\| \frac{1}{2}(2x_1^*, \ldots, 2x_n^* + r - r) - x^* \right\| = \|x^* - x^*\| = 0 \neq r$$

$$\implies \frac{1}{2}(x_1^*, \ldots, x_n^* + r) + (1 - \frac{1}{2})(x_1^*, \ldots, x_n^* - r) = x^* \notin \partial B_r(x^*).$$

$**$ The domain located outside the hyper-sphere, described by

$$S = \{x \in \mathbb{R}^n : \quad \|x - x^*\| > r\} = \mathbb{R}^n \setminus \overline{B_r(x^*)} \qquad \text{is not convex.}$$

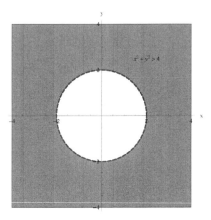

FIGURE 1.14: An unbounded open non convex set of $\mathbb{R}^2$

Indeed, we have

$$(x_1^*, \ldots, x_n^* \pm 2r) \in S \quad \text{since} \quad \|(0, \ldots, \pm 2r)\| = 2r > r$$

$$\frac{1}{2}(x_1^*, \ldots, x_n^* + 2r) + (1 - \frac{1}{2})(x_1^*, \ldots, x_n^* - 2r)$$

$$= \frac{1}{2}(2x_1^*, \ldots, 2x_n^* + 2r - 2r) = x^* \notin S.$$

For example, in the plane, the set

$$\{(x, y): \quad x^2 + y^2 > 4\} = \mathbb{R}^2 \setminus \overline{B_2((0,0))} \quad \text{is not convex.}$$

Moreover, the set is open since it is the complementary of the closed disk with center $(0,0)$ and radius 2 (see Figure 1.14). It is not bounded since for $t \geqslant 2$, the points $(0, t^2)$ belong to the set, but $\|(0, t^2)\| = t^2 \longrightarrow +\infty$ as $t \longrightarrow +\infty$.

$***$ The region located outside the hyper-sphere, including the hyper-sphere, described by

$$S = \{x \in \mathbb{R}^n: \quad \|x - x_0\| \geqslant r\} = \mathbb{R}^n \setminus B_r(x^*) \quad \text{is not convex.}$$

Example 8. "*The union of convex sets is not necessarily convex*"

$*$ The union of the disk and the line in Figure 1.9 is not convex.

$**$ The set $E = \{(x, y) \in \mathbb{R}^2: \quad xy + x - y - 1 > 0\}$, graphed in Figure 1.15, is not convex.

Indeed, we have

$$xy + x - y - 1 > 0 \iff (x-1)(y+1) > 0$$
$$\iff x > 1 \quad \text{and} \quad y > -1 \quad \text{or} \quad x < 1 \quad \text{and} \quad y < -1.$$

Thus E is the union of the sets

$$E_1 = \{(x,y) \in \mathbb{R}^2: \quad x > 1 \text{ and } y > -1\}$$
$$E_2 = \{(x,y) \in \mathbb{R}^2: \quad x < 1 \text{ and } y < -1\}$$

E_1 and E_2 are convex since they are described by linear inequalities. However, $E = E_1 \cup E_2$ is not convex since for example $(2,0)$ and $(0,-2)$ are points of E, but

$$\frac{1}{2}(2,0) + \left(1 - \frac{1}{2}\right)(0,-2) = (1,-1) \quad \text{doesn't belong to the set } E.$$

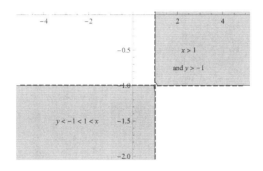

FIGURE 1.15: Union of convex sets

1.3   Functions of Several Variables

We refer the reader to any book of calculus [1], [3], [21], [23] for details on the points introduced in this section.

Definition 1.3.1 *A function f of n variables $x_1, \cdots, x_n$ is a rule that assigns to each n-vector $x = (x_1, \ldots, x_n)$ in the domain of f, denoted by D_f, a unique number $f(x) = f(x_1, \ldots, x_n)$.*

<u>Example 1</u>. Formulas may be used to model problems from different fields.

– Linear function

$$f(x_1, \ldots, x_n) = a_1 x_1 + a_2 x_2 + \ldots + a_n x_n.$$

– The body mass index is described by the function

$$B(w, h) = \frac{w}{h^2}$$

where w is the the weight in kilograms and h is the height measured in meters.

– The distance of a point $P(x, y, z)$ to a given point $P_0(x_0, y_0, z_0)$ is a function of three variables

$$d(x, y, z) = \sqrt{(x - x_0)^2 + (y - y_0)^2 + (z - z_0)^2}.$$

– The Cobb-Douglas function or the production function, describes the relationship between the output: the product Q and the inputs: $x_1, \ldots, x_n$ (capital, labor, ...) involved in the production process

$$Q(x_1, \cdots, x_n) = C x_1^{a_1} x_2^{a_2} \ldots x_n^{a_n} \quad C, a_1, \ldots, a_n \quad \text{are constants, } C > 0.$$

– The electric potential function for two positive charges, one at $(0, 1)$ with twice the magnitude as the charge at $(0, -1)$, is given by

$$\varphi(x, y) = \frac{2}{\sqrt{x^2 + (y - 1)^2}} + \frac{1}{\sqrt{x^2 + (y + 1)^2}}.$$

Example 2. When given a formula of a function, first identify its domain of definition before any other calculation.

The domains of definition of the functions given by the following formulas:

$$f(x) = \sqrt{x} \qquad\qquad g(x,y) = \sqrt{x} \qquad\qquad h(x,y,z) = \sqrt{x}$$

are

$$D_f = \{x \in \mathbb{R} / \ x \geqslant 0\}$$

$D_g = \{(x,y) \in \mathbb{R}^2 / \ x \geqslant 0\} :$ the half plane bounded by the y axis, including the axis and the points located in the 1st and 4th quadrants.

$D_h = \{(x,y,z) \in \mathbb{R}^3 / \ x \geqslant 0\} :$ the half space bounded by the plane yz, including this plane and the points with positive 1st coordinates $x \geqslant 0$.

The three domains D_f, D_g, D_h are closed, convex, unbounded subsets of $\mathbb{R}$, $\mathbb{R}^2$ and $\mathbb{R}^3$ respectively; see Figure 1.16.

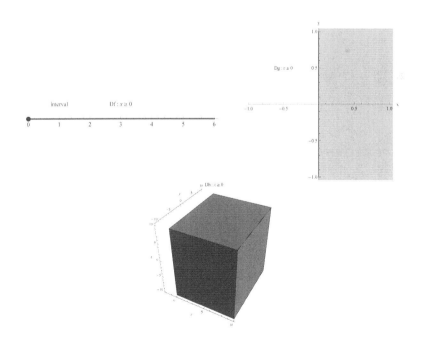

FIGURE 1.16: Domains of definition

Graphs and Level Curves

With the aid of monotony, and convexity, sketching the graph of a real function is performed by plotting few points. This is not possible in the case of dimension 3.

To get familiar with some sets in $\mathbb{R}^3$, we describe the traces' method used for plotting graphs of functions of two variables. The method consists on sketching the intersections of the graph (or surface) with well-chosen planes, usually planes that are parallel to the coordinates planes:

$$xy\text{-plane}: \ z = 0 \qquad xz\text{-plane}: \ y = 0 \qquad yz\text{-plane}: \ x = 0.$$

These intersections are called traces.

Definition 1.3.2 *The graph of a function* $f : x = (x_1, \ldots, x_n) \in D_f \subset \mathbb{R}^n \longmapsto z = f(x) \in \mathbb{R}$ *is the set*

$$G_f = \{(x, f(x)) \in \mathbb{R}^{n+1} : \quad x \in D_f\}.$$

The set of points x *in* $\mathbb{R}^n$ *satisfying* $f(x) = k$ *is called a level surface of* f.

When $n = 2$, a level surface $f(x, y) = k$ is called level curve. It is the projection of the trace $G_f \cap [z = k]$ onto the xy-plane. Drawing level curves of f is another way to picture the values of f.

The following examples illustrate how to proceed to graph some surfaces and level curves.

Example 3. A **cylinder** is a surface that consists of all lines that are parallel to a given line and that pass through a given plane curve.

Let

$$E = \{(x, y, z), \ x = y^2\}.$$

The set E cannot be the graph of a function $z = f(x, y)$ since $(1, 1, z) \in E$ for any z, and then $(1, 1)$ would have an infinite number of images. However, we can look at E as the graph of the function $x = f(y, z) = y^2$. Moreover, we have

$$E = \bigcup_{z \in \mathbb{R}} \{(x, y, z), \ x = y^2, \quad (x, y) \in \mathbb{R}^2\}.$$

This means that any horizontal plane $z = k$ (// to the xy plane) intersects the graph in a curve with equation $x = y^2$. So these horizontal traces $E \cap [z = k]$, $k \in \mathbb{R}$ are parabolas. The graph is formed by taking the parabola $x = y^2$ in the xy-plane and moving it in the direction of the z-axis. The graph is a parabolic cylinder as it can be seen as formed by parallel lines passing through the parabola $x = y^2$ in the xy-plane (see Figure 1.17).

Note that for any $k \in \mathbb{R}$, the level curve $z = k$ is the parabola $x = y^2$ in the xy plane.

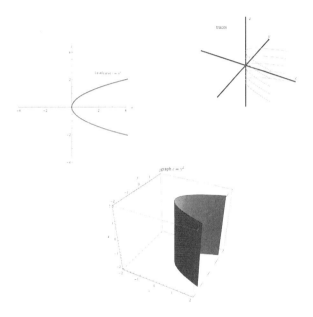

FIGURE 1.17: Parabolic cylinder

Example 4. An **Elliptic Paraboloid**, in its standard form, is the graph of the function

$$f(x, y) = z = \frac{x^2}{a^2} + \frac{y^2}{b^2} \qquad \text{with} \quad a > 0, \ b > 0.$$

The graph

$$G_f = \bigcup_{z \in [0, +\infty)} \left\{ (x, y, z), \quad \frac{x^2}{a^2} + \frac{y^2}{b^2} = z \right\}$$

can be seen as the union of ellipses $\dfrac{x^2}{a^2} + \dfrac{y^2}{b^2} = k$ in the planes $z = k$, $k \geqslant 0$.
By choosing the traces in Table 1.3, we can shape the graph in the space (see Figure 1.18 for $a = 2$, $b = 3$):

plane	trace
$xy \ (z = 0)$	$point : (0,0)$
$xz \ (y = 0)$	$parabola : z = \dfrac{x^2}{a^2}$
$yz \ (x = 0)$	$parabola : z = \dfrac{y^2}{b^2}$
$z = 1$	$ellipse : \dfrac{x^2}{a^2} + \dfrac{y^2}{b^2} = 1$

TABLE 1.3: Traces to sketch a paraboloid

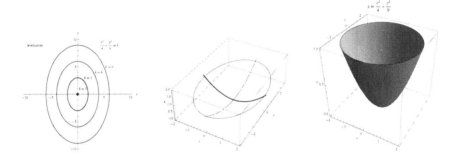

FIGURE 1.18: Elliptic paraboloid

Note that for any $k < 0$, the level curves $z = k$ are not defined. For $k > 0$, the level curves are ellipses $\dfrac{x^2}{(a\sqrt{k})^2} + \dfrac{y^2}{(b\sqrt{k})^2} = 1$ centered at the origin. For $k = 0$, the level curve is reduced to the point $(0,0)$.

<u>**Example 5**</u>. The **Elliptic Cone**, in its standard form, is described by the equation

$$z^2 = \frac{x^2}{a^2} + \frac{y^2}{b^2} \qquad \text{with} \quad a > 0, \ b > 0.$$

It is the union of the graphs of the functions $z = \pm\sqrt{\dfrac{x^2}{a^2} + \dfrac{y^2}{b^2}}$.

To sketch the cone, one can make the choice of traces in Table 1.4 (see Figure 1.19 for $a = 2, \ b = 3$):

plane	trace
$xy\ (z=0)$	$point : (0,0)$
$xz\ (y=0)$	$lines : z = \pm\dfrac{x}{a}$
$yz\ (x=0)$	$lines : z = \pm\dfrac{y}{b}$
$z = \pm 1$	$ellipse : \dfrac{x^2}{a^2} + \dfrac{y^2}{b^2} = 1$

TABLE 1.4: Traces to sketch a cone

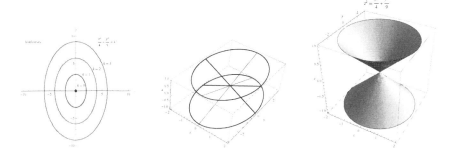

FIGURE 1.19: Elliptic cone

Note that for any $k \neq 0$, the level curves $z = \pm k$ are ellipses $\dfrac{x^2}{(|k|a)^2} + \dfrac{y^2}{(|k|b)^2} = 1$ centered at the origin. For $k = 0$, the level curve is reduced to the point $(0,0)$.

Example 6. The **Elliptic Ellipsoïd**, in its standard form, is described by the equation

$$\frac{x^2}{a^2} + \frac{y^2}{b^2} + \frac{z^2}{c^2} = 1 \qquad \text{with} \quad a > 0, b > 0, c > 0.$$

It is the union of the graphs of the functions $z = \pm c\sqrt{1 - \dfrac{x^2}{a^2} - \dfrac{y^2}{b^2}}$ that one can sketch by making the following choice of traces in Table 1.5 (see Figure 1.20 for $a = 2,\ b = 3,\ c = 4$):

plane	trace
$xy \ (z = 0)$	$ellipse : \dfrac{x^2}{a^2} + \dfrac{y^2}{b^2} = 1$
$xz \ (y = 0)$	$ellipse : \dfrac{x^2}{a^2} + \dfrac{z^2}{c^2} = 1$
$yz \ (x = 0)$	$ellipse : \dfrac{y^2}{b^2} + \dfrac{z^2}{c^2} = 1$

TABLE 1.5: Traces to sketch an ellipsoïd

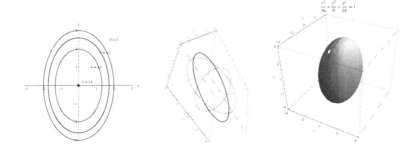

FIGURE 1.20: An elliptic ellipsoïd

For $k \in \mathbb{R}$, the level curves $z = \pm k$ are ellipses centered at the origin with vertices $\left(-a\sqrt{1 - \dfrac{k^2}{c^2}}, a\sqrt{1 - \dfrac{k^2}{c^2}} \right)$, $\left(-b\sqrt{1 - \dfrac{k^2}{c^2}}, b\sqrt{1 - \dfrac{k^2}{c^2}} \right)$ in the xy plane.

Limits and Continuity

For the local study of a function, the concept of limit is generalized to functions of several variables as follows

Definition 1.3.3 *Let $x_0 \in \mathbb{R}^n$ and let f be a function defined on $D_f \cap \left(B_r(x_0) \setminus \{x_0\} \right)$. We write* $\qquad \lim_{x \longrightarrow x_0} f(x) - L$

$$\iff \left(\forall \epsilon > 0, \ \exists \delta > 0 \ such \ that \ \forall x : 0 < \|x - x_0\| < \delta \implies |f(x) - L| < \epsilon \right).$$

Remark 1.3.1 *i) The definition above supposes that f is defined in a neighborhood of x_0, except possibly at x_0. It includes points x_0 located at the boundary of the domain of f.*

ii) One can establish, using similar tools in one dimension [2], that the standard properties of limits hold for limits of functions of n variables.

iii) If the limit of $f(x)$ fails to exist as $x \longrightarrow x_0$ along some smooth curve, or if $f(x)$ has different limits as $x \longrightarrow x_0$ along two different smooth curves, then the limit of $f(x)$ does not exist as $x \longrightarrow x_0$.

Example 7.

- $\lim\limits_{x \longrightarrow a} x_i = a_i, \quad i = 1, \cdots, n \quad a = (a_1, \cdots, a_n) \in \mathbb{R}^n.$

Indeed, for $\epsilon > 0$, choose $\delta = \epsilon > 0$. Then, we have for x satisfying

$$\|x - a\| < \delta \quad \Longrightarrow \quad |x_i - a_i| \leqslant \|x - a\| < \delta = \epsilon.$$

- Algebraic operations on limits.

$$\lim_{(x,y,z) \longrightarrow (1,2,3)} 3xy^2 + z - 5$$

$$= \lim_{(x,y,z) \longrightarrow (1,2,3)} [3xy^2] + \lim_{(x,y,z) \longrightarrow (1,2,3)} z - \lim_{(x,y,z) \longrightarrow (1,2,3)} 5$$

$$= 3[\lim_{(x,y,z) \longrightarrow (1,2,3)} x] . [\lim_{(x,y,z) \longrightarrow (1,2,3)} y]^2 + 3 - 5 = 3(1)(2)^2 + 3 - 5 = 10.$$

- The limit

$$\lim_{(x,y) \to (0,0)} \frac{2x^2 y}{x^4 + y^2} \qquad \text{does not exist.}$$

Indeed, if we consider the smooth curves C_1 and C_2 with equations $y = x^2$ and $y = x$ respectively, we find that

$$\lim_{\substack{(x,y) \to (0,0) \\ (x,y) \in C_1}} \frac{2x^2 y}{x^4 + y^2} = \lim_{x \to 0} \frac{2x^2 x^2}{x^4 + (x^2)^2} = \lim_{x \to 0} \frac{2x^4}{2x^4} = 1,$$

$$\lim_{\substack{(x,y) \to (0,0) \\ (x,y) \in C_2}} \frac{2x^2 y}{x^4 + y^2} = \lim_{x \to 0} \frac{2x^2 x}{x^4 + x^2} = \lim_{x \to 0} \frac{2x}{x^2 + 1} = 0,$$

the limits have different values along C_1 and C_2 (see Figure 1.21).

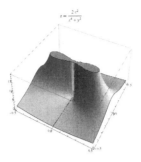

FIGURE 1.21: Behavior of $f(x,y) = \dfrac{2x^2 y}{x^4 + y^2}$ near $(0,0)$

Definition 1.3.4 *Let f be a function defined on $D_f \subset \mathbb{R}^n$. Then*

$$f \text{ is continuous at } x_0 \quad \Longleftrightarrow \quad \begin{cases} f(x_0) \text{ is defined and} \\[2mm] \displaystyle\lim_{x \longrightarrow x_0} f(x) = f(x_0). \end{cases}$$

If f is continuous at every point in an open set $\mathcal{O}$, then we say that f is continuous on $\mathcal{O}$.

Remark 1.3.2 *A function of n variables that can be constructed from continuous functions by combining the operations of addition, substraction, multiplication, division and composition is continuous wherever it is defined.*

Example 8. Give the largest region where f is continuous

$$f(x,y) = \frac{1}{e^{xy} - 1}.$$

Solution: f is continuous on its domain of definition

$$D_f = \mathbb{R}^2 \setminus \{(x,y) \in \mathbb{R}^2 / \ x = 0 \text{ or } y = 0\}.$$

More precisely, we have

  * $(x, y) \longmapsto x\,y$ is continuous on $\mathbb{R}^2$ as the product of the function $(x, y) \longmapsto x$ and the function $(x, y) \longmapsto y$

  ** $(x, y) \longmapsto \dfrac{1}{e^{xy} - 1}$ is continuous on D_f as the composition of the C^0 function $(x, y) \longmapsto xy$ on $\mathbb{R}^2$ and the C^0 function $t \longmapsto \dfrac{1}{e^t - 1}$ on $\mathbb{R} \setminus \{0\}$:

$$(x, y) \in D_f \qquad \longmapsto \qquad xy = t \in \mathbb{R} \setminus \{0\} \qquad \longmapsto \qquad \frac{1}{e^t - 1}.$$

First-order Partial Derivatives

Our purpose, now, is to generalize the concept of differentiability to functions of several variables. More precisely, we will show that the existence of a line tangent for a real differentiable function f at a point x_0 is extended to the existence of an hyperplane for a differentiable function with several variables. First, we introduce some tools:

> **Definition 1.3.5** *If* $z = f(x) = f(x_1, \cdots, x_n)$, *then the quantity*
>
> $$\frac{\partial f}{\partial x_i}(x) = \lim_{h \longrightarrow 0} \frac{f(x_1, \cdots, x_i + h, \cdots, x_n) - f(x_1, \cdots, x_i, \cdots, x_n)}{h}$$
>
> *is the partial derivative of* $f(x_1, \cdots, x_n)$ *with respect to* x_i *when all the other variables* x_j $(j \neq i, i = 1, \ldots, n)$ *are held constant.*

Remark 1.3.3 - *The partial derivative*

$$\frac{\partial f}{\partial x_i}(a) = \frac{d}{dx_i}[f(a_1, \ldots, x_i, \ldots, a_n)]\Big|_{x_i = a_i}, \qquad i = 1, \ldots, n$$

can be viewed as the slope of the line tangent to the curve $C_i : z = f(a_1, \ldots, x_i, \ldots, a_n)$ *at the point* a, *or the rate of change of* z *with respect to* x_i *along the curve* C_i *at* a.

- Other notations are :

$$\frac{\partial f}{\partial x_i} = \frac{\partial z}{\partial x_i} = f_{x_i} = z_{x_i} \qquad i = 1, \cdots, n.$$

- We call **gradient** of f the vector

$$\nabla f(x) = \langle f_{x_1}, f_{x_2}, \cdots, f_{x_n} \rangle = f'(x).$$

Example 9. Let $f(w, x, y, z) = xe^{yw} \sin z$.   Find

$$f_x(1, 2, 3, \pi/2), \quad f_y(1, 2, 3, \pi/2), \quad f_z(1, 2, 3, \pi/2) \quad \text{and} \quad f_w(1, 2, 3, \pi/2).$$

Solution: We have

$$f_x = e^{yw} \sin z \quad f_x(1, 2, 3, \pi/2) = e^{yw} \sin z \Big]_{(w,x,y,z)=(1,2,3,\pi/2)} = e^3$$

$$f_y = xwe^{yw} \sin z \quad f_y(1, 2, 3, \pi/2) = xwe^{yw} \sin z \Big]_{(w,x,y,z)=(1,2,3,\pi/2)} = 2e^3$$

$$f_z = xe^{yw} \cos z \quad f_z(1, 2, 3, \pi/2) = xe^{yw} \cos z \Big]_{(w,x,y,z)=(1,2,3,\pi/2)} = 0$$

$$f_w = xye^{yw} \sin z \quad f_w(1, 2, 3, \pi/2) = xye^{yw} \sin z \Big]_{(w,x,y,z)=(1,2,3,\pi/2)} = 6e^3.$$

Example 10. The rate of change of the (BMI) body mass index function $B(w, h) = w/h^2$ with respect of the weight w at a constant height h is

$$\frac{\partial B}{\partial w} = \frac{1}{h^2} > 0.$$

Thus, at constant height, people's BMI differs by a factor of $1/h^2$.

The rate of change of the BMI with respect of the height h at a constant weight w is

$$\frac{\partial B}{\partial h} = -\frac{2w}{h^3} < 0.$$

Therefore, with similar weight, people's BMI is a decreasing function of the height.

Higher Order Partial Derivatives

• Each partial derivative is also a function of n variables. These functions may themselves have partial derivatives, called second order derivatives. For each $i = 1, \ldots, n$, we have

$$\frac{\partial}{\partial x_j}\left(\frac{\partial f}{\partial x_i}\right) = \frac{\partial^2 f}{\partial x_j \partial x_i} = f_{x_i x_j}.$$

The n second-order partial derivatives $f_{x_i x_i}$ are called direct second-order partial; the others, $f_{x_i x_j}$ where $i \neq j$, are called mixed second-order partial. Usually these second-order partial derivatives are displayed in an $n \times n$ matrix named the Hessian

$$H_f(x) = (f_{x_i x_j})_{n \times n} = \begin{bmatrix} f_{x_1 x_1} & f_{x_1 x_2} & \cdots & f_{x_1 x_n} \\ f_{x_2 x_1} & f_{x_2 x_2} & \cdots & f_{x_2 x_n} \\ \vdots & \vdots & \vdots & \vdots \\ f_{x_n x_1} & f_{x_n x_2} & \cdots & f_{x_n x_n} \end{bmatrix}$$

• The mixed derivatives are equal in the following situation [15]

Theorem 1.3.1 *Clairaut's theorem*

Let $f(x) = f(x_1, x_2, \cdots, x_n)$. If $f_{x_i x_j}$ and $f_{x_j x_i}$, $i \neq j$ for $i, j \in \{1, \cdots, n\}$ are defined on a neighborhood of a point $a \in \mathbb{R}^n$ and are continuous at a then

$$f_{x_i x_j}(a) = f_{x_j x_i}(a).$$

• Third-order, fourth-order and higher-order partial derivatives can be obtained by successive differentiation. Clairaut's theorem reduces the steps of calculations when the continuity assumption is satisfied.

Example 11. Write the Hessian of the Cobb-Douglas function

$$Q(L, K) = cL^a K^b \qquad (c, a, b \text{ are positive constants})$$

where the two inputs are labor L and capital K.

Solution: For $L, K > 0$, we have

$$\ln Q = \ln c + a \ln L + b \ln K$$

$$\frac{\partial(\ln Q)}{\partial L} = \frac{Q_L}{Q} = \frac{a}{L} \implies Q_L = \frac{a}{L}Q$$

$$\frac{\partial(\ln Q)}{\partial K} = \frac{Q_K}{Q} = \frac{b}{K} \implies Q_K = \frac{b}{K}Q$$

$$Q_{LL} = \frac{a}{L}Q_L + \left(-\frac{a}{L^2}\right)Q = \frac{a}{L}\left(\frac{a}{L}\right)Q + \left(-\frac{a}{L^2}\right)Q = \frac{a(a-1)}{L^2}Q$$

$$Q_{KK} = \frac{b}{K}Q_K + \left(-\frac{b}{K^2}\right)Q = \frac{b}{K}\left(\frac{b}{K}\right)Q + \left(-\frac{b}{K^2}\right)Q = \frac{b(b-1)}{K^2}Q$$

$$Q_{KL} = Q_{LK} = \frac{a}{L}Q_K = \frac{ab}{LK}Q.$$

The Hessian matrix of Q is given by:

$$H_Q(L,K) = \begin{bmatrix} Q_{LL} & Q_{LK} \\ Q_{KL} & Q_{KK} \end{bmatrix} = Q \begin{bmatrix} \dfrac{a(a-1)}{L^2} & \dfrac{ab}{LK} \\ \dfrac{ab}{LK} & \dfrac{b(b-1)}{K^2} \end{bmatrix}.$$

Example 12. Laplace's equation of a function $u = u(x_1, \ldots, x_n)$ is

$$\triangle u = \frac{\partial^2 u}{\partial x_1^2} + \frac{\partial^2 u}{\partial x_2^2} + \ldots + \frac{\partial^2 u}{\partial x_n^2} = 0.$$

For which value of k, the function $u = (x_1^2 + x_2^2 + \ldots + x_n^2)^k$ satisfies Laplace's equation?

Solution: We have

$$\frac{\partial u}{\partial x_i} = 2x_i k(x_1^2 + x_2^2 + \ldots + x_n^2)^{k-1}$$

$$\frac{\partial^2 u}{\partial x_i^2} = 2k(x_1^2 + x_2^2 + \ldots + x_n^2)^{k-1} + 4x_i^2 k(k-1)(x_1^2 + x_2^2 + \ldots + x_n^2)^{k-2}$$

$$\triangle u = \frac{\partial^2 u}{\partial x_1^2} + \frac{\partial^2 u}{\partial x_2^2} + \ldots + \frac{\partial^2 u}{\partial x_n^2}$$

$$= 2kn(x_1^2 + x_2^2 + \ldots + x_n^2)^{k-1} + 4k(k-1)$$

$$\times \left(\sum_{i=1}^{n} x_i^2\right)(x_1^2 + x_2^2 + \ldots + x_n^2)^{k-2}$$

$$= 2k[n + 2(k-1)](x_1^2 + x_2^2 + \ldots + x_n^2)^{k-1}.$$

Thus $\triangle u = 0$ if $n + 2(k-1) = 0$ ie. for $k = 1 + n/2$.

Differentiability

While the existence of a derivative of a one variable function at a point guarantees the continuity of the function at this point, the existence of partial derivatives for a function of several variables doesn't. Indeed, for example

$$f(x,y) = \begin{cases} 2 & \text{if} \quad x > 0 \quad \text{and} \quad y > 0 \\ 0 & \text{if not} \end{cases}$$

has partial derivatives at $(0,0)$ since

$$f_x(0,0) = \lim_{h \to 0} \frac{f(h,0) - f(0,0)}{h} = \lim_{h \to 0} \frac{0 - 0}{h} = \lim_{h \to 0} 0 = 0,$$

$$f_y(0,0) = \lim_{h \to 0} \frac{f(0,h) - f(0,0)}{h} = 0$$

but f is not continuous at $(0,0)$ since

$$\lim_{t \to 0^+} f(t,t) = \lim_{t \to 0^+} 2 = 2 \neq 0 = f(0,0).$$

This motivates, the following definition

Definition 1.3.6 *A function of n variables is said to be differentiable at $a = (a_1, \ldots, a_n)$ provided that $f_{x_i}(a)$, $i = 1, \ldots, n$ exist and that there exists a function $\varepsilon : \mathbb{R}^+ \longrightarrow \mathbb{R}$ such that:*

$$f(x) = f(a) + f_{x_1}(a)(x_1 - a_1) + \ldots + f_{x_n}(a)(x_n - a_n) + \|x - a\| \varepsilon(\|x - a\|)$$

with

$$\lim_{x \longrightarrow a} \varepsilon(\|x - a\|) = 0.$$

Remark 1.3.4 *The definition extends the concept of differentiability of functions of one variable to functions of n variables in such a way that we preserve properties like:*

- f continuous at a;

- the values of f at points near a can be very closely approximated by the values of a linear function:

$$f(x) \approx f(a) + f_{x_1}(a)(x_1 - a_1) + \ldots + f_{x_n}(a)(x_n - a_n).$$

The next theorem provides particular conditions for a function f to be differentiable.

Theorem 1.3.2 *If all first-order partial derivatives of f exist and are continuous at a point, then f is differentiable at that point.*

If f has continuous partial derivatives of first-order in a domain D, we call f **continuously differentiable** in D. In this case, f is also called a C^1 function on D. If all partial derivatives up to order k exist and are continuous, f is called a C^k function.

Example 13. Use the linear approximation to estimate the change of the Cobb-Douglas production function

$$Q(L,K) = L^{1/3}K^{2/3} \qquad \text{from} \quad (20,10) \quad \text{to} \quad (20.6,10.3).$$

Solution: We have

$$Q_L(L,K) = \frac{1}{3L}Q, \quad Q_K(L,K) = \frac{2}{3K}Q, \quad Q(20,10) = 20^{1/3}10^{2/3} = 10(2^{1/3}),$$

$$Q_L(20,10) = \frac{1}{3(20)}Q(20,10), \qquad Q_K(20,10) = \frac{2}{3(10)}Q(20,10)$$

Thus, close to $(20,10)$, we have

$$Q(L,K) \approx Q(20,10) + Q_L(20,10)(L-20) + Q_K(20,10)(K-10)$$

$$= \left(1 + \frac{1}{60}(L-20) + \frac{2}{30}(K-10)\right)Q(20,10)$$

from which we deduce the estimate

$$Q(20.6,10.3) \approx \left(1 + \frac{1}{60}(20.6-20) + \frac{2}{30}(10.3-10)\right)Q(20,10) = 1.003\,Q(20,10).$$

Another consequence of the differentiability is the chain rule for derivation under composition.

Theorem 1.3.3 *Chain rule 1*

If f is differentiable at $x = (x_1, x_2, \ldots, x_n)$ and each $x_j = x_j(t)$, $j = 1, \ldots, n$, is a differentiable function of a variable t, then $z = f(x(t))$ is differentiable at t and

$$\frac{dz}{dt} = \frac{\partial z}{\partial x_1}\frac{dx_1}{dt} + \frac{\partial z}{\partial x_2}\frac{dx_2}{dt} + \cdots\cdots + \frac{\partial z}{\partial x_n}\frac{dx_n}{dt}.$$

Proof. Since f is differentiable at the point a, then, for $x(t)$ close to $a = x(t_0)$, we have

$$f(x(t)) - f(a) = f_{x_1}(a)(x_1(t) - a_1) + \cdots + f_{x_n}(a)(x_n(t) - a_n)$$

$$+ \|x(t) - a\|\varepsilon(\|x(t) - a\|) \qquad \text{with} \qquad \lim_{x \longrightarrow a} \varepsilon(\|x - a\|) = 0.$$

Dividing each side of the equality by $\Delta t = t - t_0$, we obtain

$$\frac{f(x(t)) - f(a)}{\Delta t} = f_{x_1}(a)\left(\frac{x_1(t) - a_1}{\Delta t}\right) + \ldots\ldots + f_{x_n}(a)\left(\frac{x_n(t) - a_n}{\Delta t}\right)$$

$$+ \left\|\frac{x(t) - a}{\Delta t}\right\|\varepsilon(\|x(t) - a\|).$$

Then letting $t \longrightarrow t_0$ and using the fact that each $x_j = x_j(t)$, $j = 1, \ldots, n$, is a differentiable function of the variable t and that $\lim_{x \longrightarrow a} \varepsilon(\|x - a\|) = 0$, then

$$\lim_{t \longrightarrow t_0} \frac{f(x(t)) - f(a)}{\Delta t} = f_{x_1}(a). \lim_{t \longrightarrow t_0} \left(\frac{x_1(t) - a_1}{\Delta t}\right) + \ldots$$

$$+ f_{x_n}(a). \lim_{t \longrightarrow t_0} \left(\frac{x_n(t) - a_n}{\Delta t}\right) + \left\|\lim_{t \longrightarrow t_0} \frac{x(t) - a}{\Delta t}\right\|. \lim_{t \longrightarrow t_0} \epsilon(\|x(t) - a\|)$$

from which we deduce that

$$\left.\frac{d(f(x(t)))}{dt}\right|_{t=t_0} = f_{x_1}(a).\frac{dx_1}{dt}(t_0) + \ldots\ldots + f_{x_n}(a).\frac{dx_n}{dt}(t_0) + \left\|\frac{dx}{dt}(t_0)\right\|.0$$

and the result follows.

In the general situation, each variable x_i is a function of m independent variables $t_1, t_2, \ldots, t_m$. Then $z = f(x(t_1, t_2, \ldots, t_m))$ is a function of $t_1, t_2, \ldots, t_m$. To compute $\dfrac{\partial z}{\partial t_j}$, we hold t_i with $i \neq j$ fixed and compute the ordinary derivative of z with respect to t_j. The result is given by the following theorem:

Theorem 1.3.4 *Chain rule 2*

If f is differentiable at $x = (x_1, x_2, \ldots, x_n)$ and each $x_j = x_j(t_1, t_2, \cdots, t_m)$, $j = 1, \cdots, n$, is a differentiable function of m variables $t_1, t_2, \ldots, t_m$, then $z = f(x(t_1, t_2, \ldots, t_m))$ is differentiable at $(t_1, t_2, \ldots, t_m)$ and

$$\frac{\partial z}{\partial t_i} = \frac{\partial z}{\partial x_1}\frac{\partial x_1}{\partial t_i} + \frac{\partial z}{\partial x_2}\frac{\partial x_2}{\partial t_i} + \ldots\ldots + \frac{\partial z}{\partial x_n}\frac{\partial x_n}{\partial t_i}.$$

Example 14. Let

$$f(x,y) = x^2 - 2xy + 2y^3, \qquad x = s\ln t, \qquad y = st.$$

Use the chain rule formula to find

$$\frac{\partial f}{\partial s}, \qquad \frac{\partial f}{\partial t}, \qquad \frac{\partial f}{\partial s}\Big|_{s=1,t=1} \qquad \text{and} \qquad \frac{\partial f}{\partial t}\Big|_{s=1,t=1}.$$

Solution: i) We have

$$\frac{\partial f}{\partial x} = 2x - 2y, \qquad\qquad \frac{\partial f}{\partial y} = -2x + 6y^2,$$

$$x = x(s,t), \qquad\qquad \frac{\partial x}{\partial s} = \ln t, \qquad\qquad \frac{\partial x}{\partial t} = \frac{s}{t},$$

$$y = y(s,t), \qquad\qquad \frac{\partial y}{\partial s} = t, \qquad\qquad \frac{\partial y}{\partial t} = s.$$

Hence the partial derivatives of f at (s,t) are:

$$\frac{\partial f}{\partial s} = \frac{\partial f}{\partial x}\cdot\frac{\partial x}{\partial s} + \frac{\partial f}{\partial y}\cdot\frac{\partial y}{\partial s} = (2x - 2y)\ln t + (-2x + 6y^2)t$$

$$= (2s\ln t - 2st)\ln t + (-2s\ln t + 6s^2t^2)t$$

$$\frac{\partial f}{\partial t} = \frac{\partial f}{\partial x}\cdot\frac{\partial x}{\partial t} + \frac{\partial f}{\partial y}\cdot\frac{\partial y}{\partial t} = (2x - 2y)\frac{s}{t} + (-2x + 6y^2)s$$

$$= (2s\ln t - 2st)\frac{s}{t} + (-2s\ln t + 6s^2t^2)s.$$

ii) When $s = 1$ and $t = 1$, we have

$$x(1,1) = (1)\ln(1) = 0, \qquad\qquad \text{and} \qquad\qquad y(1,1) = 1.$$

Thus the partial derivatives of f at $(s,t) = (1,1)$ are:

$$\frac{\partial f}{\partial s}\Big|_{s=1,t=1} = (2x(s,t) - 2y(s,t))\ln t + (-2x(s,t) + 6y^2(s,t))t\Big|_{s=1,t=1} = 6$$

$$\frac{\partial f}{\partial t}\Big|_{s=1,t=1} = (2x(s,t) - 2y(s,t))\frac{s}{t} + (-2x(s,t) + 6y(s,t)^2)s\Big|_{s=1,t=1} = 4.$$

Solved Problems

1. – Sketch the domains of definition of the functions given by the following formulas:

$$i) \quad f(x,y) = e^{2x}\sqrt{y-x^2} \qquad\qquad ii) \quad f(x,y,z) = z\sqrt{(1-x^2)(y^2-4)}$$

$$iii) \quad H(x,y,z) = \sqrt{z-x^2-y^2}.$$

Solution:

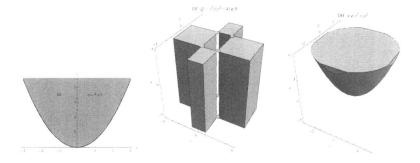

FIGURE 1.22: Domains of definitions

i) $\quad f(x,y) = e^{2x}\sqrt{y-x^2}$

$$D_f = \{(x,y) \in \mathbb{R}^2 : \quad y-x^2 \geqslant 0\}$$

the plane region located above the parabola, including the parabola.

ii) $\quad f(x,y,z) = z\sqrt{(1-x^2)(y^2-4)}$

$$D_f = \{(x,y,z) \in \mathbb{R}^3 : \quad (1-x^2)(y^2-4) \geqslant 0\}$$

x	-2	-1	1	2	
$1-x^2$	$-$	$-$	$+$	$-$	$-$
y^2-4	$+$	$-$	$-$	$-$	$+$

so

$$D_f \ = \Big\{ [-1,1] \times \big((-\infty, -2] \cup [2, +\infty) \big) \times \mathbb{R} \Big\}$$
$$\cup \Big\{ \big((-\infty, -1] \cup [1, +\infty) \big) \times [-2, 2] \times \mathbb{R} \Big\}.$$

iii) $H(x, y, z) = \sqrt{z - x^2 - y^2}$

$$D_H = \{(x, y, z) \in \mathbb{R}^3 : \quad z - x^2 - y^2 \geqslant 0\}$$

set of points bounded by the paraboloid $z = x^2 + y^2$, including the paraboloid. The three domains are illustrated in Figure 1.22.

2. – Match the functions with their graphs in Figure 1.23.

   a. $y - z^2 = 0$ *b.* $x + y + z = 0$ *c.* $4x^2 + \dfrac{y^2}{9} + z^2 = 1$

   d. $x^2 + \dfrac{y^2}{9} - z^2 = 1$ *e.* $x^2 + \dfrac{y^2}{9} = z^2$ *f.* $z - y^2 = 0$

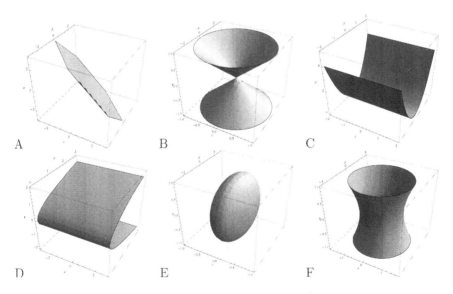

FIGURE 1.23: Surfaces in $\mathbb{R}^3$

Solution:

equation of the surface	*its graph*	*why?*
a. $\;\; y - z^2 = 0$	(D)	*parabolic cylinder in the direction of the* $x - axis,\; located\; in\; y \geqslant 0$
b. $\;\; x + y + z = 0$	(A)	*a plane*
c. $\;\; 4x^2 + \frac{y^2}{9} + z^2 = 1$	(E)	*ellipsoid centered at* $(0,0,0)$
d. $\;\; x^2 + \frac{y^2}{9} - z^2 = 1$	(F)	*the traces at* $z = -1,\, 0,\, 1$ *are ellipses*
e. $\;\; x^2 + \frac{y^2}{9} = z^2$	(B)	*elliptic cone*
f. $\;\; z - y^2 = 0$	(C)	*parabolic cylinder in the direction of the* $x - axis,\; located\; in\; z \geqslant 0$

3. – Sketch the graphs of the following functions:

$i)\; f(x,y) = \sqrt{81 - x^2}$ $\qquad ii)\; f(x,y) = 3$ $\qquad iii)\; f(x,y) = -\sqrt{x^2 + y^2}.$

Solution: i)

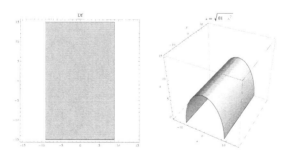

FIGURE 1.24: Domain and graph of $z = \sqrt{81 - x^2}$

Domain of f : $D_f = \{(x, y) \in \mathbb{R}^2 : 81 - x^2 \geqslant 0\} = \{(x, y) \in \mathbb{R}^2 : |x| \leqslant 9\}$

Graph of f : $G_f = \{(x, y, z) \in \mathbb{R}^3 : \exists (x, y) \in D_f \text{ such that } z = \sqrt{81 - x^2}\}$
$$= \{(x, y, z) \in \mathbb{R}^3 : \exists (x, y) \in D_f \text{ such that } x^2 + z^2 = 81, \quad z \geqslant 0\}.$$

It is the half circular cylinder located in the $z \geqslant 0$ with radius 9 and axis the y axis (see Figure 1.24).

   ii)

Domain of f : $D_f = \{(x, y) \in \mathbb{R}^2 : f(x, y) = 3 \in \mathbb{R}\} = \mathbb{R}^2$

Graph of f : $G_f = \{(x, y, z) \in \mathbb{R}^3 : \exists (x, y) \in D_f \text{ such that } z = 3\}$

It is the plane passing through $(0, 0, 3)$ with normal vector $k = \langle 0, 0, 1 \rangle$ (see Figure 1.25).

   iii)

Domain of f : $D_f = \{(x, y) \in \mathbb{R}^2 : x^2 + y^2 \geqslant 0\} = \mathbb{R}^2$

Graph of f : $G_f = \{(x, y, z) \in \mathbb{R}^3 : \exists (x, y) \in D_f \text{ such that }$
$$z = -\sqrt{x^2 + y^2}\}$$

$$= \{(x, y, z) \in \mathbb{R}^3 : \exists (x, y) \in \mathbb{R}^2 \text{ such that }$$
$$z^2 = x^2 + y^2, \ z \leqslant 0\}$$

The graph is the part of the circular cone $z^2 = x^2 + y^2$ located in the region $[z \leqslant 0]$; see Figure 1.25.

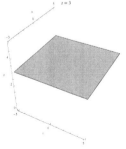
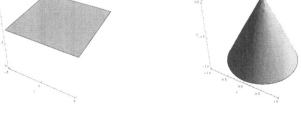

FIGURE 1.25: Graph of $z = 3$ and   graph of $z = -\sqrt{x^2 + y^2}$

▌ **4.** – Match the surfaces with the level curves in Figure 1.26.

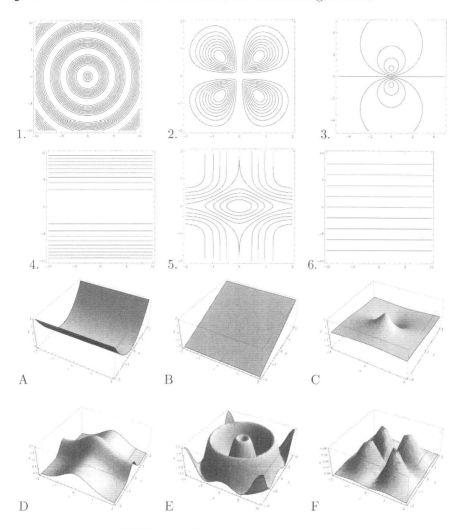

FIGURE 1.26: Surfaces and their level curves

Solution:

level curves	(1)	(2)	(3)	(4)	(5)	(6)
surface	(E)	(F)	(C)	(A)	(D)	(B)

5. – Draw a set of level curves for the following functions:

$$i) \quad z = x^2 + y \qquad\qquad ii) \quad f(x,y,z) = (x-2)^2 + y^2 + z^2.$$

Solution: i) We have: $D_f = \{(x,y) \in \mathbb{R}^2, \quad x^2 + y \in \mathbb{R}\} = \mathbb{R}^2$,

$$z = x^2 + y = k \Longleftrightarrow y - k = x^2 : \text{ parabola with vertex } (0,k)$$

and axis the line Oy; see Figure 1.27.

ii) The level curve (see the 2nd graph in Figure 1.27) $(x-2)^2 + y^2 + z^2 = k$ is reduced to

$$\begin{cases} \text{the point } (2,0,0) & \text{if} \quad k = 0 \\[2mm] \text{the sphere centered at } (2,0,0) \text{ with radius } \sqrt{k} & \text{if} \quad k > 0 \\[2mm] \text{no points} & \text{if} \quad k < 0. \end{cases}$$

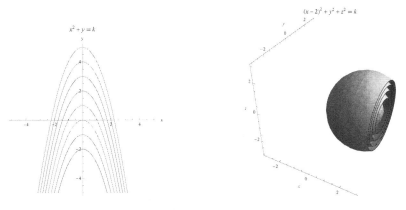

FIGURE 1.27: Level curves $x^2 + y = k$ and level surfaces $(x-2)^2 + y^2 + z^2 = k$

6. – Sketch the largest region on which the function is continuous. Explain why the function is continuous.

$$f(x,y,z) = \sqrt{y - x^2} \, \ln z.$$

Solution: f is continuous on its domain of definition

$$D_f = \{(x, y, z) \in \mathbb{R}^3 \ / \ y - x^2 \geqslant 0 \quad \text{and} \quad z > 0\}$$

because it is the product of the two continuous functions:

* $\quad u : (x, y, z) \longmapsto \ln z$ continuous on $D_1 = \{(x, y, z) : z > 0\}$
with values in $\mathbb{R}$ as the composite of the polynomial function
$(x, y, z) \in D_1 \longmapsto z \in \mathbb{R}^+ \setminus \{0\}$ and the function $t \longmapsto \ln t$ continuous on
$\mathbb{R}^+ \setminus \{0\}$; we have $(x, y, z) \in D_1 \longmapsto \quad z = t \in \mathbb{R}^+ \setminus \{0\} \quad \longmapsto \ln t$.

** $v : (x, y, z) \longmapsto \sqrt{y - x^2}$: continuous on $D_2 = \{(x, y, z) : y - x^2 \geqslant 0\}$
as the composite of the polynomial function $(x, y, z) \in D_2 \longmapsto y - x^2$
$\in \mathbb{R}^+$ and the function $t \longmapsto \sqrt{t}$ continuous on $\mathbb{R}^+$; we have
$$(x, y, z) \in D_2 \longmapsto \quad y - x^2 = t \in \mathbb{R}^+ \setminus \{0\} \quad \longmapsto \sqrt{t}.$$

*** $\quad f = u.v$ is continuous on $D_1 \cap D_2 = D_f$, the set in Figure 1.28.

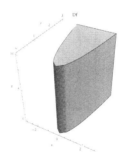

FIGURE 1.28: Domain of continuity of $f(x, y, z) = \sqrt{y - x^2} \ \ln z$

7. – Let $f(x, y, z) = x^2 y^2 - y^3 + 3x^4 + xe^{-2z} \sin(\pi y) + 5$. Find

$(a) \quad f_{xy}$ $\qquad$ $(b) \quad f_{yz}$ $\qquad$ $(c) \quad f_{xz}$ $\qquad$ $(d) \quad f_{zz}$

$(e) \quad f_{zyy}$ $\qquad$ $(f) \quad f_{xxy}$ $\qquad$ $(g) \quad f_{zyx}$ $\qquad$ $(h) \quad f_{xxyz}.$

Solution: Note that f is indefinitely differentiable. Therefore, we can change the order of differentiation with respect of the variables by using Clairaut's theorem.

$$f_x = 2xy^2 + 12x^3 + e^{-2z}\sin(\pi y)$$

$$f_y = 2x^2 y - 3y^2 + \pi x e^{-2z}\cos(\pi y) \qquad f_z = -2xe^{-2z}\sin(\pi y)$$

(a) $f_{xy} = (f_x)_y = 4xy + \pi e^{-2z}\cos(\pi y)$

(b) $f_{yz} = (f_y)_z = -2\pi x e^{-2z}\cos(\pi y)$

(c) $f_{xz} = (f_x)_z = -2e^{-2z}\sin(\pi y)$ (d) $f_{zz} = (f_z)_z = 4xe^{-2z}\sin(\pi y)$

(e) $f_{zyy} = (f_{zy})_y = (f_{yz})_y = 2\pi^2 x e^{-2z}\sin(\pi y),$

(f) $f_{xxy} = (f_x)_{xy} = (f_x)_{yx} = (f_{xy})_x = 4y$

(g) $f_{zyx} = (f_{zy})_x = (f_{yz})_x = -2\pi e^{-2z}\cos(\pi y),$

(h) $f_{xxyz} = (f_{xxy})_z = (4y)_z = 0.$

8. – Show that $u = \ln(x^2 + y^2)$ satisfies Laplace equation $\dfrac{\partial^2 u}{\partial x^2} + \dfrac{\partial^2 u}{\partial y^2} = 0.$

Show, without calculation, that: $\dfrac{\partial^2 u}{\partial x \partial y} = \dfrac{\partial^2 u}{\partial y \partial x}.$

Solution: We have

$$\frac{\partial u}{\partial x} = \frac{2x}{x^2 + y^2} \qquad\qquad \frac{\partial u}{\partial y} = \frac{2y}{x^2 + y^2}$$

$$\frac{\partial^2 u}{\partial x^2} = 2\frac{(1)(x^2 + y^2) - x(2x)}{(x^2 + y^2)^2} = 2\frac{(y^2 - x^2)}{(x^2 + y^2)^2} \qquad \frac{\partial^2 u}{\partial y^2} = 2\frac{(x^2 - y^2)}{(x^2 + y^2)^2}$$

$$\frac{\partial^2 u}{\partial x^2} + \frac{\partial^2 u}{\partial y^2} = 2\frac{(y^2 - x^2)}{(x^2 + y^2)^2} + 2\frac{(x^2 - y^2)}{(x^2 + y^2)^2} = 0.$$

Note that $\dfrac{\partial u}{\partial x}$ is a fraction. Then $\dfrac{\partial^2 u}{\partial y \partial x}$ is also a fraction. As a consequence, $\dfrac{\partial^2 u}{\partial y \partial x}$ is continuous on $\mathbb{R}^2 \setminus \{(0,0)\}$.

In the same way, $\dfrac{\partial u}{\partial y}$ is a fraction, $\dfrac{\partial^2 u}{\partial x \partial y}$ is also a fraction. Therefore, $\dfrac{\partial^2 u}{\partial x \partial y}$ is continuous on $\mathbb{R}^2 \setminus \{(0,0)\}$.

From Clairaut's Theorem, the two second derivatives u_{xy} and u_{yx} are equal on $\mathbb{R}^2 \setminus \{(0,0)\}$.

9. – Find the value $\dfrac{dw}{ds}\Big|_{s=0}$ if

$$w = x^2 e^{2y} \cos(3z); \qquad x = \cos s, \qquad y = \ln(s+2), \qquad z = s.$$

Solution: We have $x = x(s)$, $y = y(s)$, $z = z(s)$ and $w = w(x, y, z)$. Then

$$\frac{dx}{ds} = -\sin s, \qquad \frac{dy}{ds} = \frac{1}{s+2}, \qquad \frac{dz}{ds} = 1$$

$$\frac{\partial w}{\partial x} = 2x e^{2y} \cos(3z), \qquad \frac{\partial w}{\partial y} = 2x^2 e^{2y} \cos(3z), \qquad \frac{\partial w}{\partial z} = -3x^2 e^{2y} \sin(3z)$$

$$x(0) = 1, \qquad y(0) = \ln 2, \qquad z(0) = 0$$

$$\frac{dw}{ds} = \frac{\partial w}{\partial x}\frac{dx}{ds} + \frac{\partial w}{\partial y}\frac{dy}{ds} + \frac{\partial w}{\partial z}\frac{dz}{ds}$$

$$= [2x e^{2y} \cos(3z)](-\sin s) + [2x^2 e^{2y} \cos(3z)]\left(\frac{1}{s+2}\right) + [-3x^2 e^{2y} \sin(3z)]$$

$$\frac{dw}{ds}\Big|_{s=0} = e^{2\ln 2} = 4.$$

10. – Let

$$R = \ln(u^2 + v^2 + w^2), \qquad u = x + 2y, \qquad v = 2x - y, \qquad w = 2xy.$$

Find

$$\frac{\partial R}{\partial x}\Big|_{x=1, y=0} \qquad \text{and} \qquad \frac{\partial R}{\partial y}\Big|_{x=1, y=0}.$$

Solution: We have

$$\frac{\partial R}{\partial u} = \frac{2u}{u^2 + v^2 + w^2}, \qquad \frac{\partial R}{\partial v} = \frac{2v}{u^2 + v^2 + w^2}, \qquad \frac{\partial R}{\partial w} = \frac{2w}{u^2 + v^2 + w^2},$$

$$\frac{\partial u}{\partial x} = 1, \qquad \frac{\partial v}{\partial x} = 2, \qquad \frac{\partial w}{\partial x} = 2y,$$

$$\frac{\partial u}{\partial y} = 2, \qquad \frac{\partial v}{\partial y} = -1, \qquad \frac{\partial w}{\partial y} = 2x.$$

The partial derivatives of R are:

$$\frac{\partial R}{\partial x} = \frac{\partial R}{\partial u}\cdot\frac{\partial u}{\partial x} + \frac{\partial R}{\partial v}\cdot\frac{\partial v}{\partial x} + \frac{\partial R}{\partial w}\cdot\frac{\partial w}{\partial x} = \frac{2u + 4v + 4wy}{u^2 + v^2 + w^2}$$

$$\frac{\partial R}{\partial y} = \frac{\partial R}{\partial u}\cdot\frac{\partial u}{\partial y} + \frac{\partial R}{\partial v}\cdot\frac{\partial v}{\partial y} + \frac{\partial R}{\partial w}\cdot\frac{\partial w}{\partial y} = \frac{4u - 2v + 4wx}{u^2 + v^2 + w^2}.$$

When $x = 1$ and $y = 0$, we have

$$u = 1 \qquad\qquad v = 2, \qquad\qquad w = 0, \qquad\qquad u^2 + v^2 + w^2 = 5.$$

Thus

$$\frac{\partial R}{\partial x} = \frac{2(1) + 4(2) + 4(0)}{5} = 2, \qquad\qquad \frac{\partial R}{\partial y} = \frac{4(1) - 2(2) + 4(0)}{5} = 0.$$

11. – Use the linear approximation of $f(x, y, z) = x^3\sqrt{y^2 + z^2}$ at the point $(2, 3, 4)$ to estimate the number

$$(1.98)^3\sqrt{(3.01)^2 + (3.97)^2}.$$

Solution: Since f is differentiable at the point $(2, 3, 4)$, the linear approximation of $L(x, y, z)$ at the point $(2, 3, 4)$ is given by:

$$L(x, y, z) = f(2, 3, 4) + f_x(2, 3, 4)(x - 2) + f_y(2, 3, 4)(y - 3) + f_z(2, 3, 4)(z - 4).$$

We have

$$f_x = 3x^2\sqrt{y^2 + z^2}, \qquad\qquad f_y = \frac{yx^3}{\sqrt{y^2 + z^2}}, \qquad\qquad f_z = \frac{zx^3}{\sqrt{y^2 + z^2}}$$

and

$$f(2,3,4) = 40, \qquad f_x(2,3,4) = 60, \qquad f_y(2,3,4) = \frac{24}{5}, \qquad f_z(2,3,4) = \frac{32}{5}.$$

Thus

$$L(x, y, z) = 40 + \frac{12}{5}(x-2) + \frac{24}{5}(y-3) + \frac{32}{5}(z-4).$$

Using this approximation, one obtain the following estimate:

$$(1.98)^3 \sqrt{(3.01)^2 + (3.97)^2} \approx L(1.98, 3.01, 3.97)$$

$$= 40 + 60(1.98 - 2) + \frac{24}{5}(3.01 - 3) + \frac{32}{5}(3.97 - 4)3$$

$$= 40 + 60(-0.02) + \frac{24}{5}(0.01) + \frac{32}{5}(-0.03) = 38.656.$$

12. – Determine whether the limit exists. If so, find its value.

$$\lim_{(x,y)\to(0,0)} \frac{x^4 - x + y - x^3 y}{x - y}, \qquad \lim_{(x,y)\to(0,0)} \frac{\cos(xy)}{x+y}, \qquad \lim_{(x,y)\to(1,1)} \frac{x - y^4}{x^3 - y^4}.$$

Solution: We have

i) $$\lim_{(x,y)\to(0,0)} \frac{x^4 - x + y - x^3 y}{x - y} = \lim_{(x,y)\to(0,0)} \frac{x^3(x - y) - (x - y)}{x - y}$$

$$= \lim_{(x,y)\to(0,0)} x^3 - 1 = -1.$$

ii) $\displaystyle \lim_{(x,y)\to(0,0)} \frac{\cos(xy)}{x+y}$ doesn't exist since

$$\lim_{(x,y)=(t,t),t>0\to(0,0)} \frac{\cos(xy)}{x+y} = \lim_{t\to 0^+} \frac{\cos(t^2)}{2t} = +\infty.$$

and $$\lim_{(x,y)=(t,t),t<0\to(0,0)} \frac{\cos(xy)}{x+y} = \lim_{t\to 0^-} \frac{\cos(t^2)}{2t} = -\infty.$$

iii) Let C_1 and C_2 the curves $x = 1$ and $x = y$ respectively. We have

$$\lim_{(x,y)\to(1,1) \atop (x,y)\in C_1} \frac{x - y^4}{x^3 - y^4} = \lim_{y\to 1} \frac{1 - y^4}{1 - y^4} = \lim_{y\to 1} 1 = 1,$$

$$\lim_{(x,y)\to(1,1) \atop (x,y)\in C_2} \frac{y(1 - y^3)}{y^3(1 - y)} = \lim_{y\to 1} \frac{y^2 + y + 1}{y^2} = 3.$$

The limits are different along C_1 and C_2. Thus, the limit doesn't exist.

Chapter 2

Unconstrained Optimization

In this chapter, we are interested in optimizing several variables' functions $f : x = (x_1, \ldots, x_n) \longmapsto f(x) \in \mathbb{R}$ over subsets S of $\mathbb{R}^n$ with nonempty interior.

Many results are well known when dealing with functions of one variable ($n = 1$). The concept of differentiability offered useful and flexible tools to get local and global behaviors of a function. These results are generalized to functions of n variables in these notes. Indeed, we obtain, in Section 2.1, a characterization of local critical points as solutions of the vectorial equation $f'(x) = 0$ when f is regular. In Section 2.2, we use the second partial derivatives to identify the nature of the critical points. In Section 2.3, first we define the convexity-concavity property for a function of several variables, then we show how to use it to identify the global extreme points. Finally, Section 2.4 extends the extreme value theorem to continuous functions on closed bounded subsets of $\mathbb{R}^n$.

2.1 Necessary Condition

In this section, we would like to have a close look at our candidates for optimality. In other words, if we are close enough of such points (when they exist), what conditions would be satisfied? Doing so, we hope to reduce the size of the set of the candidates' points then identify among these points the extreme ones. This motivates the following definition of local extreme points.

Definition 2.1.1 *local (global) maximum (minimum)*

Let $S \subset \mathbb{R}^n$ and $f : S \longmapsto \mathbb{R}$ be a function. A point $x^* \in S$ is said to be

  – *a local maximum (resp. minimum) of f if*

$$\exists r > 0 \quad such \ that \quad f(x) \leqslant f(x^*) \quad (resp. \geqslant) \quad \forall x \in B_r(x^*) \cap S.$$

  – *a strict local maximum (resp. minimum) of f if*

$$\exists r > 0 \quad such \ that \quad f(x) < f(x^*) \quad (resp. >) \quad \forall x \in B_r(x^*) \cap S, \quad x \neq x^*.$$

  – *a global maximum (resp. minimum) of f if*

$$f(x) \leqslant f(x^*) \qquad (resp. \geqslant) \qquad \forall x \in S.$$

  – *a strict global maximum (resp. minimum) of f if*

$$f(x) < f(x^*) \qquad (resp. >) \qquad \forall x \in S, \quad x \neq x^*.$$

Remark 2.1.1 *Note that a global extreme point is also a local extreme point when S is an open set, but the converse is not always true.*

Indeed, suppose, for example, that x^* is such that

$$\min_S f(x) = f(x^*)$$

then

$$f(x) \geqslant f(x^*) \qquad \forall x \in S.$$

Because S is an open set and $x^* \in S$, there exists a ball $B_r(x^*)$ such that $B_r(x^*) \subset S$, and then, in particular,

$$f(x) \geqslant f(x^*) \qquad \forall x \in B_r(x^*)$$

which shows that x^* is a local minimum.

To show that the converse is not true, consider the function $f(x) = x^3 - 3x$. The study of the variations of f, in Table 2.1, and its graph, in Figure 2.1, show that f has a local minimum at $x = 1$ and a local maximum at $x = -1$, but none of them is a global maximum or a global minimum, as we have

$$f'(x) = 3x^2 - 3 \qquad f''(x) = 6x \qquad \lim_{x \to +\infty} f(x) = +\infty \qquad \lim_{x \to -\infty} f(x) = -\infty.$$

Now, here is a characterization of a local extreme point for a regular objective function.

x	$-\infty$		-1		0	1		$+\infty$
$f'(x)$		$+$			$-$		$+$	
$f(x)$	$-\infty$	$\nearrow$	2		$\searrow$	-2	$\nearrow$	$+\infty$
$f''(x)$		$-$		$-$		$+$		$+$
f is		*concave*		*concave*		*convex*		*convex*

TABLE 2.1: Study of $f(x) = x^3 - 3x$

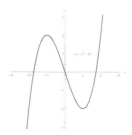

FIGURE 2.1: Local extreme points but not global ones

Theorem 2.1.1 *Necessary condition for local extreme points*

Let $S \subset \mathbb{R}^n$ and $f : S \longmapsto \mathbb{R}$ be a differentiable function at an interior point $x^* \in \overset{\circ}{S}$. Then

$$x^* \quad \text{is a local extreme point} \quad \implies \quad \nabla f(x^*) = \mathbf{0}.$$

Proof. Suppose f has a local minimum at x^*. Since f is differentiable at $x^* = (x_1^*, x_2^*, \ldots, x_n^*)$, its first derivatives exist. From the definition of the partial derivative, we have, for $j \in \{1, \ldots, n\}$

$$\frac{\partial f}{\partial x_j}(x^*) = \lim_{t \to 0} \frac{f(x_1^*, \ldots, x_j^* + t, \ldots, x_n^*) - f(x_1^*, \ldots, x_j^*, \ldots, x_n^*)}{t}$$

Because f has an interior local minimum at x^*, there is an $\epsilon > 0$ such that

$$\forall x \in B_\epsilon(x^*) \subset S \quad \implies \quad f(x) \geqslant f(x^*).$$

In particular, for $|t| < \epsilon$, we have

$$\|(x_1^*, \ldots, x_j^* + t, \ldots, x_n^*) - x^*\| = \|(x_1^*, \ldots, x_j^* + t, \ldots, x_n^*) - (x_1^*, \ldots, x_j^*, \ldots, x_n^*)\|$$

$$= \|(0,\ldots,0,t,0,\ldots,0)\| = \sqrt{0^2 + \ldots + 0^2 + t^2 + 0^2 + \ldots + 0^2} = |t| < \epsilon.$$

Thus the points $(x_1^*,\ldots,x_j^* + t,\ldots,x_n^*)$ remain inside the ball $B_\epsilon(x^*)$ and therefore satisfy

$$f(x_1^*,\ldots,x_j^* + t,\ldots,x_n^*) \geqslant f(x^*)$$

$$\Longleftrightarrow \quad f(x_1^*,\ldots,x_j^* + t,\ldots,x_n^*) - f(x_1^*,\ldots,x_j^*,\ldots,x_n^*) \geqslant 0.$$

Thus, if t is positive,

$$\frac{f(x_1^*,\ldots,x_j^* + t,\ldots,x_n^*) - f(x_1^*,\ldots,x_j^*,\ldots,x_n^*)}{t} \geqslant 0$$

and letting $t \to 0^+$, we deduce that

$$\lim_{t \to 0^+} \frac{f(x_1^*,\ldots,x_j^* + t,\ldots,x_n^*) - f(x_1^*,\ldots,x_j^*,\ldots,x_n^*)}{t} \geqslant 0.$$

In the same way, if t is negative,

$$\frac{f(x_1^*,\ldots,x_j^* + t,\ldots,x_n^*) - f(x_1^*,\ldots,x_j^*,\ldots,x_n^*)}{t} \leqslant 0$$

and letting $t \to 0^-$, we deduce that

$$\lim_{t \to 0^-} \frac{f(x_1^*,\ldots,x_j^* + t,\ldots,x_n^*) - f(x_1^*,\ldots,x_j^*,\ldots,x_n^*)}{t} \leqslant 0.$$

Because

$$\lim_{t \to 0^+} \frac{f(x_1^*,\ldots,x_j^* + t,\ldots,x_n^*) - f(x_1^*,\ldots,x_j^*,\ldots,x_n^*)}{t}$$

$$= \lim_{t \to 0^-} \frac{f(x_1^*,\ldots,x_j^* + t,\ldots,x_n^*) - f(x_1^*,\ldots,x_j^*,\ldots,x_n^*)}{t} = \frac{\partial f}{\partial x_j}(x^*)$$

we have $\dfrac{\partial f}{\partial x_j}(x^*) \geqslant 0$ and $\dfrac{\partial f}{\partial x_j}(x^*) \leqslant 0$, and we deduce that $\dfrac{\partial f}{\partial x_j}(x^*) = 0$.
This holds for each $j \in \{1,\ldots,n\}$. Hence $\nabla f(x^*) = \mathbf{0}$.

A similar argument applies if f has a local maximum at x^*.

Remark 2.1.2 *Note that a local extremum can also occur at a point where a function is not differentiable.*

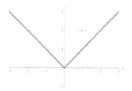

FIGURE 2.2: A minimum point where $f(x) = |x|$ is not differentiable

• For example, the one variable function $f(x) = |x|$, illustrated in Figure 2.2, has a local minimum at 0 but f is not differentiable at 0 since we have

$$\lim_{x \to 0^+} \frac{f(x) - f(0)}{x} = \lim_{x \to 0^+} \frac{x - 0}{x} = 1$$

$$\lim_{x \to 0^-} \frac{f(x) - f(0)}{x} = \lim_{x \to 0^-} \frac{-x - 0}{x} = -1.$$

Moreover 0 is a global minimum since we have

$$f(x) = |x| \geqslant 0 = f(0) \qquad \forall x \in \mathbb{R}.$$

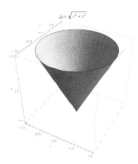
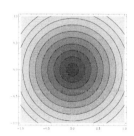

FIGURE 2.3: A minimum point where $f(x, y) = \sqrt{x^2 + y^2}$ is not differentiable

• The two variables function $f(x, y) = \sqrt{x^2 + y^2}$, graphed in Figure 2.3, attains its minimum value at $(0, 0)$ because we can see that

$$f(x, y) = \sqrt{x^2 + y^2} \geqslant 0 = f(0, 0) \qquad \forall (x, y) \in \mathbb{R}^2.$$

But f is not differentiable at $(0, 0)$ since, for example $f_x(0, 0)$ doesn't exist. Indeed, we have

$$\frac{f(0+h,0)-f(0,0)}{h}=\frac{\sqrt{h^2}-0}{h}=\frac{|h|}{h}\quad\longrightarrow\quad\begin{cases}1,&\text{if}\quad h\to 0^+\\[2mm]-1,&\text{if}\quad h\to 0^-.\end{cases}$$

The above remark leads to the following definition.

Definition 2.1.2 *Critical point*

An interior point x^* of the domain of a function f is a critical point of f if it is a stationary point where $\nabla f(x^*) = \mathbf{0}$ or a point where f is not differentiable.

Example 1. $(0,0)$ is the only stationary point for the functions f and g

$$i)\quad f(x,y)=x^2+y^2 \qquad\qquad ii)\quad g(x,y)=1-x^2-y^2.$$

It is a local and absolute minimum for f and a local and absolute maximum for g. The values of the level curves are increasing in Figure 2.4, while they are decreasing in Figure 2.5.

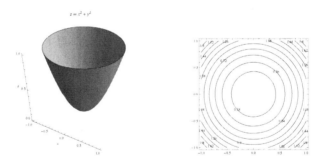

FIGURE 2.4: Local minimum point that is a global one

Indeed, we have for any $(x,y)\in\mathbb{R}^2$

$$f(x,y)=x^2+y^2\geqslant 0=f(0,0)\quad\text{and}\quad g(x,y)=1-(x^2+y^2)\leqslant 1=g(0,0).$$

Example 2. In economics, one is interested in maximizing the total profit $P(x)$ in the sale of x units of some product. If $C(x)$ is the total cost of production and $R(x)$ is the revenue function then

$$P(x)=R(x)-C(x).$$

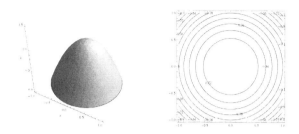

$z = -x^2 - y^2 + 1$

FIGURE 2.5: Local maximum point that is a global one

The maximum profit occurs when $P'(x) = 0$, or $R'(x) = C'(x)$. From the linear approximation, we have for $\Delta x = 1$,

$$R(x+1) - R(x) \approx R'(x)\Delta x = R'(x), \qquad C(x+1) - C(x) \approx C'(x)\Delta x = C'(x),$$

$$C(x+1) - C(x) \approx R(x+1) - R(x)$$

that is, the cost of manufacturing an additional unit of a product is approximately equal to the revenue generated by that unit. $P'(x)$, $R'(x)$, $C'(x)$ are interpreted respectively as the additional profit, revenue and cost that result from producing one additional unit when the production and sales levels are at x units.

Remark 2.1.3 *A function needs not to have a local extremum at every local critical point.*

• For example, the one variable function $f(x) = x^3$ has a local critical point since

$$f'(x) = 3x^2 = 0 \qquad \Longleftrightarrow \qquad x = 0.$$

But 0 is not a local extremum (see Figure 2.6). Indeed we have

$$f(x) = x^3 > 0 = f(0) \quad \forall x > 0 \qquad \text{and} \qquad f(x) = x^3 < 0 = f(0) \quad \forall x < 0.$$

The point 0 is called an inflection point.
• The two variables function $f(x, y) = y^2 - x^2$, graphed in Figure 2.7, has a critical point at $(0, 0)$ since we have

$$\nabla f(x, y) = \langle -2x, 2y \rangle = \langle 0, 0 \rangle \qquad \Longleftrightarrow \qquad (x, y) = (0, 0).$$

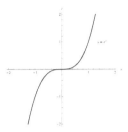

FIGURE 2.6: The critical point $x = 0$ is an inflection point for $f(x) = x^3$

However, the function f has neither a relative maximumnor a relative minimum at (0,0). Indeed, along the x and y axis, we have

$$f(x,0) = -x^2 \leqslant 0 = f(0,0) \quad \forall x \in \mathbb{R} \text{ and } f(0,y) = y^2 \geqslant 0 = f(0,0) \quad \forall y \in \mathbb{R}.$$

The point $(0,0)$ is called a saddle point. Figure 2.7 shows how the values of the level curves are increasing in one side and decreasing on the other side.

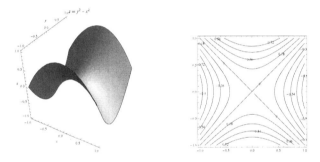

FIGURE 2.7: $(0,0)$ is a saddle point for $f(x,y) = y^2 - x^2$

Definition 2.1.3 *Saddle point*

A differentiable function $f(x)$ has a saddle point at a critical point x^* if in every open ball centered at x^* there are domain points x where $f(x) > f(x^*)$ and domain points x where $f(x) < f(x^*)$.

Remark 2.1.4 *In two dimensions, the projection of horizontal traces shows circular curves around (x^*, y^*) when it is a local extreme point, and hyperbolas when the point is a saddle point.*

Now, we give a necessary condition when the extreme point is not necessarily an interior point [5].

Theorem 2.1.2 *Necessary condition for a relative extreme point on a convex set*

Let $S \subset \Omega \subset \mathbb{R}^n$, Ω an open set, S a convex set and $f : \Omega \longmapsto \mathbb{R}$ be a differentiable function at a point $x^* \in S$. Then

$$f(x) \geqslant f(x^*) \qquad (resp. \leqslant) \qquad \forall x \in S$$

$$\implies \quad \nabla f(x^*).(x - x^*) \geqslant 0 \qquad (resp. \leqslant 0) \qquad \forall x \in S.$$

Proof. Let $x \in S$, $x \neq x^*$. Since S is convex, $\theta x + (1 - \theta)x^* = x^* + \theta(x - x^*) \in S$ for $\theta \in [0, 1]$. Suppose f has a relative minimum at x^*. Since f is differentiable at x^*, we can write

$$f(x^* + \theta(x - x^*)) - f(x^*) = \theta[\, f'(x^*).(x - x^*) + \epsilon(\theta)\,], \qquad \lim_{\theta \to 0} \epsilon(\theta) = 0.$$

If $f'(x^*).(x - x^*) < 0$, then

$$\exists \theta_0 \in (0, 1) : \quad \forall \theta \in (0, \theta_0) \qquad \implies \qquad |\epsilon(\theta)| < -\frac{1}{2}f'(x^*).(x - x^*).$$

Hence, $\qquad \forall \theta \in (0, \theta_0), \quad f(x^* + \theta(x - x^*)) - f(x^*) <$

$$\theta[\, f'(x^*).(x - x^*) - \frac{1}{2}f'(x^*).(x - x^*)\,] = \frac{\theta}{2}\, f'(x^*).(x - x^*) < 0$$

which contradicts the fact that x^* is a relative minimum. Therefore, we have $f'(x^*).(x - x^*) \geqslant 0$.
The case of a relative maximum is proved similarly.

Example 3. Consider the real function $f(x) = x^2$ with $x \in [1, 2]$; see Figure 2.8.
The interval $S = [1, 2]$ is a convex subset of $\Omega = \mathbb{R}$. f is differentiable on $\mathbb{R}$, and has no critical points on $(1, 2)$ since

$$f'(x) = 2x \neq 0 \qquad \text{on} \quad (1, 2).$$

From the theorem, if $x = a \in [1, 2]$ is a minimum point, then it must satisfy

$$f'(a)(x - a) \geqslant 0 \quad \forall x \in [1, 2] \iff 2a(x - a) \geqslant 0 \quad \forall x \in [1, 2] \implies a = 1.$$

If $x = b \in [1, 2]$ is a maximum point, then it must satisfy

$$f'(b)(x - b) \leqslant 0 \quad \forall x \in [1, 2] \iff 2b(x - b) \geqslant 0 \quad \forall x \in [1, 2] \implies b = 2.$$

Hence, $x = 1$ and $x = 2$ are the candidate points for optimality. In fact, f attains its minimum and maximum values at $x = 1$ and $x = 2$ respectively on $[1, 2]$. We can see that

$$1 \leqslant x \leqslant 2 \quad \Longrightarrow \quad 1 = 1^2 = f(1) \leqslant x^2 = f(x) \leqslant f(2) = 2^2 = 4 \quad \forall x \in [1, 2].$$

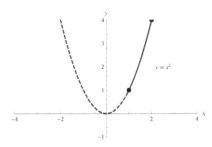

FIGURE 2.8: Extreme points of $f(x) = x^2$ on the convex $[1, 2]$

Example 4. Solve the problem

$$\min f(x_1, x_2) = x_1^2 - x_1 + x_2 + x_1 x_2 \quad \text{subject to} \quad x_1 \geqslant 0, \quad x_2 \geqslant 0.$$

Solution: The set $S = \{(x_1, x_2) : x_1 \geqslant 0, \ x_2 \geqslant 0\}$ is a convex subset of $\mathbb{R}^2$ and f is differentiable on $\Omega = \mathbb{R}^2$, and has no critical points in the interior of S, ie. $\{(x_1, x_2) : x_1 > 0, \ x_2 > 0\}$ since

$$f'(x_1, x_2) = \langle 2x_1 - 1 + x_2, 1 + x_1 \rangle = \langle 0, 0 \rangle \quad \Longleftrightarrow \quad (x_1, x_2) = (-1, 3) \notin \overset{\circ}{S}.$$

So the minimum value, if it exists, must be attained on the boundary of S. Note that

$$f(x_1, 0) = x_1^2 - x_1 = (x_1 - \frac{1}{2})^2 - \frac{1}{4} \geqslant -\frac{1}{4} = f(\frac{1}{2}, 0) \qquad \forall x_1 \geqslant 0$$

and

$$f(0, x_2) = x_2 \geqslant 0 \qquad \forall x_2 \geqslant 0.$$

Since $-1/4 < 0$, the point $(\frac{1}{2}, 0)$ is the global minimum point of f on S, as shown in Figure 2.9.

At this point

$$f'(x_1, x_2)\Big|_{x_1 = \frac{1}{2}, x_2 = 0} = \langle 2x_1 - 1 + x_2, 1 + x_1 \rangle \Big|_{x_1 = \frac{1}{2}, x_2 = 0} = \langle 0, \frac{3}{2} \rangle \neq \mathbf{0}.$$

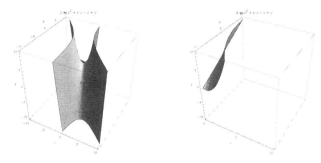

FIGURE 2.9: Min f attained at the boundary of $x_1 \geqslant 0$, $x_2 \geqslant 0$

and

$$\nabla f(\frac{1}{2}, 0).\langle x_1 - \frac{1}{2}, x_2 - 0 \rangle = \frac{3}{2}x_2 \geqslant 0 \qquad \forall (x_1, x_2) \in S = \mathbb{R}^+ \times \mathbb{R}^+.$$

Remark 2.1.5 * *Note that, it is not easy to find the candidate points by solving an inequality $\nabla f(x^*).(x - x^*) \geqslant 0$ (resp. $\leqslant 0$). However, the information gained is useful to establish other results.*

** *Solving the equation $\nabla f(x) = 0$ is not that easy either! It induces nonlinear equations or large linear systems when the number of variables is large. To overcome this difficulty, we resort to approximate methods. Newton's method is one of the well known approximate methods for approaching a root of the equation $F(x) = 0$. In Exercise 5, the method is described and applied for solving a nonlinear equation in one dimension. Steepest descent method, Conjugate gradient methods and many other methods are developed for approaching the solution [22], [5].*

∗ ∗ ∗ Finally, the following example, in dimension 2, shows the necessity of using new methods for finding the critical points. Indeed, the graph, Figure 2.10, of

$$z = f(x, y) = 10e^{-(x^2+y^2)} + 5e^{-[(x+5)^2+(y-3)^2]/10} + 4e^{-2[(x-4)^2+(y+1)^2]},$$

on the window $[-10, 8] \times [-10, 8] \times [-1, 12]$, shows three peaks. Thus, we have at least three local maxima points. These points are solution of the system

$$\begin{cases} f_x = -20xe^{-(x^2+y^2)} - 10(x+5)e^{-[(x+5)^2+(y-3)^2]/10} - 16(x-4)e^{-2[(x-4)^2+(y+1)^2]} = 0 \\ f_y = -20ye^{-(x^2+y^2)} - (y-3)e^{-[(x+5)^2+(y-3)^2]/10} - 16(y+1)e^{-2[(x-4)^2+(y+1)^2]} = 0, \end{cases}$$

a nonlinear system, for which it is not evident to find an explicit solution by algebraic manipulations. The following Maple software command searches for a solution near $(0,0)$ using an approximate method:

$$f := (x,y) -> 10 * exp(-x^2 - y^2) + 5 * exp(-((x+5)^2 + (y-3)^2) * (1/10))$$
$$+4 * exp(-2 * ((x-4)^2 + (y+1)^2))$$

$with(Optimization):$

$NLPSolve(f(x,y), x = -8..8, y = -8..8, initialpoint = x = 0, y = 0, maximize);$

The result is

$[10.1678223807097599, [x = -0.842598632890276e-2, y = 0.505559179745079e-2]].$

Thus, $(-0.084e^{-2}, 0.5e^{-2}) \approx (-0.115, 0.067)$ is an approximate critical point, where f takes the approximate local maximal value 10.1678. A search near $(-5,3)$ and $(4,-1)$ yields to the other approximate local maxima points:

$with(Optimization); NLPSolve(f(x,y), x = -8..8, y = -8..8,$
$initialpoint = x = -4, y = 3, maximize)$
$[5.00000000000001688, [x = -5.00000000010854, y = 2.99999999999990]]$
$with(Optimization): NLPSolve(f(x,y), x = -8..8, y = -8..8,$
$initialpoint = x = 4, y = -1, maximize);$
$[4.00030684298145278, [x = 3.99996531847993, y = -.999984626392930]$

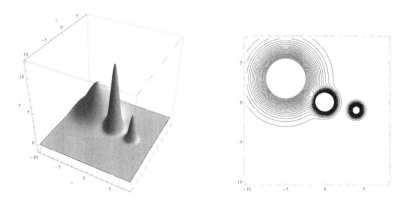

FIGURE 2.10: Location of mountains

Solved Problems

1. – *A suitable choice of the objective function.*

Find a point on the curve $y = x^2$ that is closest to the point $(3, 0)$.

Solution: *When formulating an optimization problem, sometimes, one can encounter some technical difficulties by considering an auxiliary objective function instead of considering the direct one. This situation is illustrated by the two choices below.*

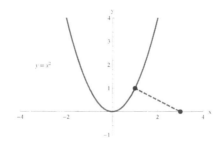

FIGURE 2.11: Closest point

- **1st choice.** Let

$$D = \text{ distance between (3,0) and any point } (x, y).$$

Since (x, y) lies on the curve $y = x^2$, the distance D must satisfy

$$D = D(x) = \sqrt{(x - 3)^2 + (y - 0)^2} = \sqrt{(x - 3)^2 + x^4}.$$

We need to solve the problem (see Figure 2.11)

$$\min_{x \in \mathbb{R}} D(x).$$

Since $\mathbb{R}$ is an open set, the minimum must occur at a critical point, i.e., since D is differentiable, at a point where

$$\frac{dD}{dx} = \frac{2(x-3) + 4x^3}{2\sqrt{(x-3)^2 + x^4}} = 0$$

$$\Longleftrightarrow \quad 2(x-3) + 4x^3 = 2(x-1)(2x^2 + 2x + 3) = 0 \quad \Longleftrightarrow \quad x = 1.$$

Since $D \in C^0(\mathbb{R})$ and

$$\lim_{x \to -\infty} D(x) = +\infty \qquad \text{and} \qquad \lim_{x \to +\infty} D(x) = +\infty,$$

the minimum exists and it must be at $x = 1$ [1]. The variations of D is given by Table 2.2.

x	$-\infty$		2		$+\infty$
$D'(x)$		$-$	0	$+$	
$D(x)$	$+\infty$	$\searrow$	$D(1)$	$\nearrow$	$+\infty$

TABLE 2.2: Variations of $D(x) = \sqrt{(x-3)^2 + x^4}$

Thus

$$\min_{x \in \mathbb{R}} D(x) = D(1) = \sqrt{5}.$$

• **2nd choice.** Note that, for any $x_0, x \in \mathbb{R}$, we have

$$0 \leqslant D^2(x_0) \leqslant D^2(x) \quad \Longleftrightarrow \quad 0 \leqslant D(x_0) = \sqrt{D^2(x_0)} \leqslant \sqrt{D^2(x)} = D(x)$$

since $\sqrt{t}$ is an increasing function on the interval $[0, +\infty)$. It suffices, then, to minimize on $\mathbb{R}$ the function

$$F(x) = D^2(x) = (x-3)^2 + x^4.$$

Since $\mathbb{R}$ is an open set, the minimum must occur at a critical point, i.e., since F is differentiable, at a point where

$$\frac{dF}{dx} = 2(x-3) + 4x^3 = 0 \quad \Longleftrightarrow \quad 2(x-1)(2x^2 + 2x + 3) = 0 \quad \Longleftrightarrow \quad x = 1.$$

Since $F \in C^0(\mathbb{R})$ and

$$\lim_{x \to -\infty} F(x) = +\infty \qquad \text{and} \qquad \lim_{x \to +\infty} F(x) = +\infty,$$

the minimum exists and it must be at $x = 1$. The variations of F is given by Table 2.3. The point $(1,1)$ is the closest point on the curve $[y = x^2]$ to the point $(3,0)$.

x	$-\infty$		1		$+\infty$
$F'(x) = 2(x-1)(2x^2+2x+3)$		$-$	0	$+$	
$F(x)$	$+\infty$	$\searrow$	$F(1)$	$\nearrow$	$+\infty$

TABLE 2.3: Variations of $F(x) = (x-3)^2 + x^4$

2. – To minimize the material in manufacturing a closed can with volume capacity of V *units*, we need to choose a suitable radius for the container. Find the radius if the container is cylindrical.

Solution: From Section 1.1, Example 1, we are lead to solve the minimization problem

$$\begin{cases} \text{minimize } A = A(r) = 2\pi r^2 + \dfrac{2V}{r} & \text{over the set } S \\[2mm] S = (0, +\infty) = \{r \in \mathbb{R} \ \ / \ \ r > 0\}. \end{cases}$$

Since S is an open set, the minimum must occur at a critical point, ie., since $A(r)$ is differentiable, at a point where

$$\frac{dA}{dr} = 4\pi r - \frac{2V}{r^2} = 0 \qquad \Longrightarrow \qquad r = \left(\frac{V}{2\pi}\right)^{1/3} \in S.$$

Since $A \in C^0(S)$ and

$$\lim_{r \to 0^+} A(r) = +\infty \qquad \text{and} \qquad \lim_{r \to +\infty} A(r) = +\infty,$$

the minimum exists and it must be on $r = \left(\dfrac{V}{2\pi}\right)^{1/3}$. Indeed the variations of A are as shown in Table 2.4.

r	0		$V^{1/3}/\sqrt[3]{2\pi}$		$+\infty$
$A'(r) = 4\pi r - \frac{2V}{r^2}$		$-$	0	$+$	
$A(r)$	$+\infty$	$\searrow$	$A((V/2\pi)^{1/3})$	$\nearrow$	$+\infty$

TABLE 2.4: Variations of $A(r) = 2\pi r^2 + \dfrac{2V}{r}$

So we should choose for the can a radius $r = (V/2\pi)^{1/3}$ and a height $h = V/(2\pi r) = (V/2\pi)^{2/3}$.

3. – Locate all absolute maxima and minima if any for each function.

$$i) \quad f(x,y) = 1 - (x+1)^2 - (y-5)^2$$

$$ii) \quad g(x,y) = 3x - 2y + 5$$

$$iii) \quad h(x,y) = x^2 - xy + y^2 - 3y.$$

Solution: i)

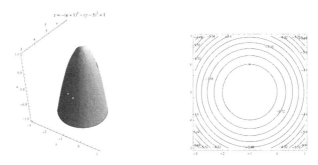

FIGURE 2.12: Graph and level curves of $z = f(x,y)$

Since f is differentiable on $\mathbb{R}^2$, its absolute extremum that are also local extremum (if they exist), are stationary points, ie. solution of

$$\nabla f = \langle -2(x+1), -2(y-5) \rangle = \langle 0,0 \rangle \qquad \Longleftrightarrow \qquad (x,y) = (-1,5).$$

So, there is only one critical point. It satisfies

$$f(-1,5) = 1 \geqslant 1 - (x+1)^2 - (y-5)^2 = f(x,y) \qquad \forall (x,y) \in \mathbb{R}^2.$$

Hence, it is a global maximum of f in $\mathbb{R}^2$; see Figure 2.12. However, f does not have a global minimum since the following hold:

$$(x+1)^2 + (y-5)^2 = \|(x,y) - (-1,5)\|^2$$

$$\|(x,y) - (-1,5)\|^2 \leqslant (\|(x,y)\| + \|(-1,5)\|)^2 = (\sqrt{x^2+y^2} + \sqrt{26})^2$$

$$\|(x,y) - (-1,5)\|^2 \geqslant \left| \|(x,y)\| - \|(-1,5)\| \right|^2 = (\sqrt{x^2+y^2} - \sqrt{26})^2.$$

Then

$$1 - (\sqrt{x^2 + y^2} + \sqrt{26})^2 \leqslant f(x,y) \leqslant 1 - (\sqrt{x^2 + y^2} - \sqrt{26})^2$$

and we deduce that

$$\lim_{\|(x,y)\| \to +\infty} f(x,y) = -\infty.$$

It suffices also to show that f takes large negative values on a subset of its domain $\mathbb{R}^2$, like

$$f(x,5) = 1 - (x+1)^2 \longrightarrow -\infty \qquad \text{as} \qquad x \longrightarrow \pm\infty.$$

ii) Since g is differentiable on $\mathbb{R}^2$, its absolute extreme points that are also

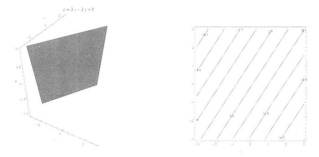

FIGURE 2.13: Graph and level curves of $g(x,y) = 3x - 2y + 5$

local extreme points (if they exist), are stationary points, ie. solution of $\nabla g = \langle 0,0 \rangle$. But

$$\nabla g = \langle 3, -2 \rangle \neq \langle 0,0 \rangle.$$

So, there is no critical point. g has no local or global extreme point. The graph $z = g(x,y)$ is a plane in $\mathbb{R}^3$ which spreads in the space taking large values when x or $y \longrightarrow \pm\infty$; see Figure 2.13. For example

$$g(0,y) = -2y + 5 \longrightarrow \mp\infty \text{ as } \qquad y \longrightarrow \pm\infty$$
$$g(x,0) = 3x + 5 \longrightarrow \pm\infty \qquad \text{as} \qquad x \longrightarrow \pm\infty.$$

iii) Since h is differentiable on $\mathbb{R}^2$, its absolute extreme points that are also local extreme points (if they exist) are stationary points, ie. solution of

$$\nabla h = \langle 2x - y, -x + 2y - 3 \rangle = \langle 0,0 \rangle \qquad \Longleftrightarrow \qquad (x,y) = (1,2).$$

So, there is only one critical point. It satisfies

$$h(1,2) = 1 - 2 + 4 - 6 = -3$$

$$h(x,y) - h(1,2) = x^2 - xy + y^2 - 3y + 3 = (x - \frac{y}{2})^2 - \frac{y^2}{4} + y^2 - 3y + 3$$

$$= (x - \frac{y}{2})^2 + \frac{3}{4}(y-1)^2 \geqslant 0 \qquad \forall(x,y) \in \mathbb{R}^2.$$

Hence, the point $(1,2)$ is a global minimum of h in $\mathbb{R}^2$. Here also, one can see that h takes large values, for example, along the x axis, we have

$$h(x,0) = x^2 \longrightarrow +\infty \qquad \text{when } x \longrightarrow \pm\infty.$$

So h has no global maximum (see Figure 2.14).

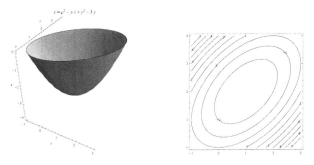

FIGURE 2.14: Graph and level curves of $h(x,y) = x^2 - xy + y^2 - 3y$

4. – Consider the problem

$$\min_{S} f(x,y) = y \qquad \text{where} \qquad S = \{(x,y): x^2 + y^2 \leqslant 1\}.$$

i) Does f have local minimum points?

ii) Where may the minimum points locate if they exist?

iii) Solve the inequality

$$\nabla f(a,b).(x - a, y - b) \geqslant 0 \qquad \forall(x,y) \in S$$

to find the candidate points (a,b) and solve the problem.

iv) Can you proceed as in iii) if $S = \{(x,y): x^2 + y^2 \geqslant 1\}$? What is the solution in this case?

Solution: i) Since f is differentiable on $\mathbb{R}^2$, a local minimum point would be a critical point, ie. solution of $\nabla f = \langle 0, 0 \rangle$. But

$$\nabla f = \langle 0, 1 \rangle \neq \langle 0, 0 \rangle \qquad \forall (x, y) \in \{(x, y) : \ x^2 + y^2 < 1\} = \overset{\circ}{S}.$$

So, there is no critical point. f has no local minimum point.

ii) If the minimum points exist, they may be on the unit circle, the boundary of S:

$$\partial S = \{(x, y) : \ x^2 + y^2 = 1\}.$$

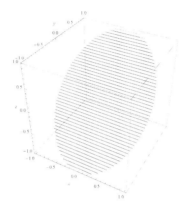

FIGURE 2.15: Graph of $f(x, y) = y$ on the set $x^2 + y^2 \leqslant 1$

iii) Since S is convex and f differentiable on $\mathbb{R}^2$, a solution (a, b) of the problem, if it exists, must satisfy

$$\begin{cases} a^2 + b^2 = 1 \\ \nabla f(a, b).(x - a, y - b) \geqslant 0 \qquad \forall (x, y) \in S \end{cases}$$

$$\Longleftrightarrow \quad \begin{cases} a^2 + b^2 = 1 \\ \langle 0, 1 \rangle.\langle x - a, y - b \rangle \geqslant 0 \qquad \forall (x, y) \in S \end{cases}$$

$$\Longleftrightarrow \quad \begin{cases} a^2 + b^2 = 1 \\ y - b \geqslant 0 \qquad \forall (x, y) \in S \end{cases} \quad \Longleftrightarrow \quad y \geqslant b \qquad \forall (x, y) \in S$$

Thus

$$b = -1 \qquad \text{and} \qquad a = 0.$$

So the only point candidate is $(a, b) = (0, 1)$. In fact, it is the minimum point (see Figure 2.15) since we have

$$f(x, y) = y \geqslant -1 = f(0, -1) \qquad \forall (x, y) \in S.$$

iv) We cannot proceed as in *iii)* because the set $S = \{(x, y) : x^2 + y^2 \geqslant 1\}$ is not convex. And because this set is not bounded, we can see that f can take large negative values. Therefore, it doesn't attain a minimum value. For example, on the negative y axis, we have

$$f(0, y) = y \quad \longrightarrow -\infty \qquad \text{as} \qquad y \longrightarrow -\infty.$$

5. – Newton's Method[2]   Let $I = [a, b]$ and let $F : I \longmapsto \mathbb{R}$ be twice differentiable on I. Suppose that

$$\exists m, M \in \mathbb{R}^+ : \quad |F'(x)| \geqslant m > 0 \quad \text{and} \quad |F''(x)| \leqslant M \quad \forall x \in I$$
$$F(a).F(b) < 0, \qquad\qquad K = M/2m.$$

Then there exists a subinterval I^* containing a root r of F such that for any $x_1 \in I^*$, the sequence x_n defined by

$$x_{n+1} = x_n - \frac{F(x_n)}{F'(x_n)} \qquad \forall n \in \mathbb{N},$$

belongs to I^* and (x_n) converges to r. Moreover

$$|x_{n+1} - r| \leqslant K|x_n - r|^2 \qquad \forall n \in \mathbb{N}.$$

Application   Let $f(x) = x^3 - 2x - 5$.

i) Show that f has a root on the interval $I = [2, 2.2]$.

ii) If $x_1 = 2$ and if (x_n) is the sequence obtained by Newton's method, show that
$$|x_{n+1} - r| \leqslant (0.7)|x_n - r|^2$$

iii) Show that x_4 is exact up to 6 decimals.

Solution: i) We have

$$f(2.2) = (2.2)^3 - 2.(2.2) - 5 = 1.248 > 0 \qquad f(2) = 8 - 4 - 5 = -1 < 0$$

f is continuous on $[2, 2.2]$ and 0 is between $f(2)$ and $f(2.2)$.

From the intermediate value theorem, there exists $x_0 \in (2, 2.2)$ such that $f(x_0) = 0$.

ii) The sequence (x_n) obtained by Newton's method is:

$$\begin{cases} x_1 = 2 \\ x_{n+1} = x_n - \dfrac{f(x_n)}{f'(x_n)} = x_n - \dfrac{x_n^3 - 2x_n - 5}{3x_n^2 - 2} = \dfrac{2x_n^3 + 5}{3x_n^2 - 2} \end{cases}$$

with

$$f'(x) = 3x^2 - 2 \qquad\qquad f''(x) = 6x.$$

Because, the functions f' and f'' are increasing on $[2, 2.2]$, we have

$$10 = f'(2) \leqslant f'(x) \leqslant f'(2.2) = 12.52$$
$$12 = f''(2) \leqslant f''(x) \leqslant f''(2.2) = 13.2.$$

In particular

$$|f'(x)| \geqslant 10 = m \qquad \text{and} \qquad |f''(x)| \leqslant 13.2 = M \qquad \forall x \in [2, 2.2].$$

We deduce that the sequence (x_n) converges to a root r of $f(x) = 0$ in $[2, 2.2]$ and satisfies

$$|x_{n+1} - r| \leqslant 0.7|x_n - r|^2 \qquad\qquad K = \frac{M}{2m} = 0.66 < 0.7.$$

iii) Denote by $e_n = x_n - r$, the approximation error of the root r, then

$$|Ke_{n+1}| \leqslant K^2 |e_n|^2 = |Ke_n|^2 \qquad \Longrightarrow \qquad |Ke_{n+1}| \leqslant |Ke_1|^{2n} \qquad \text{by induction,}$$

where

$$|e_1| = |x_1 - r| < (2.2 - 2) = 0.2 \qquad \text{since} \qquad x_1, r \in [2, 2.2].$$

Thus

$$|Ke_{n+1}| \leqslant |Ke_1|^{2n} \leqslant \left((0.7)(0.2) \right)^{2n} = (0.0196)^n$$

To obtain an accuracy up to 6 decimals, it suffices to choose n such that

$$|e_{n+1}| \leqslant \frac{(0.0196)^n}{0.66} \leqslant 10^{-6}.$$

We have

$$\frac{(0.0196)^2}{0.66} = 0.000582061 \qquad \frac{(0.0196)^3}{0.66} \approx 0.0000114084$$

$$\frac{(0.0196)^4}{0.66} \approx 0.0000002236 < 10^{-6}.$$

The desired accuracy is obtained for $n = 4$.

The approximate values of this root are:

$$x_2 = \frac{2x_1^3 + 5}{3x_1^2 - 2} = \frac{2(8) + 5}{3(2^2) - 2} = \frac{21}{10} = 2.1$$

$$x_3 = \frac{2x_2^3 + 5}{3x_2^2 - 2} = \frac{2(2.1)^3 + 5}{3(2.1)^2 - 2} = \frac{23.522}{11.23} = 2.0945681$$

$$x_4 = \frac{2x_3^3 + 5}{3x_3^2 - 2} = \frac{2(\frac{23.522}{11.23})^3 + 5}{3(\frac{23.522}{11.23})^2 - 2} \approx 2.09455148$$

$$x_5 = \frac{2x_4^3 + 5}{3x_4^2 - 2} \approx \frac{23.3782059}{11.1614377} \approx 2.0945514841.$$

We can see that x_4 is exact up to six decimals; see Figure 2.16.

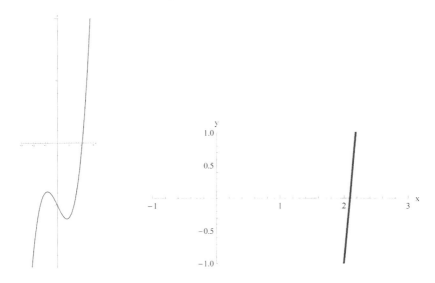

FIGURE 2.16: Approximate position of the root of $f(x) = x^3 - 2x - 5$

2.2   Classification of Local Extreme Points

For a C^2 function f of one variable, in a neighborhood of a critical point x^*, one can write by using the second order Taylor's formula:

$$f(x) = f(x^*) + \frac{f'(x^*)}{1!}(x - x^*) + (x - x^*)^2 \frac{f''(c)}{2!}$$

for some number c between x^* and x. Then, since $f'(x^*) = 0$, we have

$$f(x) = f(x^*) + (x - x^*)^2 \frac{f''(c)}{2!}.$$

Now, if we have $f''(x^*) > 0$, then by continuity of f''. we deduce that for x close to x^*, ($x \in (x^* - \epsilon, x^* + \epsilon)$), we will have

$$f''(c) > 0 \quad \Longrightarrow \quad f(x) > f(x^*) \qquad \forall x \in (x^* - \epsilon, x^* + \epsilon) \setminus \{x^*\}.$$

This means that x^* is a strict local minimum point. Similarly, we show that

$$f''(x^*) < 0 \quad \Longrightarrow \quad x^* \quad \text{is a strict local maximum point.}$$

This classification of critical points, into minima and maxima points, where the sign of the second derivative intervenes, is generalized to C^2 functions with several variables in the theorem below, following the definition:

Definition 2.2.1 *Let $H_f(x) = (f_{x_i x_j}(x))_{n \times n}$ be the Hessian of a C^2 function f. Then, the n leading minors of H_f are defined by*

$$D_k(x) = \begin{vmatrix} f_{x_1 x_1} & f_{x_1 x_2} & \cdots & f_{x_1 x_k} \\ f_{x_2 x_1} & f_{x_2 x_2} & \cdots & f_{x_2 x_k} \\ \vdots & \vdots & \vdots & \vdots \\ f_{x_k x_1} & f_{x_k x_2} & \cdots & f_{x_k x_k} \end{vmatrix}, \qquad k = 1, \ldots, n.$$

Theorem 2.2.1 *Second derivatives test - Sufficient conditions for a strict local extreme point*

Let $S \subset \mathbb{R}^n$ and $f : S \longmapsto \mathbb{R}$ be a C^2 function in a neighborhood of a critical point $x^* \in S$ $(\nabla f(x^*) = 0)$. Then

$$(i) \quad D_k(x^*) > 0, \qquad \forall k = 1, \ldots, n$$
$$\Longrightarrow \quad x^* \text{ is a strict local minimum point,}$$

(ii) $(-1)^k D_k(x^*) > 0, \qquad \forall k = 1, \ldots, n$

$\implies \qquad x^*$ is a strict local maximum point,

(iii) $D_n(x^*) \neq 0$ and neither of the conditions in (i) and (ii) are satisfied, then x^* is a saddle point.

Before proving the theorem, we will see its application through some examples.

Example 1. Profit in selling one commodity

A commodity is sold at 5\$ per unit. The total cost for producing x units is given by
$$C(x) = x^3 - 10x^2 + 17x + 66.$$
Find the most profitable level of production.

Solution: The total revenue for selling x units is $R(x) = 5x$. Thus, the profit $P(x)$ on x units is

$$P(x) = R(x) - C(x) = 5x - (x^3 - 10x^2 + 17x + 66) = -x^3 + 10x^2 - 12x - 66.$$

The profit, illustrated in Figure 2.17, will be at its maximum at points where
$$\frac{dP}{dx} = -3x^2 + 20x - 12 = -3(x-6)(x - \frac{2}{3}) = 0.$$
We deduce that we have two critical points $x = 6$ and $x = \frac{2}{3}$.

The Hessian of P is

$$H_P(x) = \left[\frac{d^2 P}{dx^2}\right] = [-6x + 20].$$
Applying the second derivatives test, we obtain

$*$ at $x = 6$,
$$(-1)^1 D_1(6) = (-1)\left(\frac{d^2 P}{dx^2}\right)(6) = (-1)(-6(6) + 20) = 16 > 0$$
Thus, $x = 6$ is a local maximum.

$**$ at $x = 2/3$,
$$D_1(2/3) = \left(\frac{d^2 P}{dx^2}\right)(\frac{2}{3}) = -6(\frac{2}{3}) + 20 = 16 > 0$$
Thus, $x = \frac{2}{3}$ is a local minimum.

Thus six units is a candidate point for optimality. We have to check that it is the point at which we have the most profitable profit. This can be done by comparing $P(x)$ and $P(6)$. Indeed, we have

$$P(x) - P(6) = -(x-6)^2(x+2) \leqslant 0 \qquad \forall x > 0, \quad x \neq 6$$

$$\implies \qquad P(x) < P(6) \qquad \forall x \in (0, +\infty) \setminus \{6\}.$$

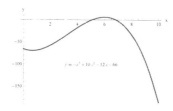

FIGURE 2.17: Graph of P and the maximum profit at $x = 6$

Example 2. Profit in selling two commodities

The cost to produce x units of a commodity A and y units of a commodity B is
$$C(x, y) = 0.2x^2 + 0.05xy + 0.05y^2 + 20x + 10y + 2500.$$

If each unit from A and B are sold for 75 and 45 respectively, find the daily production levels x and y that maximize the profit per day.

Solution: The daily profit is given by

$$P(x, y) = 75x + 45y - C(x, y) = -0.2x^2 - 0.05xy - 0.05y^2 + 55x + 35y - 2500.$$

Since P is differentiable (because it is a polynomial), the points that maximize the profit are critical ones, i.e, solutions of

$$\nabla P(x, y) = \langle -0.4x - 0.05y + 55, -0.05x - 0.1y + 35 \rangle = \langle 0, 0 \rangle \iff \begin{cases} x = 100 \\ y = 300. \end{cases}$$

We deduce that $(100, 300)$ is the only critical point of P; see Figure 2.18. Now, we apply the second derivatives test to classify that point. We have

$$H_P(x, y) = \begin{bmatrix} P_{xx} & P_{xy} \\ P_{yx} & P_{yy} \end{bmatrix} = \begin{bmatrix} -0.4 & 0 \\ 0 & -0.1 \end{bmatrix}$$

$$D_1(100, 300) = \begin{vmatrix} P_{xx} \end{vmatrix} = P_{xx} = -0.4 < 0,$$

$$D_2(100, 300) = \begin{vmatrix} P_{xx} & P_{xy} \\ P_{xy} & P_{yy} \end{vmatrix} = \begin{vmatrix} -0.4 & 0 \\ 0 & -0.1 \end{vmatrix} = 0.004 > 0.$$

So $(100, 300)$ is a local maximum point. In fact, it is a global maximum point where P attains the optimal value $P(100, 300) = 5500$. This is true because P is concave in $\mathbb{R}^2$. Indeed, we have

$$D_1(x, y) < 0 \quad \text{and} \quad D_2(x, y) > 0 \quad \forall (x, y) \in \mathbb{R}^2 \quad \text{(see next section)}.$$

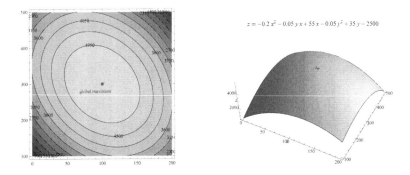

FIGURE 2.18: Profit function $P(x, y)$ and maximum point $(100, 300)$

Example 3. Several local extreme points

Find the stationary points and classify them when

$$f(x, y) = 3x - x^3 - 2y^2 + y^4.$$

Solution: Since f is a differentiable function (because it is a polynomial), the local extreme points are critical points, i.e, solutions of $\nabla f(x, y) = \langle 0, 0 \rangle$. We have

$$\nabla f(x, y) = \langle 3 - 3x^2, -4y + 4y^3 \rangle = \langle 0, 0 \rangle$$

$$\Longleftrightarrow \begin{cases} x^2 = 1 \\ \text{and} \\ y(y^2 - 1) = 0 \end{cases} \Longleftrightarrow \begin{cases} x = 1 \quad \text{or} \quad x = -1 \\ \text{and} \\ y = 0 \quad \text{or } y = 1 \text{ or } y = -1 \end{cases}$$

$$\Longleftrightarrow \begin{cases} x = 1 \\ \text{and} \\ y = 0 \end{cases} \quad \text{or} \quad \begin{cases} x = 1 \\ \text{and} \\ y = 1 \end{cases} \quad \text{or} \quad \begin{cases} x = 1 \\ \text{and} \\ y = -1 \end{cases}$$

$$\text{or} \quad \begin{cases} x = -1 \\ \text{and} \\ y = 0 \end{cases} \quad \text{or} \quad \begin{cases} x = -1 \\ \text{and} \\ y = 1 \end{cases} \quad \text{or} \quad \begin{cases} x = -1 \\ \text{and} \\ y = -1. \end{cases}$$

We deduce that $(1, 0)$, $(1, 1)$, $(1, -1)$, $(-1, 0)$, $(-1, 1)$ and $(-1, -1)$ are the critical points of f. The level curves, graphed in Figure 2.19, show the nature of these points.

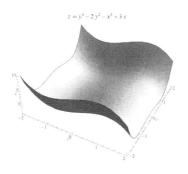
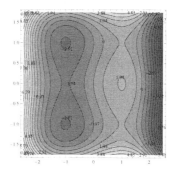

FIGURE 2.19: Local extreme points of $f(x, y) = 3x - x^3 - 2y^2 + y^4$

* *Classification of the critical points:* We have

$$f_{xx}(x, y) = -6x, \qquad f_{yy}(x, y) = 12y^2 - 4, \qquad f_{xy}(x, y) = 0,$$

$$H_f(x, y) = \begin{bmatrix} f_{xx} & f_{xy} \\ f_{yx} & f_{yy} \end{bmatrix} = \begin{bmatrix} -6x & 0 \\ 0 & 12y^2 - 4 \end{bmatrix}$$

$$D_1(x, y) = \left| \; f_{xx} \; \right| = f_{xx} = -6x,$$

$$D_2(x, y) = \begin{vmatrix} f_{xx} & f_{xy} \\ f_{xy} & f_{yy} \end{vmatrix} = \begin{vmatrix} -6x & 0 \\ 0 & 12y^2 - 4 \end{vmatrix} = -24x[3y^2 - 1].$$

Applying the second derivative test, we obtain:

(x, y)	$D_1(x, y)$	$D_2(x, y)$	type
$(1, 0)$	-6	$\begin{vmatrix} -6 & 0 \\ 0 & -4 \end{vmatrix} = 24$	local maximum
$(\;1, 1)$	6	$\begin{vmatrix} 6 & 0 \\ 0 & 8 \end{vmatrix} = 48$	local minimum
$(-1, -1)$	6	$\begin{vmatrix} 6 & 0 \\ 0 & 8 \end{vmatrix} = 48$	local minimum
$(1, 1)$	-6	$\begin{vmatrix} -6 & 0 \\ 0 & 8 \end{vmatrix} = -48$	saddle point
$(1, -1)$	-6	$\begin{vmatrix} -6 & 0 \\ 0 & 8 \end{vmatrix} = -48$	saddle point
$(-1, 0)$	6	$\begin{vmatrix} 6 & 0 \\ 0 & -4 \end{vmatrix} = -24$	saddle point

TABLE 2.5: Critical points' classification for $f(x, y) = 3x - x^3 - 2y^2 + y^4$

The proof of Theorem 2.2.1 uses Taylor's formula for a function of several variables and a characterization of symmetric quadratic forms (see the end of this section). Taylor's formula will be used several times through out the next chapters. It is therefore important to understand its proof.

Theorem 2.2.2 2nd order Taylor's formula for a function of n variables

Suppose f is C^2 in an open set of $\mathbb{R}^n$ containing the line segment $[x^*, x^* + h]$. Then

$$f(x^* + h) = f(x^*) + \sum_{i=1}^{n} \frac{\partial f}{\partial x_i}(x^*)h_i + \frac{1}{2}\sum_{i=1}^{n}\sum_{j=1}^{n} \frac{\partial^2 f}{\partial x_i \partial x_j}(x^* + c\,h)h_i h_j$$

or

$$f(x^* + h) = f(x^*) + \nabla f(x^*).h + \frac{1}{2}{}^t h H_f(x^* + ch)h$$

for some $c \in (0,1)$ and where $x^* = \begin{bmatrix} x_1^* \\ \vdots \\ x_n^* \end{bmatrix}$, $h = \begin{bmatrix} h_1 \\ \vdots \\ h_n \end{bmatrix}$, ${}^t h = \begin{bmatrix} h_1 & \cdots & h_n \end{bmatrix}$. Here, we identified the column vector $x^* + th$ with the point $(x_1^* + th_1, \ldots, x_n^* + th_n)$, $t \in \mathbb{R}$.

Proof. Define the function

$$g(t) = f(x_1^* + th_1, \ldots, x_n^* + th_n) = f(x^* + th).$$

Note that

$$g(t) = f(x_1(t), x_2(t), \ldots, x_n(t)) \qquad \text{with} \qquad x_j(t) = x_j^* + th_j \quad j = 1, \ldots, n.$$

Since the real functions x_j, $j = 1, \ldots, n$, are differentiable with $x_j'(t) = h_j$, then g is differentiable and we have by the chain rule formula

$$g'(t) = \frac{\partial f}{\partial x_1}\frac{\partial x_1}{\partial t} + \frac{\partial f}{\partial x_2}\frac{\partial x_2}{\partial t} + \cdots\cdots + \frac{\partial f}{\partial x_n}\frac{\partial x_n}{\partial t}$$

$$- \left[f_{x_1}(x^* + th)\right]h_1 + \left[f_{x_2}(x^* + th)\right]h_2 + \cdots\cdots + \left[f_{x_n}(x^* + th)\right]h_n$$

$$= \left[\nabla f(x^* + th)\right].h.$$

Because f is C^2, then g is also C^2, and we have

$$g''(t) = \left[\frac{d}{dt}f_{x_1}(x^*+th)\right]h_1 + \left[\frac{d}{dt}f_{x_2}(x^*+th)\right]h_2 + \ldots\ldots + \left[\frac{d}{dt}f_{x_n}(x^*+th)\right]h_n.$$

For each $i = 1, \ldots, n$, we have

$$f_{x_i}(x^* + th) = f_{x_i}(x_1(t), x_2(t), \ldots, x_n(t)).$$

Then

$$\frac{d}{dt}f_{x_i}(x^* + th) = \frac{\partial f_{x_i}}{\partial x_1}\frac{\partial x_1}{\partial t} + \frac{\partial f_{x_i}}{\partial x_2}\frac{\partial x_2}{\partial t} + \ldots\ldots + \frac{\partial f_{x_i}}{\partial x_n}\frac{\partial x_n}{\partial t}$$

$$= \left[f_{x_i x_1}(x^* + th)\right]h_1 + \left[f_{x_i x_2}(x^* + th)\right]h_2 + \ldots\ldots + \left[f_{x_i x_n}(x^* + th)\right]h_n$$

$$= \sum_{j=1}^{n}\left[f_{x_i x_j}(x^* + th)\right]h_j.$$

Hence

$$g''(t) = \sum_{i=1}^{n}\sum_{j=1}^{n}\left[f_{x_i x_j}(x^* + th)\right]h_i h_j.$$

Now, since f is defined on the segment $[x^*, x^* + h]$, g is defined on the interval $[0, 1]$ and by using the 2nd order Taylor's formula for real functions [1], [2], we get

$$g(1) = g(0) + \frac{g'(0)}{1!}(1 - 0) + \frac{g''(c)}{2!}(1 - 0)^2$$

$$= g(0) + g'(0) + \frac{1}{2}g''(c) \qquad \text{for some } c \in (0, 1),$$

or equivalently

$$f(x^* + h) = f(x^*) + \sum_{i=1}^{n}f_{x_i}(x^*)h_i + \frac{1}{2}\sum_{i=1}^{n}\sum_{j=1}^{n}f_{x_i x_j}(x^* + ch)h_i h_j.$$

Proof. (Theorem 2.2.1) Since x^* is an interior point of S and is a local stationary point of f then

$$\nabla f(x^*) = 0.$$

For $h \in \mathbb{R}^n$ such that $x^* + h \in S$, we have from the 2nd order Taylor's formula

$$f(x^* + h) = f(x^*) + \frac{1}{2}\, {}^t h H_f(x^* + ch)h \qquad \text{for some } c \in (0, 1).$$

Situation (i)   Suppose that $D_k(x^*) > 0$ for all $k = 1, \ldots, n$.

By continuity of the second-order partial derivatives of f, there exists $r > 0$ such that

$$D_k(x) > 0 \qquad \forall x \in B_r(x^*) \qquad \forall k = 1, \ldots, n.$$

As a consequence, the quadratic form

$$Q(h)(x) = \sum_{i=1}^{n} \sum_{j=1}^{n} f_{x_i x_j}(x)h_i h_j = {}^t h H_f(x)h$$

with the associated symmetric matrix $H_f(x) = \left[f_{x_i x_j}(x)\right]_{n \times n}$ is definitely positive.

Since $x^* + ch \in B_r(x^*)$, then

$$Q(h)(x^* + ch) = {}^t h H_f(x^* + ch)h > 0.$$

Therefore, we have for $x^* + h \in B_r(x^*)$

$$f(x^* + h) - f(x^*) = \frac{1}{2}Q(h)(x^* + ch) > 0$$

which shows that the stationary point x^* is a strict local minimum point for f in S.

Situation (ii)   Suppose that $(-1)^k D_k(x^*) > 0$ for all $k = 1, \ldots, n$.

By continuity of the second-order partial derivatives of f, there exists $r > 0$ such that

$$(-1)^k D_k(x) > 0 \qquad \forall x \in B_r(x^*) \qquad \forall k = 1, \ldots, n.$$

From the property of determinants, we can write

$$(-1)^k D_k(x^*) = \begin{vmatrix} (-f)_{x_1 x_1} & (-f)_{x_1 x_2} & \cdots & (-f)_{x_1 x_k} \\ (-f)_{x_2 x_1} & (-f)_{x_2 x_2} & \cdots & (-f)_{x_2 x_k} \\ \vdots & \vdots & \vdots & \vdots \\ (-f)_{x_k x_1} & (-f)_{x_k x_2} & \cdots & (-f)_{x_k x_k} \end{vmatrix}$$

As a consequence, the quadratic form

$$^t h H_{-f}(x) h = \sum_{i=1}^{n} \sum_{j=1}^{n} (-f)_{x_i x_j}(x) h_i h_j$$

with the associated symmetric matrix $H_{-f}(x) = \left[(-f)_{x_i x_j}(x) \right]_{n \times n}$ is definite positive.

Therefore, we have for $x^* + h \in B(x^*, r)$

$$(-f)(x^* + h) - (-f)(x^*) = (\frac{1}{2})^t h H_{-f}(x^* + ch) h > 0$$

$$\implies \quad (-f)(x^* + h) > (-f)(x^*) \quad \Longleftrightarrow \quad f(x^*) > f(x^* + h)$$

which shows that the stationary point x^* is a strict local maximum point for f in S.

Situation (iii) Assume $D_n(x^*) \neq 0$ and neither of the conditions $i)$ and $ii)$ hold.

Note that situation (i) (resp. (ii)) means also that the matrix $A = \left[f_{x_i x_j}(x^*) \right]_{n \times n}$ is definite positive (resp. negative), which is equivalent to each of its eigen value λ_i to be positive (resp. negative). So, if neither (i) or (ii) hold, there exist $i_0, j_0 \in \{1, \ldots, n\}$ such that

$$D_n(x^*) = \prod_{i=1}^{n} \lambda_i \neq 0 \qquad \text{with} \qquad \lambda_{i_0} > 0 \qquad \text{and} \qquad \lambda_{j_0} < 0.$$

Now, since A is symmetric, there exists an orthogonal matrix $O = (p_{ij})_{n \times n}$ $(O^{-1} =^t O)$ such that

$$A = O D^t O \qquad\qquad D = \begin{bmatrix} \lambda_1 & \cdots & 0 \\ 0 & \ddots & 0 \\ 0 & \cdots & \lambda_n \end{bmatrix}$$

Then the quadratic form $Q(h)$ can be written as

$$Q(h)(x^*) =^t h A h =^t [^t O h] D [^t O h] = \sum_{i=1}^{n} \lambda_i \left(\sum_{j=1}^{n} p_{ji} h_j \right)^2.$$

Choose h_s and h_s' such that for $s > 0$,

$$^t O h_s = \frac{s}{\sqrt{\lambda_{i_0}}} e_{i_0} + \frac{2s}{\sqrt{-\lambda_{j_0}}} e_{j_0} \qquad\qquad ^t O h_s' = \frac{2s}{\sqrt{\lambda_{i_0}}} e_{i_0} + \frac{s}{\sqrt{-\lambda_{j_0}}} e_{j_0},$$

which is possible since ^tO is invertible. Then we have

$$Q(h_s)(x^*) = \lambda_{i_0}\left(\frac{s}{\sqrt{\lambda_{i_0}}}\right)^2 + \lambda_{j_0}\left(\frac{2s}{\sqrt{-\lambda_{j_0}}}\right)^2 = s^2 - 4s^2 = -3s^2 < 0$$

$$Q(h'_s)(x^*) = \lambda_{i_0}\left(\frac{2s}{\sqrt{\lambda_{i_0}}}\right)^2 + \lambda_{j_0}\left(\frac{s}{\sqrt{-\lambda_{j_0}}}\right)^2 = 4s^2 - s^2 = 3s^2 > 0.$$

We deduce, by continuity of $Q(h)(x)$ the existence of $\delta > 0$ such that $\forall s \in (0, \delta)$

$$f(x^* + h_s) - f(x^*) = \frac{1}{2}Q(h_s)(x^* + ch_s) < 0$$

$$f(x^* + h'_s) - f(x^*) = \frac{1}{2}Q(h'_s)(x^* + ch'_s) > 0.$$

f takes values greater and less than $f(x^*)$ in the neighborhood of x^*. Therefore x^* is a saddle point.

The following theorem shows that the Hessian matrix of a C^2 function at a local maximum (resp. minimum) point is necessarily positive (resp. negative) semi definite. However, this condition is not sufficient as we can see it in a suggested exercise where the origin is neither a local minimum, nor a local maximum.

Theorem 2.2.3 *Necessary conditions for a local extreme point*

Let $S \subset \mathbb{R}^n$ and $f : S \longmapsto \mathbb{R}$ be a C^2 function in a neighborhood of a critical point $x^* \in \overset{\circ}{S}$ $(\nabla f(x^*) = 0)$. Then

(i) x^* is a local minimum point $\implies$ $\triangle_k(x^*) \geqslant 0$ $\forall k = 1, n$

(ii) x^* is a local maximum point $\implies$ $(-1)^k \triangle_k(x^*) \geqslant 0$ $\forall k = 1, n$

where $\triangle_k(x^*)$ is the principal minor of order k of the Hessian matrix $H_f(x^*)$; that is the determinant of a matrix obtained by deleting $n - k$ rows and $n - k$ columns such that if the ith row (column) is selected, then so is the ith column (row).

Proof. **(i)** Suppose that x^* is an interior local minimum point for f. There exists $r > 0$ such that

$$f(x^*) \leqslant f(x) \qquad \forall x \in B_r(x^*).$$

In particular, for $t \in (-r, r)$ and $h \in \mathbb{R}^n$ with $\|h\| = 1$, we have $x^* + th \in B_r(x^*)$ since

$$\|x^* + th - x^*\| = |t|\|h\| = |t| < r.$$

Then

$$g(0) = f(x^*) \leqslant f(x^* + th) = g(t) \qquad \forall t \in (-r, r).$$

So g is a one variable function that has an interior local minimum at $t = 0$. Consequently, it satisfies

$$g'(0) = 0 \qquad \text{and} \qquad g''(0) \geqslant 0.$$

From previous calculations, we have

$$g''(t) = \sum_{i=1}^{n} \sum_{j=1}^{n} f_{x_i x_j}(x^* + th) h_i h_j.$$

Hence

$$g''(0) = \sum_{i=1}^{n} \sum_{j=1}^{n} f_{x_i x_j}(x^*) h_i h_j =^t h H_f(x^*) h \geqslant 0.$$

The above inequality remains true for $h = 0$ and for $h \neq 0$. It suffices to consider for this last case $h/\|h\|$ which is a unit vector. Hence the Hessian matrix of f at x^* is positive semi definite by the result below from Algebra (see [10]).

(ii) is proved similarly.

Quadratic forms

Consider the quadratic form in n variables

$$Q(h) = \sum_{i=1}^{n} \sum_{j=1}^{n} a_{ij} h_i h_j =^t h A h \qquad {}^t h = \begin{bmatrix} h_1 & \cdots & h_n \end{bmatrix}$$

associated to the symmetric matrix

$$A = (a_{ij})_{i,j=1,\dots,n} = \begin{bmatrix} a_{11} & \cdots & a_{1n} \\ \vdots & \ddots & \vdots \\ a_{n1} & \cdots & a_{nn} \end{bmatrix} \qquad (a_{ij} = a_{ji}).$$

Definition.

Q is positive (resp. negative) definite if $Q(h) > 0$ (resp. < 0) for all $h \neq 0$.

Q is positive (resp. negative) semi definite if $Q(h) \geqslant 0$ (resp. $\leqslant 0$) for all $h \in \mathbb{R}^n$.

We have the following necessary and sufficient condition for a quadratic form Q to be positive (negative), definite or semi definite.

Theorem.

$$Q \text{ is positive definite} \iff D_r > 0 \quad r = 1, \ldots, n$$
$$Q \text{ is negative definite} \iff (-1)^r D_r > 0 \quad r = 1, \ldots, n$$

where D_r is the leading principal minor of order r of the matrix A;

$$D_r = \begin{vmatrix} a_{11} & \cdots & a_{1n} \\ \vdots & \ddots & \vdots \\ a_{n1} & \cdots & a_{nn} \end{vmatrix} \quad \text{for} \quad r = 1, \ldots, n.$$

Theorem.

$$Q \text{ is positive semi definite} \iff \Delta_r \geqslant 0 \quad r = 1, \ldots, n$$
$$Q \text{ is negative semi definite} \iff (-1)^r \Delta_r \geqslant 0 \quad r = 1, \ldots, n$$

where Δ_r is the principal minor of order r of the matrix A; that is the determinant of the matrix obtained from the matrix A by deleting $n - r$ rows and $n - r$ columns such that if the ith row (column) is selected, then so is the ith column (row).

Solved Problems

1. – Use the following functions to show that the positivity or negativity semi definite of the Hessian of the objective function at a critical point is not a necessary condition for local optimality.

$$f(x,y) = x^4 + y^4, \qquad g(x,y) = -(x^4 + y^4), \qquad h(x,y) = x^4 - y^4.$$

Solution:

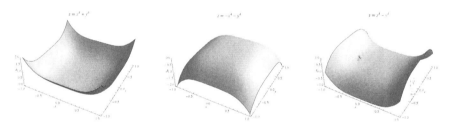

FIGURE 2.20: Graphs of $f,\ g,\ h$

We have

$$\nabla f(x,y) = \langle 4x^3, 4y^3 \rangle, \quad \nabla g(x,y) = \langle -4x^3, -4y^3 \rangle, \quad \nabla h(x,y) = \langle 4x^3, -4y^3 \rangle.$$

So $(0,0)$ is the only stationary point for f, g and h. But, we cannot conclude anything about its nature by using the second derivatives test since the Hessian matrix at $(0,0)$ of each function is equal to the zero matrix.

$$H_f = \begin{bmatrix} 12x^2 & 0 \\ 0 & 12y^2 \end{bmatrix}, \qquad H_g = -H_f, \qquad H_h = \begin{bmatrix} 12x^2 & 0 \\ 0 & -12y^2 \end{bmatrix}$$

$$H_f(0,0) = H_g(0,0) = H_h(0,0) = \begin{bmatrix} 0 & 0 \\ 0 & 0 \end{bmatrix}$$

$$\Delta_1^1(0,0) = \Delta_1^2(0,0) = \Delta_2(0,0) = 0$$

where Δ_1^l is the principal minor of order l obtained by removing the l^{eme} row and l^{eme} column $l = 1,2$. Thus the Hessian matrices of f, g and h are positive and negative semi definite at $(0,0)$. However, this doesn't imply that $(0,0)$ is a local minimum or maximum point. Indeed, by looking at the functions directly, we can classify the point. The three situations are shown in Figure 2.20.

First, note that $(0,0)$ is a global maximum for f since we have

$$f(x,y) = x^4 + y^4 \geqslant 0 = f(0,0) \qquad\qquad \forall (x,y) \in \mathbb{R}^2.$$

Next, note that $(0,0)$ is a global minimum for g. Indeed, we have

$$g(x,y) = -(x^4 + y^4) \leqslant 0 = g(0,0) \qquad\qquad \forall (x,y) \in \mathbb{R}^2.$$

Finally, $(0,0)$ is a saddle point for h since we have

$$h(x,0) = x^4 \geqslant 0 = h(0,0) \qquad\qquad \forall x \in \mathbb{R}$$
$$h(0,y) = -y^4 \leqslant 0 = h(0,0) \qquad\qquad \forall y \in \mathbb{R}.$$

Thus, for any disk centered at $(0,0)$, h takes values greater and lower than $h(0,0)$.

2. – Classify the stationary points of

$$f(x_1, x_2, x_3, x_4) = 20x_2 + 48x_3 + 6x_4 + 8x_1 x_2 - 4x_1^2 - 12x_3^2 - x_4^2 - 4x_2^3.$$

Does f attain its global extreme values on $\mathbb{R}^4$?

Solution: Since the function f is differentiable (because it is a polynomial), the local extreme points are critical points, i.e, solutions of

$$\nabla f(x_1, x_2, x_3, x_4) = \langle 8x_2 - 8x_1 ,\ 20 + 8x_1 - 12x_2^2,\ 48 - 24x_3 ,\ 6 - 2x_4 \rangle = \mathbf{0}_{\mathbb{R}^4}$$

$$\Longleftrightarrow \quad \begin{cases} x_2 = x_1 \\[2mm] x_3 = 2 \end{cases} \qquad \begin{aligned} & 5 + 2x_1 - 3x_2^2 = 0 \\[2mm] & x_4 - 3. \end{aligned}$$

We deduce that $(-1, -1, 2, 3)$ and $(\frac{5}{3}, \frac{5}{3}, 2, 3)$ are the critical points of f.

- *Classification of the critical points:* The Hessian matrix of f is

$$H_f(x_1, x_2, x_3, x_4) = \begin{bmatrix} -8 & 8 & 0 & 0 \\ 8 & -24x_2 & 0 & 0 \\ 0 & 0 & -24 & 0 \\ 0 & 0 & 0 & -2 \end{bmatrix}$$

The leading principal minors at the point $(-1, -1, 2, 3)$ are

$$D_1 = -8 < 0, \quad D_2 = -256 < 0, \quad D_3 = -24D_2 > 0 \text{ and } D_4 = -2D_3 < 0.$$

Then, $(-1, -1, 2, 3)$ is a saddle point.

The leading principal minors at the point $(\frac{5}{3}, \frac{5}{3}, 2, 3)$ are

$$D_1 = -8 < 0, \quad D_2 = 256 > 0, \quad D_3 = -24D_2 < 0 \text{ and } D_4 = -2D_3 > 0.$$

Then, $(\frac{5}{3}, \frac{5}{3}, 2, 3)$ is a local maximum point.

- *Global optimal points:* Note that

$$f(0, x_2, 0, 0) = 20x_2 - 4x_2^3 \longrightarrow \mp\infty \quad \text{as} \quad x_2 \longrightarrow \pm\infty.$$

Thus f takes large negative and positive values. Therefore f doesn't attain its global optimal values on $\mathbb{R}^2$.

3. – Let $f(x, y) = \ln(1 + x^2 y)$.

   i) Find and sketch the domain of definition of f.

   ii) Find the stationary points and show that the second-derivatives test is inconclusive at these points.

   iii) Describe the behavior of f at these points.

Solution: i) The domain of f is given by:

$$D_f = \{(x, y) \in \mathbb{R}^2 : \quad 1 + x^2 y > 0\}$$
$$= \{(0, y) : \quad y \in \mathbb{R}\} \cup \{(x, y) \in \mathbb{R}^* \times \mathbb{R} : \quad y > -\frac{1}{x^2}\}.$$

The domain of f is the region located above the curve $y = -\dfrac{1}{x^2}$, including the y axis; see Figure 2.21.

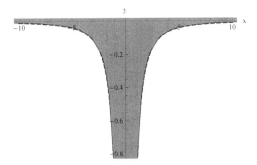

FIGURE 2.21: Domain of $f(x, y) = \ln(1 + x^2 y)$

ii) f is differentiable on its open domain D_f because

$$f = v o u \qquad \text{with} \qquad v(t) = \ln t, \qquad u(x, y) = 1 + x^2 y$$

u is differentiable in $\mathbb{R}^2$ then, in particular, in D_f

$u(D_f) \subset \mathbb{R}^{*+}$

v is differentiable in $\mathbb{R}^{*+}$.

The stationary points are solutions of

$$\nabla f(x, y) = \langle \frac{2xy}{1 + x^2 y}, \frac{x^2}{1 + x^2 y} \rangle = \langle 0, 0 \rangle$$

$$\Longleftrightarrow \begin{cases} xy = 0 \\ x^2 = 0 \end{cases} \qquad \Longleftrightarrow \qquad \begin{cases} x = 0 \quad \text{or} \quad y = 0 \\ x = 0 \end{cases}$$

$$\Longleftrightarrow \begin{cases} x = 0 \quad \text{and} \quad x = 0 \\ \text{or} \\ y = 0 \quad \text{and} \quad x = 0 \end{cases} \qquad \Longleftrightarrow \qquad x = 0, \qquad y \in \mathbb{R}.$$

We deduce that the points located on the y axis are the critical points of f.

The Hessian matrix of f is

$$H_f(x, y) = \frac{1}{(1 + x^2 y)^2} \begin{bmatrix} 1 - x^2 y & 2x \\ 2x & -x^4 \end{bmatrix}.$$

At the stationary points, we have

$$H_f(0, y) = \begin{bmatrix} 1 & 0 \\ 0 & 0 \end{bmatrix}.$$

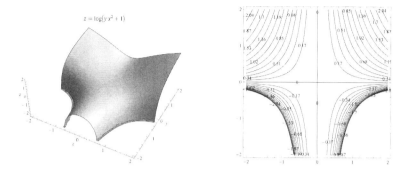

FIGURE 2.22: Graph and level curves of $f(x,y) = \ln(1 + x^2 y)$

The leading minor $D_2(0, y) = det(H_f(0, y)) = 0$, then the second derivatives test fails at these points. The behaviour of the function is illustrated in
Figure 2.22.

Classification of these points:

• The points $(0, y_0)$ with $y_0 > 0$ are local minimum points for f. Indeed, since
the logarithm function is increasing, we have

$$f(x, y) = \ln(1 + x^2 y) \geqslant \ln(1) = \ln(1 + 0^2 y_0) = 0 = f(0, y_0)$$

$$\forall x \in \mathbb{R}, \qquad \forall y \in (y_0 - \frac{y_0}{2}, y_0 + \frac{y_0}{2}) = (\frac{y_0}{2}, \frac{3y_0}{2}).$$

Thus, f takes values greater than $f(0, y_0)$ in a neighborhood of $(0, y_0)$ with
$y_0 > 0$.

• The points $(0, y_0)$ with $y_0 < 0$ are local maximum points for f. Indeed, since
ln is an increasing function, we have

$$f(x, y) = \ln(1 + x^2 y) \leqslant \ln(1) = \ln(1 + 0^2 y_0) = 0 = f(0, y_0)$$

$$\forall y \in (y_0 - \frac{-y_0}{2}, y_0 + \frac{-y_0}{2}) = (\frac{3y_0}{2}, \frac{y_0}{2}), \quad \forall x \quad \text{such that} \quad 0 < 1 + x^2 y.$$

f takes values lower than $f(0, y_0)$ in a neighborhood of $(0, y_0)$ with $y_0 < 0$.

• The point $(0, 0)$ is a saddle point for f. Indeed, we have

$$f(x, y) = \ln(1 + x^2 y) \geqslant \ln(1) = 0 = f(0, 0) \qquad\qquad \forall y \in \mathbb{R}^+$$

$f(x, y) = \ln(1 + x^2 y) \leqslant \ln(1) = 0 = f(0,0) \quad \forall y \in \mathbb{R}^-$ such that $0 < 1 + x^2 y$.

For any disk centered at $(0,0)$, f takes values greater and lower than $f(0,0)$.

4. – Find and classify all stationary points of $f(x, y) = x^2 y + y^3 x - x y$. Are there global minimum and maximum values of f on $\mathbb{R}^2$?

Solution:

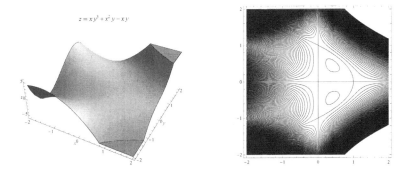

FIGURE 2.23: Graph and level curves of $f(x, y) = x^2 y + y^3 x - xy$

The function f is differentiable on its open domain $\mathbb{R}^2$ since it is a polynomial. So, the local extreme points are critical, i.e, solutions of

$$\nabla f(x,y) = \langle 2xy + y^3 - y, x^2 + 3y^2 x - x \rangle = \langle 0,0 \rangle \Longleftrightarrow$$

$$\begin{cases} y(y^2 + 2x - 1) = 0 \\ x(3y^2 + x - 1) = 0 \end{cases} \Longleftrightarrow \begin{cases} y = 0 \quad \text{or} \quad y^2 + 2x - 1 = 0 \\ x = 0 \quad \text{or} \quad 3y^2 + x - 1 = 0 \end{cases}$$

$$\Longleftrightarrow \begin{cases} y = 0 \quad \text{and} \quad x = 0 \\ \text{or} \quad [y = 0 \quad \text{and} \quad 3y^2 + x - 1 = 0] \\ \text{or} \quad [y^2 + 2x - 1 = 0 \quad \text{and} \quad x = 0] \\ \text{or} \quad [y^2 + 2x - 1 = 0 \quad \text{and} \quad 3y^2 + x - 1 = 0] \end{cases} \Longleftrightarrow$$

$$\left\{ \begin{array}{l} y = 0 \quad \text{and} \quad x = 0 \\[2mm] \text{or} \quad [y = 0 \quad \text{and} \quad x = 1] \end{array} \right. \left\{ \begin{array}{l} \text{or} \quad [y^2 - 1 = 0 \quad \text{and} \quad x = 0] \\[2mm] \text{or} \quad [y^2 = \dfrac{1}{5} \quad \text{and} \quad x = \dfrac{2}{5}]. \end{array} \right.$$

We deduce that $(0,0)$, $(1,0)$, $(0,1)$, $(0,-1)$, $(\frac{2}{5}, \frac{1}{\sqrt{5}})$ and $(\frac{2}{5}, -\frac{1}{\sqrt{5}})$ are the critical points of f. Reading the level curves in Figure 2.23, one can locate four saddle points and two local extrema.

Classification of the critical points: Applying the second derivatives test, we obtain:

critical point	$D_1(x,y)$	$D_2(x,y)$	classification
$(0,0)$	0	-1	saddle point
$(1,0)$	0	-4	saddle point
$(0,1)$	2	-4	saddle point
$(0,-1)$	-2	-4	saddle point
$(\frac{2}{5}, \frac{1}{\sqrt{5}})$	$\frac{2}{\sqrt{5}}$	$\frac{4}{5}$	local minimum point
$(\frac{2}{5}, -\frac{1}{\sqrt{5}})$	$-\frac{2}{\sqrt{5}}$	$\frac{4}{5}$	local maximum point

TABLE 2.6: Critical points' classification for $f(x,y) = x^2 y + y^3 x - xy$

where

$$f_{xx}(x,y) = 2y, \qquad f_{yy}(x,y) = 6yx, \qquad f_{xy}(x,y) = 2x + 3y^2 - 1$$

$$H_f(x,y) = \begin{bmatrix} 2y & 2x + 3y^2 - 1 \\ 2x + 3y^2 - 1 & 6xy \end{bmatrix}, \quad D_1(x,y) = \mid f_{xx} \mid = f_{xx} = 2y$$

$$\left| \begin{array}{cc} f_{xx} & f_{xy} \\ f_{xy} & f_{yy} \end{array} \right| = \left| \begin{array}{cc} 2y & 2x + 3y^2 - 1 \\ 2x + 3y^2 - 1 & 6yx \end{array} \right| = 12xy^2 - [2x + 3y^2 - 1]^2.$$

Finally, note that f takes large positive and negative values since we have

$$f(1, y) = y^3 \longrightarrow \pm\infty \qquad \text{as} \qquad y \longrightarrow \pm\infty.$$

Therefore, f doesn't attain a global maximal value nor a minimal one.

5. – A power substation must be located at a point closest to three houses located at the points $(0,0)$, $(1,1)$, $(0,2)$. Find the optimal location by minimizing the sum of the squares of the distances between the houses and the substation.

Solution: Let (x, y) be the position of the power substation. Then, we have

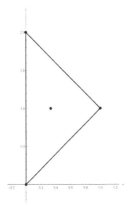

FIGURE 2.24: The closest power station to three houses

to look for (x, y) as the point that minimize the function

$$f(x, y) = d^2((x, y), (0, 0)) + d^2((x, y), (1, 1)) + \ldots + d^2((x, y), (0, 2))$$

which can be written as

$$f(x, y) = [(x - 0)^2 + (y - 0)^2] + [(x - 1)^2 + (y - 1)^2] + [(x - 0)^2 + (y - 2)^2].$$

Because f is polynomial, it is differentiable on the open set $\mathbb{R}^2$. Thus a global minimum point is also a local one. Therefore, it is a solution of

$$\nabla f(x,y) = \langle 2x + 2(x-1) + 2x, 2y + 2(y-1) + 2(y-2) \rangle = \langle 0, 0 \rangle$$

$$\iff \quad \begin{cases} 6x - 2 = 0 \\ \\ 6y - 6 = 0 \end{cases} \quad \iff \quad (x, y) = (\frac{1}{3}, 1).$$

Thus, we have one critical point and by applying the second derivatives test, we obtain:

$$H_f(x,y) = \begin{bmatrix} 6 & 0 \\ 0 & 6 \end{bmatrix} \qquad D_1(\frac{1}{3}, 1) = 6 > 0 \qquad D_2(\frac{1}{3}, 1) = \begin{vmatrix} 6 & 0 \\ 0 & 6 \end{vmatrix} = 36 > 0.$$

So $(\frac{1}{3}, 1)$ is a local minimum; see Figure 2.24 for the position of the point and the three houses.

To show that it is the point that minimizes f globally, we proceed by comparing the values of f and completing squares:

$$f(x,y) - f(\frac{1}{3}, 1) = 3x^2 - 2x + 1 + 3y^2 - 6y + 5 - (\frac{2}{3} + 2)$$

$$= 3(x - \frac{1}{3})^2 + 3(y-1)^2 \geqslant 0 \qquad \forall (x,y) \in \mathbb{R}^2.$$

6. – Based on the level curves that are visible in Figures 2.25 and 2.26, identify the approximate position of the local maxima, local minima and saddle points.

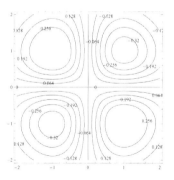

FIGURE 2.25: Level curves of $f(x,y) = -xye^{-(x^2+y^2)/2}$ on $[-2.2] \times [-2, 2]$

Solution: i) From the level curves' plotting, one can locate:

- a saddle point at $(0, 0)$

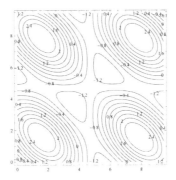

FIGURE 2.26: Level curves of $g(x,y) = \sin(x) + \sin(y) - \cos(x+y)$ for x, y in $[0, 3\pi]$

 - two local maxima at $(-1, 1)$, $(1, -1)$

 - two local minima at $(-1, -1)$, $(1, 1)$.

Using Maple software, one can check these observations by applying the second derivatives test using the coding:

$with(Student[MultivariateCalculus])$
$LagrangeMultipliers(-x*y*exp(-(x^2+y^2)*(1/2)), [], [x,y], output = detailed)$
$[x = 0, y = 0, -x*y*exp(-(1/2)*x^2 - (1/2)*y^2) = 0],$
$[x = 1, y = 1, -x*y*exp(-(1/2)*x^2 - (1/2)*y^2) = -exp(-1)],$
$[x = 1, y = -1, -x*y*exp(-(1/2)*x^2 - (1/2)*y^2) = exp(-1)],$
$[x = -1, y = 1, -x*y*exp(-(1/2)*x^2 - (1/2)*y^2) = exp(-1)],$
$[x = -1, y = -1, -x*y*exp(-(1/2)*x^2 - (1/2)*y^2) = -exp(-1)]$
$SecondDerivativeTest(-x*y*exp(-(x^2+y^2)*(1/2)),\quad [x,y] = [0,0])$
$LocalMin = [],\quad LocalMax = [],\quad Saddle = [[0,0]]$
$SecondDerivativeTest(-x*y*exp(-(x^2+y^2)*(1/2)),\quad [x,y] = [1,1])$
$LocalMin = [[1,1]],\quad LocalMax = [],\quad Saddle = []$

$\vdots$

ii) For the second figure, the exact points found, using Maple, are:

 - 5 saddle points $(\frac{3\pi}{2}, \frac{3\pi}{2})$, $(\frac{\pi}{2}, \frac{3\pi}{2})$, $(\frac{3\pi}{2}, \frac{\pi}{2})$, $(\frac{7\pi}{2}, \frac{\pi}{2})$, $(\frac{\pi}{2}, \frac{7\pi}{2})$

 - 4 local maxima at $(\frac{\pi}{2}, \frac{5\pi}{2})$, $(\frac{5\pi}{2}, \frac{\pi}{2})$, $(\frac{\pi}{2}, \frac{\pi}{2})$, $(\frac{5\pi}{2}, \frac{5\pi}{2})$

 - 2 local minima at $(\frac{11\pi}{6}, \frac{11\pi}{6})$, $(\frac{7\pi}{2}, \frac{7\pi}{2})$.

2.3 Convexity/Concavity and Global Extreme Points

In dimension 1, when a C^2 function f is convex on its domain D_f and x^* is a local minimum of f, then x^* is a global minimum. Indeed, the convexity of f is characterized by $f''(x) \geqslant 0$ [2], [1]. Then, using Taylor's formula, the values $f(x)$ and $f(x^*)$ can be compared as follows:

$$f(x) = f(x^*) + (x - x^*)f'(x^*) + \frac{(x - x^*)^2}{2}f''(c) \quad \text{for some } c \text{ between } x^* \text{ and } x.$$

Because $f'(x^*) = 0$, then

$$f(x) - f(x^*) = \frac{(x - x^*)^2}{2}f''(c) \geqslant 0.$$

As x is arbitrarily chosen in the domain of f, then

$$f(x) \geqslant f(x^*) \qquad \forall x \in D_f$$

which shows that x^* is a global minimum point for f.

In this section, we want to generalize the convexity property to functions of several variables in order to establish, later, results of global optimality.

2.3.1 Convex/Concave Several Variable Functions

Definition 2.3.1 *Let S be a convex set of $\mathbb{R}^n$ and let f be a real function*

$$f : \qquad S \longrightarrow \mathbb{R}$$
$$x = (x_1, \cdots, x_n) \longmapsto f(x).$$

Then,

$$f \text{ is convex} \iff f(ta + (1-t)b) \leqslant tf(a) + (1-t)f(b).$$

$$f \text{ is strictly convex} \iff f(ta + (1-t)b) < tf(a) + (1-t)f(b)$$
$$a \neq b, \quad t \neq 0, 1.$$

$$f \text{ is concave} \iff f(ta + (1-t)b) \geqslant tf(a) + (1-t)f(b).$$

$$f \text{ is strictly concave} \iff f(ta + (1-t)b) > tf(a) + (1-t)f(b)$$
$$a \neq b, \quad t \neq 0, 1.$$

These equivalences must hold $\forall a, b \in S, \quad \forall t \in [0, 1]$.

• Using the definition, one can check that the functions

$$i) \quad f(x) = ax + b \qquad\qquad ii) \quad f(x, y) = ax + by + c$$

are simultaneously concave and convex in $\mathbb{R}$ and $\mathbb{R}^2$ respectively. Their respective graphs represent a line $y = ax + b$ and a plane $z = ax + by + c$.

• A convex/concave function is not necessarily differentiable at every point.

$$i) \quad f(x) = |x| \qquad\qquad ii) \quad f(x, y) = \sqrt{x^2 + y^2} = \|(x, y)\|.$$

Each function is not differentiable at the origin and represents the Euclidean distance in $\mathbb{R}$ and $\mathbb{R}^2$ respectively. We use the triangular inequality to verify that they are convex.

• One can form new convex/concave functions using algebraic operations. For Example [25],

if f, g are functions defined on a convex set $S \subset \mathbb{R}^n$ and $s, t \geqslant 0$, then:

$$f \text{ and } g \text{ are concave (resp. convex)} \implies$$

$$\left\|\begin{array}{l} sf + tg \quad \text{is concave (resp. convex)} \\[2mm] \min(f, g) \quad (\text{resp. } \max(f, g)) \quad \text{is concave (resp. convex).} \end{array}\right.$$

Remark 2.3.1 *The geometrical interpretation of the convexity of f expresses that the graph of f remains under the line segment $[AB]$ joining any two points $A(a, f(a))$ and $B(b, f(b))$ of the graph of f. Indeed,*

$$[A, B] = \Big\{ (x, y) \in \mathbb{R}^n \times \mathbb{R} : \quad x = a + t(b - a),$$
$$y = f(a) + t(f(b) - f(a)) \quad t \in [0, 1] \Big\}$$

is located above the part of the graph of f

$$\Big\{ (x, y) \in \mathbb{R}^n \times \mathbb{R} : \quad x = a + t(b - a), \quad y = f(a + t(b - a)) \quad t \in [0, 1] \Big\}$$

since we have $\quad \forall t \in [0, 1]$

$$f(a + t(b - a)) = f(tb + (1 - t)a) \leqslant tf(b) + (1 - t)f(a) = f(a) + t(f(b) - f(a)).$$

Similarly, the geometrical interpretation of the concavity of f expresses that the graph of f remains above the line segment $[AB]$ joining any two points $A(a, f(a))$ and $B(b, f(b))$ of the graph of f; see Figure 2.27.

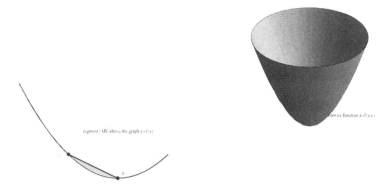

FIGURE 2.27: Shape of convex functions

Remark 2.3.2 *There is a connection with a convexity/concavity of a function f defined on a convex set $S \subset \mathbb{R}^n$ and the convexity/concavity of particular sets described by f [25]. Indeed, we have*

$$f \text{ is convex} \iff \text{the set } \left\{(x,y) \in S \times \mathbb{R} : y \geqslant f(x)\right\} \text{ is convex}$$

$$f \text{ is concave} \iff \text{the set } \left\{(x,y) \in S \times \mathbb{R} : y \leqslant f(x)\right\} \text{ is convex}$$

2.3.2 Characterization of Convex/Concave C^1 Functions

When $n = 1$, the following theorem expresses that the graph of a convex (resp. concave) C^1 function remains above (resp. below) its tangent lines.

Theorem 2.3.1 *Let S be a convex open set of $\mathbb{R}^n$ and let $f : S \longrightarrow \mathbb{R}$ be C^1. Then, for any $x, a \in S$, the following inequalities hold*

$$f \text{ is convex in } S \iff f(x) - f(a) \geqslant \left[\nabla f(a)\right].(x - a)$$

$$f \text{ is strictly convex in } S \iff f(x) - f(a) > \left[\nabla f(a)\right].(x - a), \ x \neq a$$

> f is concave in S $\iff$ $f(x) - f(a) \leqslant \left[\nabla f(a)\right].(x - a)$
>
> f is strictly concave in S $\iff$ $f(x) - f(a) < \left[\nabla f(a)\right].(x - a), \; x \neq a.$

Proof. We prove the first assertion. The other assertions can be established similarly.

$\implies)$ If f is convex in S, then, by definition, we have for $a, b \in S$,

$$f(tb + (1 - t)a) \leqslant tf(b) + (1 - t)f(a) \qquad \forall t \in [0, 1]$$

from which we deduce

$$f(b) - f(a) \geqslant \frac{f(tb + (1 - t)a) - f(a)}{t} = \frac{f(a + t(b - a)) - f(a)}{t} \quad \forall t \in (0, 1].$$

Since $f \in C^1(S)$, we obtain

$$f(b) - f(a) \geqslant \lim_{t \to 0^+} \frac{g(t) - g(0)}{t - 0} = g'(0)$$

where

$$g(t) = f(a + t(b - a)) \quad g'(t) = f'(a + t(b - a)).(b - a), \quad g'(0) = f'(a).(b - a).$$

Indeed,

$$g(t) = f(a_1 + t(x_1 - a_1), \ldots, a_n + t(x_n - a_n)) = f(x_1(t), x_2(t), \ldots, x_n(t)).$$

Each function $x_j(t) == a_j + t(b_j - a_j)$, $j = 1, \ldots, n$, is differentiable with $x'_j(t) = b_j - a_j$. So g is differentiable and we obtain, by the chain rule formula,

$$g'(t) = \frac{\partial f}{\partial x_1}\frac{\partial x_1}{\partial t} + \frac{\partial f}{\partial x_2}\frac{\partial x_2}{\partial t} + \ldots + \frac{\partial f}{\partial x_n}\frac{\partial x_n}{\partial t}$$

$$= \left[f_{x_1}(a + t(b - a))\right](b_1 - a_1) + \left[f_{x_2}(a + t(b - a))\right](b_2 - a_2) + \ldots$$

$$\ldots + \left[f_{x_n}(a + t(b - a))\right](b_n - a_n) = \left[(\nabla f)(a + t(b - a))\right].(b - a).$$

$\impliedby)$ Assume that

$$f(x) - f(u) \geqslant \left[\nabla f(u)\right].(x - u) \qquad \forall\, x, u \in S.$$

Let $a, b \in S$ and $t \in [0, 1]$. Choosing $x = a$ and $u = ta + (1 - t)b$ in the above inequality, we obtain

$$f(a) - f(ta + (1 - t)b) \geqslant \left[\nabla f(ta + (1 - t)b)\right].(a - [ta + (1 - t)b])$$

$$= (1 - t)\left[\nabla f(ta + (1 - t)b)\right].(a - b). \tag{$*$}$$

Now, choose $x = b$ and $u = ta + (1 - t)b$ in the same inequality. We get

$$f(b) - f(ta + (1 - t)b) \geqslant \left[\nabla f(ta + (1 - t)b)\right].(b - [ta + (1 - t)b])$$

$$= -t\left[\nabla f(ta + (1 - t)b)\right].(a - b). \tag{$**$}$$

Multiply the inequality $(*)$ by $t > 0$ and the inequality $(**)$ by $(1 - t) > 0$, then add the resulting inequalities. This gives

$$tf(a) + (1 - t)f(b) - (t + (1 - t))f(ta + (1 - t)b)$$

$$\geqslant [t(1 - t) - (1 - t)t]\nabla f(ta + (1 - t)b).(a - b) = 0.$$

Therefore f is convex.

Example 1. Show that $f(x, y) = x^2 + y^2$ is convex on $\mathbb{R}^2$.

Solution: We have

$$f(x, y) - f(s, t) - \nabla f(s, t).\begin{bmatrix} x - s \\ y - t \end{bmatrix}$$

$$= x^2 + y^2 - (s^2 + t^2) - \begin{bmatrix} 2s & 2t \end{bmatrix}.\begin{bmatrix} x - s \\ y - t \end{bmatrix}$$

$$= x^2 + y^2 - (s^2 + t^2) - 2s(x - s) - 2t(y - t)$$

$$= (s - x)^2 + (t - y)^2 \geqslant 0 \qquad \forall\, (x, y),\ (s, t) \in \mathbb{R}^2.$$

Thus f is convex on $\mathbb{R}^2$. Note that by taking $(s, t) = (0, 0)$, the critical point of f, we deduce that $f(x, y) - f(0, 0) \geqslant 0 \quad \forall\, (x, y) \in \mathbb{R}^2$. Hence, $(0, 0)$ is a global minimum of f.

As, we can expect, from the above example, it will not always be easy to check the convexity or concavity of a function through solving inequalities. Next, we show a more practical characterization, but requiring more regularity on the function.

2.3.3   Characterization of Convex/Concave C^2 Functions

Theorem 2.3.2 *Strict convexity/concavity*

Let S be a convex open set of $\mathbb{R}^n$ and let $f : S \longrightarrow \mathbb{R}, \quad f \in C^2(S)$. Then

(i) $D_k(x) > 0 \quad \forall x \in S, \quad k = 1, \ldots, n \quad \Longrightarrow \quad f$ *is strictly convex in S.*

(ii) $(-1)^k D_k(x) > 0 \quad \forall x \in S, \; k = 1, \ldots, n \Longrightarrow f$ *is strictly concave in S.*

$D_k(x), \; k = 1, \ldots, n$ *are the n leading minors of the Hessian matrix* $H_f(x) = (f_{x_i x_j}(x))_{n \times n}$ *of f.*

Proof. **i)** For $a, b \in S, a \neq b$, and $t \in [0, 1]$, define the function

$$g(t) = f(tb + (1-t)a) = f(a + t(b-a)) = f(x_1(t), \ldots, x_n(t))$$

with $x_j(t) = a_j + t(b_j - a_j) \quad j = 1, \ldots, n.$
By the chain rule theorem, we have

$$g'(t) = \left[(\nabla f)(a + t(b-a)) \right].(b-a).$$

Since f is C^2, g is also C^2 and we have

$$g''(t) = \left[\frac{d}{dt} f_{x_1}(a + t(b-a)) \right](b_1 - a_1) + \ldots + \left[\frac{d}{dt} f_{x_n}(a + t(b-a)) \right](b_n - a_n).$$

For each $i = 1, \ldots, n$, we have

$$f_{x_i}(a + t(b-a)) = f_{x_i}(x_1(t), x_2(t), \ldots, x_n(t)).$$

Then

$$\frac{d}{dt} f_{x_i}(y + t(x - y)) = \frac{\partial f_{x_i}}{\partial x_1} \frac{\partial x_1}{\partial t} + \frac{\partial f_{x_i}}{\partial x_2} \frac{\partial x_2}{\partial t} + \ldots + \frac{\partial f_{x_i}}{\partial x_n} \frac{\partial x_n}{\partial t}$$

$$= \left[f_{x_i x_1}(a + t(b-a)) \right](b_1 - a_1) + \ldots + \left[f_{x_i x_n}(a + t(b-a)) \right](b_n - a_n)$$

$$= \sum_{j=1}^{n} \left[f_{x_i x_j}(a + t(b-a)) \right](x_j - y_j).$$

Hence

$$g''(t) = \sum_{i=1}^{n} \sum_{j=1}^{n} [f_{x_i x_j}(a + t(b - a))](b_i - a_i)(b_j - a_j).$$

Now, by assumption, we have $D_k(z) > 0$ for all $z \in S$ and for all $k = 1, \ldots, n$, then the quadratic form

$$Q(h) = \sum_{i=1}^{n} \sum_{j=1}^{n} \left[f_{x_i x_j}(a + t(b - a)) \right] h_i h_j$$

with the associated symmetric matrix $\left[f_{x_i x_j}(a + t(b - a)) \right]_{n \times n}$ is positive definite. As a consequence, $g''(t) > 0$ and g is strictly convex. In particular

$$f(tb + (1-t)a) = g(t) = g(t.1 + (1-t)0) < tg(1) + (1-t)g(0) = tf(b) + (1-t)f(a)$$

and the strict convexity of f follows.

ii) Under the assumptions $ii)$, the quadratic form

$$Q^*(h) = \sum_{i=1}^{n} \sum_{j=1}^{n} \left[(-f)_{x_i x_j}(a + t(b - a)) \right] h_i h_j =^t hH_{-f}(a + t(b - a))h$$

$$=^t h \begin{bmatrix} (-f)_{x_1 x_1} & (-f)_{x_1 x_2} & \cdots & (-f)_{x_1 x_k} \\ (-f)_{x_2 x_1} & (-f)_{x_2 x_2} & \cdots & (-f)_{x_2 x_k} \\ \vdots & \vdots & \vdots & \vdots \\ (-f)_{x_k x_1} & (-f)_{x_k x_2} & \cdots & (-f)_{x_k x_k} \end{bmatrix} h$$

is positive definite by assumption. As a consequence, $(-g)''(t) > 0$ and $-g$ is strictly convex. In particular

$$-f(tb + (1 - t)a) = (-g)(t) = -g(t.1 + (1 - t)0)$$
$$< t(-g)(1) + (1 - t)(-g)(0) = t(-f)(b) + (1 - t)(-f)(a)$$
$$\Longleftrightarrow \quad f(tb + (1 - t)a) > tf(b) + (1 - t)f(a) \qquad \forall a \neq b \qquad \forall t \in [0, 1]$$

and the strict concavity of f follows.

We also have the following characterization

Theorem 2.3.3 *Convexity/concavity*

Let S be a convex open set of $\mathbb{R}^n$ and let $f : S \longrightarrow \mathbb{R}$ be C^2. Then

$$f \text{ is convex in } S \iff \Delta_k(x) \geqslant 0 \quad \forall x \in S \quad \forall k = 1, \ldots, n.$$

$$f \text{ is concave in } S \iff (-1)^k \Delta_k(x) \geqslant 0 \quad \forall x \in S \quad \forall k = 1, \ldots, n.$$

A principal minor $\Delta_r(x)$ of order r in the Hessian $[f_{x_i x_j}(x)]$ of f is the determinant obtained by deleting $n - r$ rows and the $n - r$ columns with the same numbers (if the ith row (column) is selected, then so is the ith column (row)).

Proof. We prove only the first assertion. The second one is established by replacing f by $-f$.

$\impliedby$) We proceed as in the proof of the previous theorem. We conclude that $Q(h)$ is positive semi definite. As a consequence, $g''(t) \geqslant 0$ and g is convex. In particular

$$f(tb+(1-t)a) = g(t) = g(t.1+(1-t)0) \leqslant tg(1)+(1-t)g(0) = tf(b)+(1-t)f(a)$$

and the convexity of f follows.

$\implies$) Suppose f convex in S. It suffices to show that the quadratic form $Q(h)$ satisfies

$$Q(h) = \sum_{i=1}^{n} \sum_{j=1}^{n} f_{x_i x_j}(a) h_i h_j \geqslant 0 \qquad \forall a \in S.$$

So, let $a \in S$. Since S is an open set, there exists $\epsilon > 0$ such that $B_\epsilon(a) \subset S$. In particular for $h \in \mathbb{R}^n$, $h \neq \mathbf{0}$, we have

$$a + th \in B_\epsilon(a) \iff \|a + th - a\| = |t|\|h\| < \epsilon \iff |t| < \frac{\epsilon}{\|h\|} = \alpha.$$

So, for $t \in (-\alpha, \alpha)$, the function $u(t) = f(a + th)$ is well defined. We claim that u is convex. Indeed, we have for $\lambda \in [0, 1]$ and $t, s \in (-\alpha, \alpha)$,

$$u(\lambda t + (1 - \lambda)s) = f(a + [\lambda t + (1 - \lambda)s]h)$$

$$= f(\lambda a + (1 - \lambda)a + [\lambda t + (1 - \lambda)s]h)$$

$$= f(\lambda[a + th] + (1 - \lambda)[a + sh])$$

$$\leqslant \lambda f(a + th) + (1 - \lambda)f(a + sh) \quad \text{since } f \text{ is convex}$$

$$= \lambda u(t) + (1 - \lambda)u(s).$$

Hence $u''(t) \geqslant 0$ for all $t \in (-\alpha, \alpha)$. But

$$u''(t) = \sum_{i=1}^{n} \sum_{j=1}^{n} \big[f_{x_i x_j}(a + th) \big] h_i h_j$$

and for $t = 0$, we obtain the semi definite positivity of the quadratic form Q.

Example 2. Show that $f(x, y) = x^4 + y^4$ is convex on $\mathbb{R}^2$.

Solution: We have
$$\nabla f(x, y) = \langle 4x^3, 4y^3 \rangle.$$

The Hessian matrix of f is

$$H_f(x, y) = \begin{bmatrix} f_{xx} & f_{xy} \\ f_{yx} & f_{yy} \end{bmatrix} = \begin{bmatrix} 12x^2 & 0 \\ 0 & 12y^2 \end{bmatrix}.$$

The leading principal minors are

$$D_1(x, y) = 12x^2 \geqslant 0 \qquad D_2(x, y) = \begin{vmatrix} 12x^2 & 0 \\ 0 & 12y^2 \end{vmatrix} = 144x^2 y^2 \geqslant 0.$$

We cannot conclude about the strict convexity of f on $\mathbb{R}^2$ since $D_1(x, y) = 0$ if $x = 0$ and $D_2(x, y) = 0$ if $x = 0$ or $y = 0$. However, since f is C^∞, then

$$f \quad \text{is convex} \quad \Longleftrightarrow \quad H_f \quad \text{is semi-definite positive}.$$

We have $\quad \Delta_1^{11}(x, y) = 12y^2 \geqslant 0$,

$$\Delta_1^{22}(x, y) = 12x^2 \geqslant 0, \quad \text{and} \quad \Delta_2(x, y) = 144x^2 y^2 \geqslant 0.$$

Thus, f is convex on $\mathbb{R}^2$.

2.3.4   Characterization of a Global Extreme Point

From the previous characterizations of convex/concave functions, we deduce the following:

> **Theorem 2.3.4** *Let* $S \subset \Omega \subset \mathbb{R}^n$, Ω *an open set,* S *a convex set and* $x^* \in S$. *Let* $f : \Omega \longrightarrow \mathbb{R}$ *be a* C^1 *function on* Ω, *concave (resp. convex) on* S, *then*
>
> x^* *is a global maximum*
> $$\Longleftrightarrow \nabla f(x^*).(x - x^*) \leqslant 0 \quad \forall x \in S \quad (resp. \geqslant).$$
> *(resp. minimum) point*
>
> *Moreover, if* $x^* \in \overset{\circ}{S}$, *then*
>
> x^* *is a global maximum (resp. minimum) point for* f *on* $S \Longleftrightarrow \nabla f(x^*) = 0$.

Proof. • Without the convexity assumption of f, the implication

$$x^* \text{ is a global minimum point for } f \text{ on } S \quad \Longrightarrow \quad \nabla f(x^*).(x - x^*) \geqslant 0$$

is established in theorem 2.1.2. Now, Suppose that f is convex on S and that

$$\nabla f(x^*).(x - x^*) \geqslant 0 \qquad \forall x \in S,$$

then, because f is a C^1 function on Ω, we have (see proof of Theorem 2.3.1 $\Rightarrow$)

$$f(x) - f(x^*) \geqslant \nabla f(x^*).(x - x^*) \qquad \forall x \in S$$

from which we deduce that

$$f(x) \geqslant f(x^*) \qquad \forall x \in S$$

and conclude that x^* is a global minimum point for f on S.

• If x^* is an interior global minimum point for f, then it is a stationary point with no need to the convexity of f. Conversely, suppose x^* is a stationary point for f and f be $C^1(\Omega)$ and convex in S, then we have

$$f(x) - f(x^*) \leqslant \nabla f(x^*).(x - x^*).$$

Because, $\nabla f(x^*) = 0$, we deduce that $f(x) - f(x^*) \leqslant 0$, and then $f(x) \leqslant f(x^*)$ $\quad \forall x \in S$.

The case f concave is established similarly.

Example 3. Suppose that x units of a commodity are sold at $160 - 0.01x$ cents per unit and that the total cost of production, in cents, is given by

$$C(x) = 40x + 20000.$$

Find the most profitable level of production if $7000 \leqslant x \leqslant 100000$ by applying the theorem above.

Solution: Since the total revenue for selling x units is $R(x) = x(160 - 0.01x)$, the profit $P(x)$ on x units will be

$$P(x) = R(x) - C(x) = x(160 - 0.01x) - (40x + 20000) = -0.01x^2 + 120x - 20000.$$

So, we have to find

$$\max_{x \in S} P(x) \qquad \text{on the convex set} \qquad S = [7000, 100000].$$

From

$$\frac{dP}{dx} = -0.02x + 120$$

we have

$$\frac{dP}{dx} = 0 \qquad \Longleftrightarrow \qquad x = 6000.$$

The only stationary point is 6000 and cannot be the maximum point since it is not in S. Let us then explore the concavity of P. We have

$$P(x) - P(a) - \frac{dP}{dx}(a)(x - a) = -0.01(x - a)^2 \leqslant 0 \qquad \forall x, a \in S, \quad x \neq a.$$

Thus, P is strictly concave on S. Therefore, the maximum point x^* (that exists by the extreme value theorem) must satisfy

$$\frac{dP}{dx}(x^*).(x - x^*) \leqslant 0 \quad \forall x \in S \qquad \Longleftrightarrow \qquad (-0.02x^* + 120)(x - x^*) \leqslant 0 \quad \forall x \in S.$$

Since $(-0.02x^* + 120) < 0$ on S, we must have

$$x - x^* \geqslant 0 \quad \forall x \in [7000, 100000] \qquad \Longleftrightarrow \qquad x^* = 7000.$$

7000 units should be produced to attain maximum profit.

Theorem 2.3.5 *Let S be a convex set of $\mathbb{R}^n$ and $x^* \in \overset{\circ}{S}$. Let $f : S \longrightarrow \mathbb{R}$ be a C^2 concave (resp. convex) function on S, then*

x^* *is a global maximum (resp. minimum) point for f on $S \Longleftrightarrow \nabla f(x^*) = 0$.*

Proof. **i)** Suppose f to be concave.

$\implies$) Suppose x^* is a global maximum point for f. Since x^* is an interior point, it is a stationary point, that is $\nabla f(x^*) = 0$.

$\impliedby$) Suppose $\nabla f(x^*) = 0$, then, since f is C^2, we have, for some $t \in [0,1]$,

$$f(x) - f(x^*) = \nabla f(x^*).(x - x^*) +^t (x - x^*) H_f(x^* + t(x - x^*))(x - x^*),$$

then, because f is concave in S, we have

$$f(x) - f(x^*) =^t (x - x^*) H_f(x^* + t(x - x^*))(x - x^*) \leqslant 0.$$

Hence

$$f(x) \leqslant f(x^*) \qquad \forall x \in S.$$

Thus x^* is is a global maximum point for f.

ii) If f is convex in S, then $-f$ is concave in S. From i), we deduce that

$$x^* \text{ is a global minimum point for } f \text{ in } S$$
$$\iff \quad x^* \text{ is a global maximum point for } (-f) \text{ in } S$$
$$\iff \quad \nabla(-f)(x^*) = 0 \quad \iff \quad \nabla f(x^*) = -\nabla(-f)(x^*) = 0.$$

Example 4. Find the global maxima and minima points if any of f defined by

$$f(x, y, z, t) = 24x + 32y + 48z + 72t - (x^2 + y^2 + 2z^2 + 3t^2).$$

Solution: A global extreme point of f is also a local extreme one since f is defined on $\mathbb{R}^4$ which is open. Therefore, it is a stationary point since f is C^∞ (because it is a polynomial). We have

$$\nabla f(x, y, z, t) = \langle 24 - 2x, 32 - 2y, 48 - 4z, 72 - 6t \rangle = \langle 0, 0, 0, 0 \rangle$$

$$\iff \quad (x, y, z, t) = (12, 16, 12, 12).$$

The only stationary point is $(12, 16, 12, 12)$. The Hessian matrix is

$$H_f(x, y, z, t) = \begin{bmatrix} -2 & 0 & 0 & 0 \\ 0 & -2 & 0 & 0 \\ 0 & 0 & -4 & 0 \\ 0 & 0 & 0 & -6 \end{bmatrix}$$

The leading principal minors are

$$D_1 = -2, \quad D_2 = \begin{vmatrix} -2 & 0 \\ 0 & -2 \end{vmatrix} = 4, \quad D_3 = \begin{vmatrix} -2 & 0 & 0 \\ 0 & -2 & 0 \\ 0 & 0 & -4 \end{vmatrix} = -16,$$

$$D_4(x, y, z, t) = \begin{vmatrix} -2 & 0 & 0 & 0 \\ 0 & -2 & 0 & 0 \\ 0 & 0 & -4 & 0 \\ 0 & 0 & 0 & -6 \end{vmatrix} = -6(-16) = 96,$$

and satisfy

$$(-1)^k D_k(x, y, z, t) > 0 \qquad \forall (x, y, z, t) \in \mathbb{R}^4 \qquad \text{for} \qquad k = 1, 2, 3, 4.$$

Therefore, f is strictly concave on $\mathbb{R}^4$ and the point $(12, 16, 12, 12)$ is the only global maximum point.

Note that $\min_{\mathbb{R}^4} f$ doesn't exist since f takes large negative values. Indeed, we have, for example,

$$f(x, 0, 0, 0) = 24x - x^2 \longrightarrow -\infty \qquad \text{as} \qquad x \longrightarrow \pm\infty.$$

Solved Problems

1. – A power substation must be located at a point closest to n houses located at m distinct points (x_1, y_1), (x_2, y_2), ... , (x_m, y_m). Find the optimal location by minimizing the sum of the squares of the distances between the houses and the substation.

Solution: Let (x, y) be the position of the power substation. Then, we have to look for (x, y) as the point that minimizes the function

$$f(x, y) = d^2((x, y), (x_1, y_1)) + d^2((x, y), (x_2, y_2)) + \ldots + d^2((x, y), (x_m, y_m))$$

which can be written as

$$f(x, y) = [(x-x_1)^2 + (y-y_1)^2] + [(x-x_2)^2 + (y-y_2)^2] + \ldots + [(x-x_m)^2 + (y-y_m)^2].$$

Because f is polynomial, it is differentiable on the open set $\mathbb{R}^2$. Thus a global minimum point is also a local one. Therefore, it is a solution of

$$\nabla f(x, y) = \langle 2(x - x_1) + 2(x - x_2) + \ldots + 2(x - x_m),$$
$$2(y - y_1) + 2(y - y_2) + \ldots + 2(y - y_m) \rangle = \langle 0, \ldots, 0 \rangle$$

$$\iff \begin{cases} m.x - \sum_{k=1}^{m} x_k = 0 \\ \\ m.y - \sum_{k=1}^{m} y_k = 0 \end{cases} \iff x = \frac{1}{m} \sum_{k=1}^{m} x_k \quad \text{and} \quad y = \frac{1}{m} \sum_{k=1}^{m} y_k.$$

We have only one critical point. The Hessian matrix of f is

$$H_f(x, y) = \begin{bmatrix} f_{xx} & f_{xy} \\ f_{yx} & f_{yy} \end{bmatrix} = \begin{bmatrix} m & 0 \\ 0 & m \end{bmatrix}.$$

The leading principal minors satisfy

$$D_1(x, y) = m > 0 \qquad\qquad D_2(x, y) = \begin{vmatrix} m & 0 \\ 0 & m \end{vmatrix} = m^2 > 0.$$

So f is strictly convex on $\mathbb{R}^2$. Then, the critical point is the global minimum of f and describes the optimal location of the substation.

> **2.** – Let f be a function of two variables given by
>
> $$f(x, y) = x^2 + y^4 - 4xy \qquad\qquad \text{for all } x \text{ and } y.$$
>
> i) Calculate the first and second order partial derivatives of f.
>
> ii) Find all the stationary points of f and classify them by means of the second derivatives test.
>
> iii) Does f have any global extreme points?
>
> iv) Use a software to graph f.

Solution: i) and **ii)** Since the function f is differentiable (because it is a polynomial), the local extreme points are critical, i.e, solutions of

$$\nabla f(x, y) = \langle 2x - 4y, 4y^3 - 4x \rangle = \langle 0, 0 \rangle$$

$$\Longleftrightarrow \quad \begin{cases} 2x - 4y = 0 \\ \\ 4y^3 - 4x = 0 \end{cases} \quad\Longleftrightarrow\quad \begin{cases} x = 2y \\ \\ y^3 - 2y = 0 \end{cases} \quad\Longleftrightarrow\quad y(y^2 - 2) = 0$$

$$\Longleftrightarrow \quad \begin{cases} x = 2y \\ \\ y = 0 \end{cases} \quad \text{or} \quad \begin{cases} x = 2y \\ \\ y = \sqrt{2} \end{cases} \quad \text{or} \quad \begin{cases} x = 2y \\ \\ y = -\sqrt{2} \end{cases}$$

We deduce that $(0,0)$, $(2\sqrt{2}, \sqrt{2})$ and $(-2\sqrt{2}, -\sqrt{2})$ are the critical points.

Classification of the critical points:

The Hessian matrix of f is

$$H_f(x, y) = \begin{bmatrix} f_{xx} & f_{xy} \\ f_{yx} & f_{yy} \end{bmatrix} = \begin{bmatrix} 2 & -4 \\ -4 & 12y^2 \end{bmatrix}$$

(x,y)	$D_1(x,y)$	$D_2(x,y)$	type
$(0,0)$	2	-16	saddle point
$(2\sqrt{2}, \sqrt{2})$	2	32	local minimum
$(-2\sqrt{2}, \sqrt{2})$	2	32	local minimum

TABLE 2.7: Critical points classification of $f(x,y) = x^2 + y^4 - 4xy$

The leading principal minors are

$$D_1(x,y) = 2 \qquad\qquad D_2(x,y) = \begin{vmatrix} 2 & -4 \\ -4 & 12y^2 \end{vmatrix}.$$

An application of the second derivatives test gives the characterization in Table 2.7.

iii) and iv) The first graphing in Figure 2.28 shows a form of a saddle. On the second graphing, there are two families of circulaire curves and a hyperbola which confirm the previous classification of the critical points.

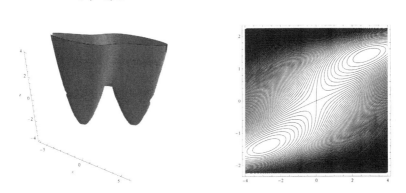

$z = y^4 - 4xy + x^2$

FIGURE 2.28: Graph and level curves of f

Global extreme points.

We cannot conclude about the concavity/convexity of f on $\mathbb{R}^2$ since the signs of the principal minors of the Hessian are as follows:

$$\Delta_1^{11}(x,y) = 12y^2 \geqslant 0, \qquad \Delta_1^{22}(x,y) = 2 \geqslant 0, \qquad \Delta_2(x,y) = 24y^2 - 16$$

and Δ_2 depends on y. Thus, f is neither convex, nor concave on $\mathbb{R}^2$.

However, we remark that, on the y axis, we have

$$f(0, y) = y^4 \longrightarrow +\infty \qquad \text{as} \qquad y \longrightarrow \pm\infty.$$

So f cannot attain a maximum value in $\mathbb{R}^2$.

Moreover, by completing the squares, we compare the values of f with its value at the local minima points $f((2\sqrt{2}, \sqrt{2}) = f((-2\sqrt{2}, -\sqrt{2}) = -4$ and obtain

$$f(x, y) + 4 = (x - 2y)^2 + (y^2 - 2)^2 \geqslant 0 \qquad \forall(x, y) \in \mathbb{R}^2.$$

Thus, f attains its global minimal value -4 at these two points.

3. – Let $f(x, y) = x^2$.

   i) Show that f has infinitely many critical points and that the second derivatives test fails for these points.

   ii) Show that f is convex on $\mathbb{R}^2$.

   iii) What is the minimum value of f? Give the minima points.

   iv) Does f have any local or global maxima? Justify your answer.

Solution:

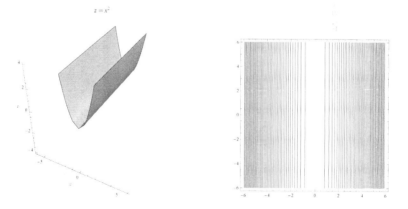

FIGURE 2.29: Graph and level curves of f

i) Since f is a differentiable function (because it is a polynomial), the local extreme points are critical ones, i.e, solutions of

$$\nabla f(x, y) = \langle 2x, 0 \rangle = \langle 0, 0 \rangle \qquad \Longleftrightarrow \qquad x = 0.$$

We deduce that the points on the y axis are all critical points of f.

Classification of the critical points:

The Hessian matrix of f is

$$H_f(x,y) = \begin{bmatrix} f_{xx} & f_{xy} \\ f_{yx} & f_{yy} \end{bmatrix} = \begin{bmatrix} 2 & 0 \\ 0 & 0 \end{bmatrix}$$

The leading principal minors at the critical points $(0,y)$ of f $(y \in \mathbb{R})$ are

$$D_1(0,y) = 2 > 0 \qquad\qquad D_2(0,y) = \begin{vmatrix} 2 & 0 \\ 0 & 0 \end{vmatrix} = 0.$$

So the second derivative test is inconclusive.

ii) The principal minors are

$$\Delta_1^{11}(x,y) = 0 \qquad \Delta_1^{22}(x,y) = 2 \qquad \Delta_2(x,y) = \begin{vmatrix} 2 & 0 \\ 0 & 0 \end{vmatrix} = 0.$$

and satisfy $\Delta_k(x,y) \geqslant 0$ for $k = 1,2$ $x,y \in \mathbb{R}^2$. Therefore f is convex in $\mathbb{R}^2$.

iii) Note that

$$f(x,y) = x^2 \geqslant 0 = f(0,y) \qquad\qquad \forall (x,y) \in \mathbb{R}^2.$$

We deduce that the critical points are global minimum for f in $\mathbb{R}^2$.

iv) Since f is infinitely differentiable (because it is polynomial) in the open set $\mathbb{R}^2$, an absolute maximum of f would be a local maximum, and therefore a critical point. But, all the critical points are minima points for f. Hence, f has no local nor absolute maxima; see Figure 2.29. In fact, on the x-axis, we have

$$f(x,0) = x^2 \longrightarrow +\infty \qquad \text{as} \qquad x \longrightarrow +\infty.$$

So f cannot attain a maximum value M in $\mathbb{R}^2$. Indeed, if not, we would have

$$f(x,y) \leqslant M \qquad \forall (x,y) \in \mathbb{R}^2.$$

Then, we have

$$f(x,0) = x^2 \leqslant M \qquad \forall x \in \mathbb{R}$$

which is not possible. For example

$$f(M,0) = M^2 > M \qquad \forall M > 1.$$

4. – Discuss the convexity/concavity of f on $\mathbb{R}^2$

$$f(x, y) = 4xy - x^2 - y^2 - 6x.$$

Are there global extreme points?

Solution:

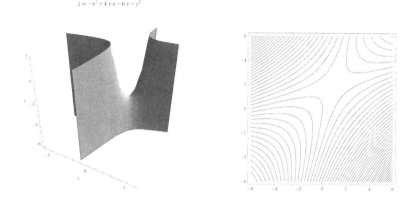

FIGURE 2.30: Graph and level curves of f

We have

$$\nabla f(x, y) = \langle 4y - 2x - 6, 4x - 2y \rangle.$$

Since f is C^∞, then

$$f \quad \text{is convex} \quad \Longleftrightarrow \quad H_f \quad \text{is semi-definite positive}$$

where the Hessian matrix of f is

$$H_f(x, y) = \begin{bmatrix} f_{xx} & f_{xy} \\ f_{yx} & f_{yy} \end{bmatrix} = \begin{bmatrix} -2 & 4 \\ 4 & -2 \end{bmatrix}$$

The principal minors of H_f are

$$\Delta_1^{11}(x, y) = -2, \qquad \Delta_1^{22}(x, y) = -2 \quad \text{and} \quad \Delta_2(x, y) = \begin{vmatrix} -2 & 4 \\ 4 & -2 \end{vmatrix} = -12.$$

So f is not convex, nor concave on $\mathbb{R}^2$; see Figure 2.30.

Remark that

$$f(0, y) = -y^2 \quad \longrightarrow -\infty \quad \text{as} \quad y \longrightarrow \pm\infty.$$

f takes large negative values and doesn't attain its minimal value.

On the other hand, when looking for the critical points of f, we obtain

$$\nabla f(x,y) = \langle 4y - 2x - 6, 4x - 2y \rangle = \langle 0,0 \rangle \quad \Longleftrightarrow \quad x = 1 \quad \text{and} \quad y = 2x = 2.$$

This point is a saddle point. It will help us to find a direction of increase of values of f. Indeed, by completing the squares, we obtain

$$f(x,y) - f(1,2) = -(2x - y)^2 + 3(x - 1)^2$$

from which we deduce that

$$f(x,2x) = f(1,2) + 3(x - 1)^2 \quad \longrightarrow +\infty \quad \text{as} \quad x \longrightarrow \pm\infty.$$

So f takes large positive values and doesn't attain its maximal value either.

5. – Let f be the function defined by:

$$f(x,y) = x^4 - 2x^2 + y^2 - 6y.$$

i) Find the critical points of f.

ii) Use the second derivative test to classify the critical points of f.

iii) Find the global minimum value of f on $\mathbb{R}^2$ by completing squares.

iv) Is there a global maximum value of f on $\mathbb{R}^2$?

v) Show that f is convex on each of the open convex sets

$$S_1 = \{(x,y): x < -1/\sqrt{3}\} \quad \text{and} \quad S_2 = \{(x,y): x > 1/\sqrt{3}\}.$$

vi) Sketch these sets and plot the critical points.

vii) Find $\min\limits_{S_1} f(x,y)$ and $\min\limits_{S_2} f(x,y)$ (justify)

viii) Set $S = S_1 \cup S_2$. Find $m_0 = \min\limits_{\mathbb{R}^2 \setminus S} (x^4 - 2x^2)$

ix) Use $(x^4 - 2x^2) \geqslant m_0$ on $\mathbb{R}^2 \setminus S$ to deduce $\min\limits_{\mathbb{R}^2 \setminus S} f(x,y)$.

Solution: The shape of the surface, in Figure 2.31, shows that the function is neither convex, nor concave.

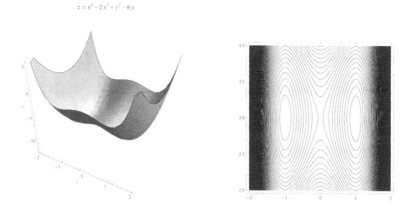

FIGURE 2.31: Graph and level curves of f

i) Since f is a differentiable function (because it is a polynomial), the local extreme points are critical points solution of

$$\nabla f(x, y) = \langle 4x^3 - 4x, 2y - 6 \rangle = \langle 0, 0 \rangle \quad \Longleftrightarrow \quad \begin{cases} 4x(x+1)(x-1) = 0 \\ \\ 2(y-3) = 0 \end{cases}$$

$$\Longleftrightarrow \quad \begin{cases} x = 0 \quad \text{or} \quad x+1 = 0 \quad \text{or} \quad x-1 = 0 \\ \\ y = 3 \end{cases}$$

$$\Longleftrightarrow \quad \begin{cases} x = 0 \quad \text{and} \quad y = 3 \\ \\ \text{or} \quad [x = -1 \quad \text{and} \quad y = 3] \\ \\ \text{or} \quad [x = 1 \quad \text{and} \quad y = 3]. \end{cases}$$

We deduce that $(-1, 3)$, $(0, 3)$ and $(1, 3)$ are the critical points of f.

ii) *Classification of the critical points:*

The Hessian matrix of f is

$$H_f(x, y) = \begin{bmatrix} f_{xx} & f_{xy} \\ f_{yx} & f_{yy} \end{bmatrix} = \begin{bmatrix} 12x^2 - 4 & 0 \\ 0 & 2 \end{bmatrix}$$

The leading principal minors are

$$D_1(x, y) = 12x^2 - 4 \qquad\qquad D_2(x, y) = \begin{vmatrix} 12x^2 - 4 & 0 \\ 0 & 2 \end{vmatrix}.$$

(x,y)	$D_1(x,y)$	$D_2(x,y)$	type
$(-1,3)$	8	$\begin{vmatrix} 8 & 0 \\ 0 & 2 \end{vmatrix} = 16$	local minimum
$(0,3)$	-4	$\begin{vmatrix} -4 & 0 \\ 0 & 2 \end{vmatrix} = -8$	saddle point
$(1,3)$	8	$\begin{vmatrix} 8 & 0 \\ 0 & 2 \end{vmatrix} = 16$	local minimum

TABLE 2.8: Classifying critical points of $f(x,y) = x^4 - 2x^2 + y^2 - 6y$

The second derivative test gives the following characterization of the points in Table 2.8.

iii) *Global minimum value of f:* We have

$$f(x,y) = (x^2 - 1)^2 + (y-3)^2 - 10 \geqslant -10 = f(1,3) = f(-1,3) \quad \forall (x,y) \in \mathbb{R}^2$$

Thus

$$\min_{(x,y)\in\mathbb{R}^2} f(x,y) = -10 = f(1,3) = f(-1,3).$$

iv) *Global maximum value of f:* We have

$$f(x,3) = (x^2 - 1)^2 - 10 \qquad \text{and} \qquad \lim_{x\to\pm\infty} f(x,3) = +\infty.$$

So f doesn't attain its maximum value on $\mathbb{R}^2$.

v) The principal minors of the Hessian of f are

$$\Delta_1^{11} = |\ f_{yy}\ | - 2 \geqslant 0, \quad \Delta_1^{22} = |\ f_{xx}\ | = 12x^2 - 4 \geqslant 0 \iff |x| \geqslant \frac{1}{\sqrt{3}}$$

$$\Delta_2 = \begin{vmatrix} 12x^2 - 4 & 0 \\ 0 & 2 \end{vmatrix} = 8(3x^2 - 1) \geqslant 0 \iff |x| \geqslant \frac{1}{\sqrt{3}}$$

So

$$\Delta_k \geqslant 0 \qquad k = 1, 2 \qquad \Longleftrightarrow \qquad |x| \geqslant \frac{1}{\sqrt{3}}.$$

Hence, H_f is semi definite positive on each open convex set S_1 and S_2. Hence

$$f \qquad \text{is convex on each of the open convex sets} \qquad S_1 \text{ and } S_2.$$

vi) Sketch of the sets S_1 and S_2 in Figure 2.32.

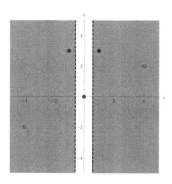

FIGURE 2.32: The convex sets S_1, and S_2

vii) Since f is convex on $S_1 = [x < -\frac{1}{\sqrt{3}}]$ and the critical point $(-1, 3)$ is in S_1 with $f(-1, 3) = -10$, then

$$\min_{S_1} \ f(x, y) = f(-1, 3) = -10.$$

f is also convex on $S_2 = [x > \frac{1}{\sqrt{3}}]$ and the critical point $(1, 3)$ is in S_2 with $f(1, 3) = -10$, then

$$\min_{S_2} \ f(x, y) = f(1, 3) = -10.$$

viii) We have

$$\varphi(x) = x^4 - 2x^2 \qquad\qquad \varphi'(x) = 4x^3 - 4x = 4x(x - 1)(x + 1)$$

Using Table 2.9, we find that

$$\min_{x \in [-\frac{1}{\sqrt{3}}, \frac{1}{\sqrt{3}}]} \varphi(x) = \varphi(-\frac{1}{\sqrt{3}}) = \varphi(\frac{1}{\sqrt{3}}) = -\frac{5}{9}.$$

x	-1		0		1
$\varphi'(x)$		$+$	0	$-$	
$\varphi(x)$	-1	$\nearrow$	0	$\searrow$	-1

TABLE 2.9: Variations of $\varphi(x) = x^4 - 2x^2$

ix) We deduce that

$$x^4 - 2x^2 \geqslant -\frac{5}{9} \qquad \forall x \in [-\frac{1}{\sqrt{3}}, \frac{1}{\sqrt{3}}]$$

$$f(x,y) \geqslant -\frac{5}{9} + y^2 - 6y = (y-3)^2 - 9 - \frac{5}{9} \geqslant -9 - \frac{5}{9} = f(\pm\frac{1}{\sqrt{3}}, 3) \ \forall (x,y) \in \mathbb{R}^2 \backslash S.$$

Hence,

$$\min_{\mathbb{R}^2 \backslash S} f(x,y) = f(-\frac{1}{\sqrt{3}}, 3) = f(\frac{1}{\sqrt{3}}, 3) = -9 - \frac{5}{9} = -\frac{86}{9}.$$

Remark. Note that

$$f(x,y) \geqslant -9 - \frac{5}{9} \geqslant -10 \qquad \forall (x,y) \in \mathbb{R}^2 \setminus S.$$

We also have

$$f(x,y) \geqslant -10 = f(-1,3) = f(1,3) \qquad \forall (x,y) \in S.$$

Hence,

$$\min_{\mathbb{R}^2} f(x,y) = f(-1,3) = f(1,3) = -10.$$

2.4 Extreme Value Theorem

The first main result of this section is

Theorem 2.4.1 *Extreme value theorem*

Let S be a closed bounded set of $\mathbb{R}^n$. Let $f \in C^0(S)$. Then f attains both its maximal and minimal values in S; that is, $\max_S f$ and $\min_S f$ exist.

The proof of the extreme value theorem uses the fact that the image of a closed bounded set S of $\mathbb{R}^n$ by a real valued continuous function $f : S \longrightarrow \mathbb{R}$ is a closed bounded set of $\mathbb{R}$ [18]. Thus $f(S)$ is a closed bounded interval $[a, b]$. Therefore

$$\exists x_m, \ x_M \in S \qquad \text{such that} \qquad f(x_m) = a, \qquad f(x_M) = b.$$

Since $f(S) = [f(x_m), f(x_M)]$, then

$$f(x_m) \leqslant f(x) \leqslant f(x_M) \qquad \forall x \in S.$$

Therefore,

$$f(x_m) = \min_S f(x) \qquad \text{and} \qquad f(x_M) = \max_S f(x).$$

Remark 2.4.1 *When f is a continuous function on a closed and bounded set S, then the extreme value theorem guarantees the existence of an absolute maximum and an absolute minimum of f on S. These absolute extreme points can occur either on the boundary of S or in the interior of S. As a consequence, to look for these points, we can proceed as follows:*

- *find the critical points of f that lie in the interior of S*

- *find the boundary points where f takes its absolute values on the boundary*

- *compare the values of f taken at the critical and boundary points found. The largest of the values of f at these points is the absolute maximum and the smallest is the absolute minimum.*

Example 1. Find the extreme values of $f(x) = \frac{1}{3}x^3 - \frac{1}{2}x^2 - 2x + 3$ on the intervals $[-1, 1]$ and $[-2, 2]$.

Solution: We have $f'(x) = x^2 - x - 2 = (x - 2)(x + 1)$ and

$$f'(x) = 0 \quad \Longleftrightarrow \quad x = 2 \quad \text{or} \quad x = -1.$$

We deduce that 2 and -1 are the critical points of f; see Figure 2.33.

• The values $\max\limits_{x \in [-1,1]} f(x)$ and $\min\limits_{x \in [-1,1]} f(x)$ exist by the extreme value theorem because f is continuous on the closed bounded interval $[-1, 1]$. Now, since there is no critical points in the interior of the interval $(-1, 1)$, these values must be in $\{f(-1),\, f(1)\}$. Comparing these two values, we conclude that

$$\max\limits_{x \in [-1,1]} f(x) = f(-1) = \frac{25}{6} \qquad \text{and} \qquad \min\limits_{x \in [-1,1]} f(x) = f(1) = \frac{5}{6}.$$

• The values $\max\limits_{x \in [-2,2]} f(x)$ and $\min\limits_{x \in [-2,2]} f(x)$ exist by the extreme value theorem because f is continuous on the closed bounded interval $[-2, 2]$. The critical point -1 is in the interior of the interval $(-2, 2)$, the absolute values must be in $\{f(-2),\, f(-1),\, f(2)\} = \{\frac{7}{3}, \frac{25}{6}, -\frac{1}{3}\}$. Comparing these three values, we conclude that

$$\max\limits_{x \in [-2,2]} f(x) = f(-1) = \frac{25}{6} \qquad \text{and} \qquad \min\limits_{x \in [-2,2]} f(x) = f(2) = -\frac{1}{3}.$$

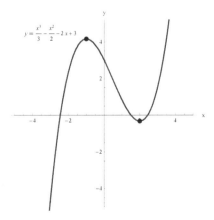

FIGURE 2.33: Absolute values on a closed interval

Example 2. Find the absolute maximum and minimum values of

$$f(x,y) = 4xy - x^2 - y^2 - 6x$$

on the closed triangle $S = \{(x,y) \; : \; 0 \leqslant x \leqslant 2, \quad 0 \leqslant y \leqslant 3x\}$.

Solution: f is continuous (because it is a polynomial) on the triangle S, which is a bounded and closed subset of $\mathbb{R}^2$. So f attains its absolute extreme points on S at the stationary points lying at the interior of S or on points located at the boundary of S (see Figure 2.34).

∗ *Interior stationary points of f:* We have

$$\nabla f = \langle 4y - 2x - 6, 4x - 2y \rangle = \langle 0,0 \rangle \qquad \Longleftrightarrow \qquad (x,y) = (1,2).$$

The point $(1,2)$ is the only critical point of f and $f(1,2) = -3$.

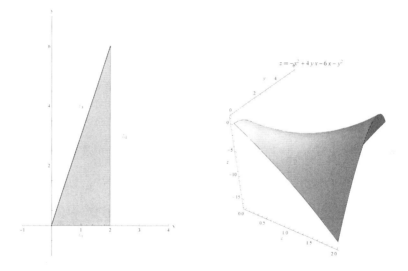

FIGURE 2.34: Extreme values of f on the triangular plane region S

∗ *Extreme values of f at the boundary of S:*

Let L_1, L_2 and L_3 be the three sides of the triangle, defined by:

$$L_1 = \{(x,0), \; 0 \leqslant x \leqslant 2\} \qquad L_2 = \{(2,y), \; 0 \leqslant y \leqslant 6\}$$
$$L_3 = \{(x,3x), \; 0 \leqslant x \leqslant 2\}.$$

– On L_1, we have: $f(x,0) = -x^2 - 6x = g(x)$, $\quad g'(x) = -2x - 6$. We deduce from the monotony of g (see Table 2.10) that

x	0	2
$g'(x)$		$-$
$g(x)$	0	$\searrow$ -16

TABLE 2.10: Variations of $g(x) = -x^2 - 6x$ on $[0, 2]$

$$\max_{L_1} f = f(0,0) = 0 \quad \text{and} \quad \min_{L_1} f = f(2,0) = -16.$$

– On L_2, we have: $f(2, y) = -y^2 + 8y - 16 = h(y), \quad h'(y) = -2y + 8.$

y	0	4	6
$h'(y)$		$+$ 0 $-$	
$h(y)$	-16	$\nearrow$ 0 $\searrow$	-4

TABLE 2.11: Variations of $h(y) = -y^2 + 8y - 16$ on $[0, 6]$

Then, from Table 2.11, we obtain

$$\max_{L_2} f = f(2,4) = 0 \quad \text{and} \quad \min_{L_2} f = f(2,0) = -16.$$

– On L_3, we have: $f(x, 3x) = 2x^2 - 6x = l(x), \quad l'(x) = 4x - 6.$

x	0	$\frac{3}{2}$	2
$l'(x)$		$-$ 0 $+$	
$l(x)$	0	$\searrow$ $-\frac{9}{2}$ $\nearrow$	-4

TABLE 2.12: Variations of $l(x) = 2x^2 - 6x$ on $[0, 2]$

Using Table 2.12, we deduce that

$$\max_{L_3} f = f(0,0) = 0 \quad \text{and} \quad \min_{L_3} f = f(3/2, 9/2) = -9/2.$$

∗ *Conclusion*: We list, in Table 2.13, the values of f at the interior critical points and at the boundary points where an absolute extreme value occurs on the considered side of the boundary. We conclude that the absolute maximum value of f is $f(0,0) = f(2,4) = 0$ and the absolute minimum value is $f(2,0) = -16$.

Now, here is a version of an extreme value theorem for a continuous function on an unbounded domain.

(x,y)	$(1,2)$	$(0,0)$	$(2,0)$	$(2,4)$	$(3/2, 9/2)$
$f(x,y)$	-3	0	-16	0	$-9/2$

TABLE 2.13: Values of f at the points

Theorem 2.4.2

Let $f(x)$ be a continuous function on an unbounded set S of $\mathbb{R}^n$ such that

$$\lim_{\|x\| \to +\infty} f(x) = +\infty \qquad\qquad (resp. \ -\infty).$$

Then, there exists an element $x^* \in S$ such that

$$f(x^*) = \min_{x \in S} f(x) \qquad\qquad (resp. \ \max_{x \in S} f(x)).$$

Proof. Let $x_0 \in S$. There exists $R_0 > 0$ such that

$$\forall x \in S: \qquad \|x\| > R_0 \qquad \Longrightarrow \qquad f(x) > f(x_0).$$

So the optimization problem $\inf_{x \in S} f(x)$ is equivalent to

$$\min_{x \in S_0} f(x) \qquad\qquad \text{with} \qquad S_0 = S \cap \{x \in \mathbb{R}^n: \quad \|x\| \leqslant R_0\}.$$

Indeed, we have

$$S_0 \subset S \qquad \Longrightarrow \qquad \min_{x \in S_0} f(x) = \inf_{x \in S_0} f(x) \geqslant \inf_{x \in S} f(x).$$

Moreover, we have

$$f(x) > f(x_0) \qquad\qquad \text{if} \quad x \in S \setminus S_0$$

$$f(x) \geqslant \min_{z \in S_0} f(z) \qquad\qquad \text{if} \quad x \in S_0$$

then

$$f(x) \geqslant \max\left(f(x_0), \min_{z \in S_0} f(z)\right) \geqslant \min_{z \in S_0} f(z) \qquad \forall x \in S.$$

Hence

$$\inf_{x \in S} f(x) \geqslant \min_{z \in S_0} f(z).$$

Note that the minimum $\min\limits_{z \in S_0} f(z)$ is attained by the extreme value theorem since S_0 is a bounded closed set of $\mathbb{R}^n$. Therefore

$$\exists x^* \in S_0 \qquad \text{such that} \qquad \min_{x \in S_0} f(x) = f(x^*).$$

Now, since, we have $\inf\limits_{x \in S} f(x) \leqslant f(x^*)$, we deduce that

$$f(x^*) = \inf_{x \in S} f(x) = \min_{x \in S_0} f(x).$$

Example 3. Let $f(x) = 3x^4 + 4x^3 - 12x^2 + 2$.

  i)   Show that f has an absolute minimum on $\mathbb{R}$.

  ii)   Find the minimal value of f on $\mathbb{R}$.

Solution: The graphing in Figure 2.35 shows three local extrema.

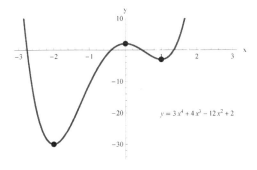

$$y = 3x^4 + 4x^3 - 12x^2 + 2$$

FIGURE 2.35: Absolute minimum of f

i) f is continuous on $\mathbb{R}$ since f is polynomial. Moreover, we have

$$\lim_{|x| \to +\infty} f(x) = +\infty.$$

Then f attains its minimum value at some point $x^* \in \mathbb{R}$.

ii) Since $\mathbb{R}$ is an open set and $x^* \in \mathbb{R}$, then x^* must be a critical point of f. We have

$$f'(x) = 12x^3 + 12x^2 - 24x = 12x(x+2)(x-1) \qquad \text{and}$$

$$f'(x) = 0 \qquad \Longleftrightarrow \qquad x = 0, \quad x = -2 \quad \text{or} \quad x = 1.$$

We deduce that

$$\min_{x \in \mathbb{R}} f(x) = \min \left\{ f(0), f(-2), f(1) \right\} = \min \left\{ 2, -30, -3 \right\} = -30 = f(-2).$$

Example 4. Let $f(x) = p(x) = x^n + +a_{n-1}x^{n-1} + \cdots + a_1 x + a_0$ be a polynomial with $n \geqslant 1$.

If n is odd, then

$$\lim_{x \to +\infty} p(x) \qquad \text{and} \qquad \lim_{x \to -\infty} p(x)$$

have opposite signs (one is $+\infty$ and the other is $-\infty$), so f has no absolute extreme points.

If n is even, then the limits above have the same sign. When they are both equal to $+\infty$, f has an absolute minimum but no absolute maximum. When the limits are both equal to $-\infty$, f has an absolute maximum but no absolute minimum.

Solved Problems

1. – Define the function $f(x, y) = \dfrac{1}{4}x^2 - \dfrac{1}{9}y^2$ on the closed unit disk. Find

i) the critical points

ii) the local extreme values

iii) the absolute extreme values.

Solution:

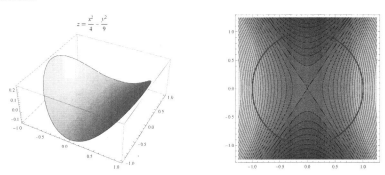

FIGURE 2.36: Graph of f on the unit disk and level curves

i) Since f is differentiable, the critical points are solution of $\nabla f(x, y) = \langle 0, 0 \rangle$. That is

$$\nabla f(x, y) = \langle \frac{x}{2}, -\frac{2y}{9} \rangle = \langle 0, 0 \rangle \quad \Longleftrightarrow \quad (x, y) = (0, 0).$$

So $(0, 0)$, the origin of the unit disk, is the unique critical point of f.

ii) *Nature of the local extreme point.* We have

$$f_{xx} = \frac{1}{2}, \qquad\qquad f_{yy} = -\frac{2}{9}, \qquad\qquad f_{xy} = 0.$$

Then $\quad D_2(0,0) = [f_{xx}f_{yy} - f_{xy}^2](0,0) = -\frac{1}{9} < 0 \quad$ and $(0,0)$ is a saddle point; see Figure 2.36.

iii) *Global extreme points.*
Since the unit disk is a bounded closed subset of $\mathbb{R}^2$, f attains its global extreme points on this set since it is continuous (because it is a polynomial function). These extreme points are interior critical points or points on the boundary of the disk.

∗ *Extreme values of f on the boundary of the disk:*

On the unit disk, f takes the values (see Table 2.14)

$$f(\cos t, \sin t) = \frac{1}{4}\cos^2 t - \frac{1}{9}\sin^2 t = g(t), \qquad\qquad t \in [0, 2\pi].$$

We have

$$g'(t) = -\frac{1}{2}\cos t \sin t - \frac{2}{9}\sin t \cos t = -\frac{13}{18}\sin t \cos t$$

θ	0		$\frac{\pi}{2}$		π		$\frac{3\pi}{2}$		2π
$\sin t$		$+$		$+$		$-$		$-$	
$\cos t$		$+$		$-$		$-$		$+$	
$g'(t)$		$-$		$\mathbin{\vert}$		$-$		$+$	
$g(t)$	$\frac{1}{4}$	$\searrow$	$-\frac{1}{9}$	$\nearrow$	$\frac{1}{4}$	$\searrow$	$-\frac{1}{9}$	$\nearrow$	$\frac{1}{4}$

TABLE 2.14: Variations of $g(t) = \frac{1}{4}\cos^2 t - \frac{1}{9}\sin^2 t$

∗ *Conclusion:*
We list, in Table 2.15, the values of f at the critical point and at the boundary points where f attains its absolute values on that boundary.

(x,y)	$(0,0)$	$(1,0)$	$(0,1)$	$(-1,0)$	$(0,-1)$
$f(x,y)$	0	$\frac{1}{4}$	$-\frac{1}{9}$	$\frac{1}{4}$	$-\frac{1}{9}$

TABLE 2.15: Values of $f(x,y) = \frac{1}{4}x^2 - \frac{1}{9}y^2$ at candidate points

The absolute maximal value of f on the disk is $\dfrac{1}{4}$ and is attained on the points $(1,0)$ and $(-1,0)$.

The absolute minimal value of f on the disk is $-\dfrac{1}{9}$ and is attained on the points $(0,1)$ and $(0,-1)$.

2. – Find the absolute extreme points of the function

$$f(x,y) = (4x - x^2)\cos y$$

on the rectangular region $1 \leqslant x \leqslant 3, \quad -\pi/4 \leqslant y \leqslant \pi/4.$

Solution:

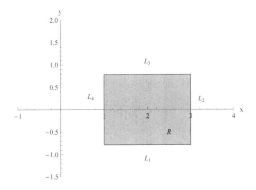

FIGURE 2.37: The plane region R

f is continuous (because it is the product of a polynomial function and the cosine function) on the rectangle $R = [1,3] \times [-\pi/4, \pi/4]$ (see Figure 2.37), which is a closed bounded set of $\mathbb{R}^2$, then f attains its absolute extreme points on R. These points are attained at the critical points of f located at the interior of R or on points located on ∂R.

* *Interior stationary points of f.* We have

$$\nabla f = \langle (4 - 2x)\cos y, \, -(4x - x^2)\sin y \rangle = \langle 0, 0 \rangle$$

$$\Longleftrightarrow \begin{cases} x = 2 \quad \text{or} \quad \cos y = 0 \\[2mm] x = 0 \quad \text{or} \quad x = 4 \quad \text{or} \quad \sin y = 0 \end{cases} \Longleftrightarrow (x,y) = (2,0).$$

The point $(2,0)$ is the only critical point of f, as shown in Figure 2.38, and $f(2,0) = 4$.

* *Extreme values of f at the boundary of R:*

Let L_1, L_2, L_3 and L_4 the four sides of the rectangle R, defined by:

$$L_1 = \{(x, -\tfrac{\pi}{4}), \quad 1 \leqslant x \leqslant 3\}, \qquad L_2 = \{(3, y), \quad -\tfrac{\pi}{4} \leqslant y \leqslant \tfrac{\pi}{4}\}$$

$$L_3 = \{(x, \tfrac{\pi}{4}), \quad 1 \leqslant x \leqslant 3\}, \qquad L_4 = \{(1, y), \quad -\tfrac{\pi}{4} \leqslant y \leqslant \tfrac{\pi}{4}\}.$$

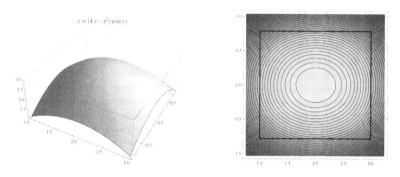

FIGURE 2.38: Values of f on R and level curves

– *On L_1, we have:* $f(x, -\tfrac{\pi}{4}) = \tfrac{\sqrt{2}}{2}(4x - x^2) = g(x)$, $\quad g'(x) = \sqrt{2}(2 - x)$.

x	1		2		3
$g'(x)$		$+$		$-$	
$g(x)$	$\tfrac{3}{2}\sqrt{2}$	$\nearrow$	$2\sqrt{2}$	$\searrow$	$\tfrac{3}{2}\sqrt{2}$

TABLE 2.16: Variations of $g(x) = \tfrac{\sqrt{2}}{2}(4x - x^2)$

We deduce from the monotony of g, described in Table 2.16, that

$$\max_{L_1} f = f(2, -\frac{\pi}{4}) = 2\sqrt{2} \qquad \min_{L_1} f = f(1, -\frac{\pi}{4}) = f(3, -\frac{\pi}{4}) = \frac{3\sqrt{2}}{2}.$$

– **On L_2,** we have: $f(3, y) = 3\cos y = h(y)$, $\quad h'(y) = -3\sin y$.
From the monotony of h (see Table 2.17), we have

$$\max_{L_2} f = f(3, 0) = 3 \qquad \min_{L_2} f = f(3, -\frac{\pi}{4}) = f(3, -\frac{\pi}{4}) = \frac{3\sqrt{2}}{2}.$$

y	$-\frac{\pi}{4}$		0		$\frac{\pi}{4}$
$h'(y)$		+		−	
$h(y)$	$\frac{3}{2}\sqrt{2}$	↗	3	↘	$\frac{3}{2}\sqrt{2}$

TABLE 2.17: Variations of $h(y) = 3\cos y$

— **On** L_3, we have: $f(x, \frac{\pi}{4}) = \frac{\sqrt{2}}{2}(4x - x^2) = l(x)$, $l'(x) = \sqrt{2}(2 - x)$.

x	1		2		3
$l'(x)$		+		−	
$l(x)$	$\frac{3}{2}\sqrt{2}$	↗	$2\sqrt{2}$	↘	$\frac{3}{2}\sqrt{2}$

TABLE 2.18: Variations of $l(x) = \frac{\sqrt{2}}{2}(4x - x^2)$

As a consequence of Table 2.18, we have

$$\max_{L_3} f = f(2, \frac{\pi}{4}) = 2\sqrt{2} \qquad \min_{L_3} f = f(1, \frac{\pi}{4}) = f(3, \frac{\pi}{4}) = \frac{3\sqrt{2}}{2}.$$

— **On** L_4, we have: $f(1, y) = 3\cos y = m(y)$, $m'(y) = -3\sin y$.

y	$-\frac{\pi}{4}$		0		$\frac{\pi}{4}$
$m'(y)$		+		−	
$m(y)$	$\frac{3}{2}\sqrt{2}$	↗	3	↘	$\frac{3}{2}\sqrt{2}$

TABLE 2.19: Variations of $m(y) = 3\cos y$

From the behaviour described in Table 2.19, we deduce that

$$\max_{L_4} f = f(1, 0) = 3 \qquad \min_{L_4} f = f(1, -\frac{\pi}{4}) = f(1, -\frac{\pi}{4}) = \frac{3\sqrt{2}}{2}.$$

∗ *Conclusion:* We list the particular points found above in Table 2.20.
The maximal value of f on R is 4 and it is attained at the point $(2, 0)$, which is an interior critical point.
The minimal value of f on R is $\frac{3\sqrt{2}}{2}$ and it is attained at the points $(1, -\frac{\pi}{4})$, $(1, \frac{\pi}{4})$, $(3, -\frac{\pi}{4})$ and $(3, \frac{\pi}{4})$.

(x,y)	$(2,0)$	$(2,\pm\frac{\pi}{4})$	$(1,\pm\frac{\pi}{4})$	$(3,\pm\frac{\pi}{4})$	$(3,0)$	$(1,0)$
$f(x,y)$	4	$2\sqrt{2}$	$\frac{3\sqrt{2}}{2}$	$\frac{3\sqrt{2}}{2}$	3	3

TABLE 2.20: Values of $f(x,y)=(4x-x^2)\cos y$ at candidate points

▌ **3.** – Find the points on the surface $z^2 = xy+4$ that are closer to the origin.

Solution: The distance of a point (x,y,z) to the origin is given by $d = \sqrt{x^2+y^2+z^2}$. The problem is equivalent to minimize $d^2 = x^2+y^2+z^2$ on the set $z^2 = xy+4$ or equivalently to look for

$$\min_{S=\mathbb{R}^2} x^2 + y^2 + (xy+4) = f(x,y).$$

Note that the function f is continuous on the unbounded set $\mathbb{R}^2$ and satisfies

$$f(x,y) \geqslant x^2 + y^2 - \frac{1}{2}(x^2+y^2) + 4 = \frac{1}{2}(x^2+y^2) + 4 = \frac{1}{2}\|(x,y)\| + 4$$

since

$$|xy| \leqslant \frac{1}{2}(x^2+y^2).$$

Thus

$$\lim_{\|(x,y)\|\to+\infty} f(x,y) = +\infty.$$

Hence, the minimization problem has a solution.

Note that a global minimum of the problem is also a local minimum, i.e., solution of

$$\nabla f = \langle 2x+y, 2y+x\rangle = \langle 0,0\rangle \iff (x,y)=(0,0) \quad \text{since} \quad \begin{vmatrix} 2 & 1 \\ 1 & 2 \end{vmatrix} \neq 0.$$

The point $(0,0)$ is the only critical point of f and $f(0,0)=4$.
Since the global minimum exists, then $(0,0)$ is the global minimum and the corresponding points on the surface $z^2 = 4+xy$ where the distance is closer to $(0,0,0)$ are: $(0,0,\pm2)$.

We can also verify that $(0,0)$ is a local minimum by applying the second derivatives test. Indeed, we have

$$f_{xx} = 2, \qquad f_{xy} = 1, \qquad f_{yy} = 2$$

$$D_1(x,y) = f_{xx} = 2 \qquad D_2(x,y) = \begin{vmatrix} f_{xx} & f_{xy} \\ f_{xy} & f_{yy} \end{vmatrix} = \begin{vmatrix} 2 & 1 \\ 1 & 2 \end{vmatrix} = 3.$$

Since $D_1(0,0) > 0$ and $D_2(0,0) > 0$, then $(0,0)$ is a strict local minimum.

4. – i) Find the quantities x, y that should be produced to maximize the total profit function

$$f(x,y) = x + 4y$$

subject to

$$2x + 3y \leqslant 19, \qquad\qquad -3x + 2y \leqslant 4$$
$$x + y \leqslant 8, \qquad\qquad 0 \leqslant x \leqslant 6, \qquad\qquad y \geqslant 0.$$

ii) Use level curves to solve the problem geometrically.

Solution:

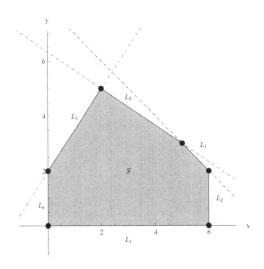

FIGURE 2.39: Hexagonal plane region S

i) Set

$$S - \{(x,y): \ 2x + 3y \leqslant 19, \ -3x + 2y \leqslant 4, \ x + y \leqslant 8, \ 0 \leqslant x \leqslant 6, \ y \geqslant 0\}.$$

The set S is the region of the plan xy, located in the first quadrant and bounded by the lines $2x+3y = 19$, $\quad -3x+2y = 4$, $\quad x+y = 8$; see Figure 2.39. It is a closed bounded convex of $\mathbb{R}^2$. Since f is continuous (because it is a polynomial), it attains its absolute extreme points on S at the stationary points lying at the interior of S or on points located at the boundary of S.

∗ *Interior stationary points of f.* We have

$$\nabla f = \langle 1, 4 \rangle \neq \langle 0, 0 \rangle \qquad \forall (x, y) \in \mathbb{R}^2.$$

There is no critical point of f.

∗ *Extreme values of f at the boundary of S:*

Let $L_1, \cdots, L_6$ be the six sides of the hexagon S, defined by:

$$L_1 = \{(x, 0), \quad 0 \leqslant x \leqslant 6\}, \qquad L_2 = \{(6, y), \quad 0 \leqslant y \leqslant 2\}$$

$$L_3 = \{(x, 8 - x), \quad 5 \leqslant x \leqslant 6\}, \qquad L_4 = \{(x, \frac{19 - 2x}{2}), \quad 2 \leqslant x \leqslant 5\},$$

$$L_5 = \{(x, \frac{4 + 3x}{2}), \quad 0 \leqslant x \leqslant 2\}, \qquad L_6 = \{(0, y), \quad 0 \leqslant y \leqslant 2\}.$$

On L_1, we have: $f(x, 0) = x$,

$$\max_{L_1} f = f(6, 0) = 6 \qquad\qquad \min_{L_1} f = f(0, 0) = 0.$$

On L_2, we have: $f(6, y) = 6 + 4y$,

$$\max_{L_2} f = f(6, 2) = 10 \qquad\qquad \min_{L_2} f = f(6, 0) = 6.$$

On L_3, we have: $f(x, 8 - x) = x + 4(8 - x) = 32 - 3x$,

$$\max_{L_3} f = f(5, 3) = 17 \qquad\qquad \min_{L_3} f = f(6, 2) = 10.$$

On L_4, we have: $f(x, \frac{19-2x}{3}) = x + \frac{4}{3}(19 - 2x) = \frac{1}{3}(76 - 5x)$,

$$\max_{L_4} f = f(2, 5) = 22 \qquad\qquad \min_{L_4} f = f(5, 3) = 17.$$

On L_5, we have: $f(x, \frac{4+3x}{2}) = x + 2(4 + 3x) = 8 + 7x$,

$$\max_{L_5} f = f(2, 5) = 22 \qquad\qquad \min_{L_5} f = f(0, 2) = 8.$$

On L_6, we have: $f(0, y) = 4y$,

$$\max_{L_6} f = f(0, 2) = 8 \qquad\qquad \min_{L_6} f = f(0, 0) = 0.$$

* *Conclusion:*
We list, in Table 2.21 below, the values of f at the boundary points where f takes absolute values on each side of the set S. We conclude that the absolute maximum value of f is $f(2, 5) = 22$ and the absolute minimum value is $f(0, 0) = 0$.

(x, y)	$(0, 0)$	$(6, 0)$	$(6, 2)$	$(5, 3)$	$(2, 5)$	$(0, 2)$
$f(x, y)$	0	6	10	17	22	8

TABLE 2.21: Values of $f(x, y) = x + 4y$ at candidate points

ii) To solve the problem geometrically, we sketch the level curves $x + 4y = k$. The profit k is attained if the line has common points with the region S. The profit 0 is attained at the point $(0, 0)$ at the level curve $x + 4y = 0$. When the profit k increases, the lines $x + 4y = k$ are parallel and move out farther to reach the point $(2, 5)$ where the highest profit is attained; see Figure 2.40.

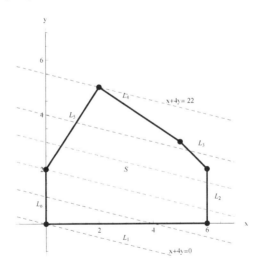

FIGURE 2.40: Level curve of highest profit

Remark 2.4.2 *Note that the points that appear in the above table are the vertices of the hexagon S. The extreme points are attained at two of these vertices. This is true in the more general problem*

$$\min_{S} p.x = p_1 x_1 + \cdots + p_n x_n \qquad or \qquad \max_{S} p.x$$

with

$$S = \{x = (x_1, \cdots, x_n) \in \mathbb{R}^{+^n} \ : \ Ax \leqslant b\}$$

where

$$A = (a_{ij})_{1 \leqslant i \leqslant m, \ 1 \leqslant j \leqslant n}, \qquad b =^t (b_1, \cdots, b_n) \qquad and \qquad p =^t (p_1, \cdots, p_n).$$

We look for the extreme points on the polyhedra

$$U = \{x = (x_1, \cdots, x_n) \in \mathbb{R}+^n \ : \ Ax = b\}.$$

We establish, when it exists, that an extreme point is at least a corner of the polyhedra U. However, when m and n take large values, the number of corners is very important, and linear programming develops various methods to approach these optimal values of the objective function [19], [5], [29].

Chapter 3

Constrained Optimization-Equality Constraints

In this chapter, we are interested in optimizing functions $f : \Omega \subset \mathbb{R}^n \longrightarrow \mathbb{R}$ over subsets described by equations

$$g(\mathbf{x}) = (g_1(\mathbf{x}), g_2(\mathbf{x}), \dots, g_m(\mathbf{x})) = \mathbf{c}_{\mathbb{R}^m} \quad with \quad m < n \quad \mathbf{x} \in \mathbb{R}^n.$$

Denote the set of the constraints by

$$S = [g(\mathbf{x}) = \mathbf{c}] = [g_1(\mathbf{x}) = c_1] \cap [g_2(\mathbf{x}) = c_2] \cap \dots \cap [g_m(\mathbf{x}) = c_m].$$

In dimension $n = 3$, when $m = 1$, the equation $g_1(x, y, z) = \mathbf{c}_{\mathbb{R}^3}$ may describe a surface, while when $m = 2$, the equations

$$g_1(x, y, z) = c_1 \quad and \quad g_2(x, y, z) = c_2$$

may describe a curve as the intersection of two surfaces. Thus, the set $[g = \mathbf{c}]$ can be seen as a set of dimensions less than 3. For $m = 3$, the set $[g = \mathbf{c}]$ may be reduced to some points or to the empty set. For this reason, we will not consider these situations and assume always $m < n$.

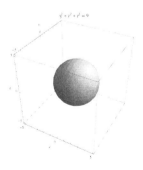

FIGURE 3.1: $S = [g_1 = 9]$, $n = 3$, $m = 1$

Example. $*$ $S = [g_1(x, y, z) = x^2 + y^2 + z^2 = 9]$ is a surface (the sphere centered at the origin with radius 3; see Figure 3.1). Here $(n = 3, \quad m = 1)$.

** $S = [g_1(x, y, z) = x^2 + y^2 + z^2 = 9] \cap [g_2(x, y, z) = z = 2]$ is the intersection of the previous sphere with the plan $z = 2$; see Figure 3.2.

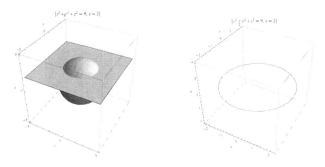

FIGURE 3.2: $S = [g_1 = 9] \cap [g_2 = 2]$, $n = 3$, $m = 2$

** $S = [g_1(x, y, z) = x^2 + y^2 + z^2 = 9] \cap [g_2(x, y, z) = z = 2] \cap [g_3(x, y, z) = y = 1] = \{(2, 1, 2), (-2, 1, 2)\}$ is the intersection of the sphere with the two planes $z = 2$ et $y = 1$. It is reduced to two points; see Figure 3.3.

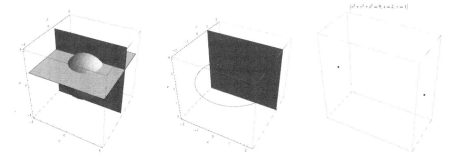

FIGURE 3.3: $S = [g_1 = 9] \cap [g_2 = 2] \cap [g_3 = 1]$, $n = 3$, $m = 3$

As in the case of unconstrained optimization, we will need to reduce our set of searches of the extreme points by looking for some necessary conditions. A local study for such points x^* cannot be done by considering balls centered at these points because the points $x^* + th$, with $|h|$ small, do not remain necessarily inside the set $[g = c]$. This situation prevents us from comparing the values $f(x^* + th)$ with $f(x^*)$. In order to remain close to x^* through points of the set $[g = c]$, an idea is to consider all the curves passing through x^* included in the constraint set. We will consider curves $t \longmapsto x(t)$ such that the set $\{x(t) : t \in [-a, a], x(0) = x^*\}$, for some $a > 0$, are included in $[g = c]$. So, if x^* is a local maximum of f, then we have

$$f(x(t)) \leqslant f(x^*) \qquad \forall t \in [-a, a].$$

Thus, 0 is local maximum point for the function $t \longmapsto f(x(t))$. *Hence*

$$\frac{d}{dt}\Big[f(x(t))\Big]\Big|_{t=0} = f'(x(t)).x'(t)\Big|_{t=0} = 0 \qquad \Longrightarrow \qquad f'(x^*).x'(0) = 0$$

$x'(0)$ *is a tangent vector to the curve* $x(t)$ *at the point* $x(0) = x^*$. *This equality musn't depend on a particular curve* $x(t)$. *So, we must have*

$$f'(x^*).x'(0) = 0 \qquad \text{for any curve } x(t) \text{ such that } g(x(t)) = \mathbf{c}.$$

In this chapter, first, we will characterize, in Section 3.1, the set of tangent vectors to such curves, then establish in Section 3.2, the equations satisfied by a local extreme point x^*. *In Section 3.3, we identify the candidates' points for optimality, and in Section 3.4, we explore the global optimality of a constrained local candidate point. Finally, we establish, in Section 3.5, the dependence of the optimal function with respect to certain of its parameters.*

3.1   Tangent Plane

Let

$$x^* \in S = [g(x) = \mathbf{c}].$$

Definition 3.1.1 *The set defined by*

$$T = \{\ x'(0) : \quad t \longmapsto x(t) \in S, \quad x \in C^1(-a, a), \quad a > 0, \quad x(0) = x^*\ \}$$

of all tangent vectors at x^* *to differentiable curves included in* S, *is called* ***tangent plane*** *at* x^* *to the surface* $[g = \mathbf{c}]$.

We have the following characterization of the tangent plane T at a regular point x^* of S.

Definition 3.1.2 *A point $x^* \in S = [g = c]$ is said to be **a regular point** of the constraints if the gradient vectors $\nabla g_1(x^*), \ldots, \nabla g_m(x^*)$ are linearly independent (LI). That is, the $m \times n$ matrix*

$$g'(x^*) = \begin{bmatrix} \frac{\partial g_1}{\partial x_1} & \frac{\partial g_1}{\partial x_2} & \cdots & \frac{\partial g_1}{\partial x_n} \\[2mm] \frac{\partial g_2}{\partial x_1} & \frac{\partial g_2}{\partial x_2} & \cdots & \frac{\partial g_2}{\partial x_n} \\[2mm] \vdots & \vdots & \vdots & \vdots \\[2mm] \frac{\partial g_m}{\partial x_1} & \frac{\partial g_m}{\partial x_2} & \cdots & \frac{\partial g_m}{\partial x_n} \end{bmatrix} \qquad \textit{has rank } m.$$

$$v_1, \ldots, v_m \in \mathbb{R}^n \text{ are LI} \iff \left(\alpha_1 v_1 + \ldots + \alpha_m v_m = 0 \implies \alpha_1 = \ldots = 0 \right).$$

The rank of a matrix $=$ rank of its transpose [10].

Theorem 3.1.1 *At a regular point $x^* \in S = [g = c]$, where g is C^1 in a neighborhood of x^*, the tangent plane T is equal to the subspace*

$$M = \{y \in \mathbb{R}^n : \quad g'(x^*)y = 0\}.$$

The proof of this theorem is an application of the implicit function theorem.

Proof. We have

$\mathbf{T} \subset \mathbf{M}$: Indeed, let $y \in \mathbf{T}$, then

$$\exists x \in C^1(-a, a) \text{ such that } g(x(t)) = c \quad \forall t \in (-a, a) \text{ for some } a > 0,$$
$$x(0) = x^*, \qquad\qquad x'(0) = y.$$

Differentiating the relation $g(x(t)) = c$, we obtain

$$g'(x(t))x'(t) = 0 \quad \forall t \in (-a, a) \implies g'(x(0))x'(0) = 0 \iff g'(x^*)y = 0.$$

Hence $y \in \mathbf{M}$.

$\mathbf{M} \subset \mathbf{T}$: $*$ Indeed, let $y \in \mathbf{M} \setminus \{0\}$ and consider the vectorial equation

$$F(t, u) = g(x^* + ty +{}^t g'(x^*)u) - c = 0,$$

where for fixed t, the vector $u \in \mathbb{R}^m$ is the unknown.

Note that F is well defined on an open subset of $\mathbb{R} \times \mathbb{R}^m$. Indeed, if g is C^1 on $B_\delta(x^*) \subset \mathbb{R}^n$, then

$$\forall (t,u) \in (-\delta_0, \delta_0) \times B_{\delta_0}(\mathbf{0}) \quad \text{with} \quad \delta_0 = \min\left(\frac{\delta}{2\|y\|}, \frac{\delta}{2\|g'(x^*)\|}\right)$$

$$\|(x^* + ty +{}^t g'(x^*)u) - x^*\| \leqslant |t|\|y\| + \|u\|\|g'(x^*)\|$$

$$< \frac{\delta}{2\|y\|}\|y\| + \frac{\delta}{2\|g'(x^*)\|}\|g'(x^*)\| = \frac{\delta}{2} + \frac{\delta}{2} = \delta$$

$$\implies \quad [x^* + ty +{}^t g'(x^*)u] \in B_\delta(x^*).$$

We have

$$F(t,u) = g(X(t,u)) - c \qquad\qquad X(t,u) = x^* + ty +{}^t g'(x^*)u$$

$$X_j(t,u) = x_j^* + ty_j + \sum_{l=1}^{m} \frac{\partial g_l}{\partial x_j}(x^*)u_l \qquad\qquad \frac{\partial X_j}{\partial u_i} = \frac{\partial g_i}{\partial x_j}(x^*)$$

$$\frac{\partial F_k}{\partial u_i}(t,u) = \sum_{j=1}^{n} \frac{\partial g_k}{\partial X_j}\frac{\partial X_j}{\partial u_i} = \sum_{j=1}^{n} \frac{\partial g_k}{\partial x_j}(X(t,u))\frac{\partial g_i}{\partial x_j}(x^*)$$

$$\left[\frac{\partial F_k}{\partial u_i}(t,u)\right]_{k,i=1,\cdots,m} = g'(X(t,u))\left({}^t g'(x^*)\right).$$

By hypotheses, we have

- F is a C^1 function in the open set $A = (-\delta_0, \delta_0) \times B_{\delta_0}(\mathbf{0})$

- $F(0,\mathbf{0}) = g(x^*) - c = 0$

- $(0,\mathbf{0}) \in (-\delta_0, \delta_0) \times B(\mathbf{0}, \delta_0)$, so $(0,\mathbf{0})$ is an interior point

- $det(\nabla_u F(0,\mathbf{0})) = \dfrac{\partial(F_1,\cdots,F_m)}{\partial(u_1,\cdots,u_m)} = det\left[g'(x^*)\left({}^t g'(x^*)\right)\right] \neq 0$ \quad as $rank\, g'(x^*) = m$.

Then, by the implicit function theorem, there exists open balls

$$B_\epsilon(0) \subset (-\delta_0, \delta_0), \quad B_\eta(\mathbf{0}) \subset B_{\delta_0}(\mathbf{0}), \quad \epsilon, \eta > 0 \quad \text{with} \quad B_\epsilon(0) \times B_\eta(\mathbf{0}) \subseteq A,$$

and such that

$$det(\nabla_u F(t, u)) \neq 0 \qquad \text{in} \qquad B_\epsilon(0) \times B_\eta(\mathbf{0})$$

$$\forall t \in B_\epsilon(0), \quad \exists! u \in B_\eta(\mathbf{0}): \qquad F(t, u) = 0$$

$$u : (-\epsilon, \epsilon) \longrightarrow B_\eta(\mathbf{0}); \qquad t \longmapsto u(t) \quad \text{is a } C^1 \text{ function.}$$

The curve
$$x(t) = X(t, u(t)) = x^* + ty +^t g'(x^*)u(t)$$
is thus, by construction, a curve on S. By differentiating both sides of

$$F(t, u(t)) = g(x(t)) - c = g(X(t, u(t))) - c = 0$$

with respect to t, we get

$$0 = \frac{d}{dt}g(x(t)) = \sum_{j=1}^{n} \frac{\partial g}{\partial X_j}\frac{\partial X_j}{\partial t}$$

$$X_j(t, u) = x_j^* + ty_j + \sum_{l=1}^{m} \frac{\partial g_l}{\partial x_j}(x^*)u_l \qquad \frac{\partial X_j}{\partial t} = y_j + \sum_{l=1}^{m} \frac{\partial g_l}{\partial x_j}(x^*)\frac{\partial u_l}{\partial t}$$

$$0 = \frac{d}{dt}g(x(t))\Big]_{t=0} = \sum_{j=1}^{n} \frac{\partial g}{\partial x_j}(X(t, u))\left[y_j + \sum_{l=1}^{m} \frac{\partial g_l}{\partial x_j}(x^*)\frac{\partial u_l}{\partial t}\right]_{t=0}$$

$$= g'(x^*)y + g'(x^*)^t g'(x^*)u'(0).$$

Since $y \in \mathbf{M} \setminus \{0\}$, we have $g'(x^*).y = \mathbf{0}$. Moreover, since $g'(x^*)^t g'(x^*)$ is nonsingular, we conclude that $u'(0) = \mathbf{0}$. Hence

$$x'(0) = y +^t g'(x^*)u'(0) = y + 0 = y$$

and y is a tangent vector to the curve $x(t)$ included in S, so $y \in \mathbf{T}$.

** If $y = 0$, the constant curve $x(t) = x^*$ is included in S and $x'(0) = 0 = y$, so $0 \in \mathbf{T}$.

It is easy to show that $\mathbf{M}$ is a subspace of $\mathbb{R}^n$. Indeed, $\mathbf{0} \in \mathbf{M}$ and for $y_1, y_2 \in \mathbf{M}, \kappa \subset \mathbb{R}$, we have

$$g'(x^*)(y_1 + \kappa y_2) = g'(x^*)y_1 + \kappa g'(x^*)y_2 = 0.$$

Theorem 3.1.2 *Implicit function theorem [15] [20]*

Let A in $\mathbb{R}^n \times \mathbb{R}^m$ be an open set. Let $F = (F_1, \ldots, F_m)$ be a $C^1(A)$ function. Consider the vector equation

$$F(x, y) = 0.$$

If

$$\exists (x^0, y^0) \in \overset{\circ}{A} = A, \qquad F(x^0, y^0) = 0 \quad and \quad det\left(F'_y(x^0, y^0)\right) \neq 0,$$

then, $\exists \epsilon, \eta > 0$ such that

$$det\left(F'_y(x, y)\right) \neq 0 \qquad \forall (x, y) \in B_\epsilon(x^0) \times B_\eta(y^0) \subset A$$

$$\forall x \in B_\epsilon(x^0), \quad \exists! y \in B_\eta(y^0): \qquad F(x, y) = 0$$

$$\varphi : B_\epsilon(x^0) \longrightarrow B_\eta(y^0); \qquad x \longmapsto \varphi(x) = y \quad is \ C^1(B_\epsilon(x^0))$$

$$\varphi'(x) = -\left(F'_y(x, y)\right)^{-1} F'_x(x, y)$$

where

$$F'_y(x, y) = \nabla_y F(x, y) = \begin{bmatrix} \frac{\partial F_1}{\partial y_1} & \cdots & \frac{\partial F_1}{\partial y_m} \\ \vdots & \vdots & \vdots \\ \frac{\partial F_m}{\partial y_1} & \cdots & \frac{\partial F_m}{\partial y_m} \end{bmatrix} \quad \text{gradient of } F \text{ with respect to } y$$

$$det(F'_y(x, y)) = \frac{\partial(F_1, \ldots, F_m)}{\partial(y_1, \ldots, y_m)} \qquad \text{Jacobian of } F \text{ with respect to } y.$$

Remark 3.1.1 *Denote by $\mathbf{T}(x^*)$ the translation of $\mathbf{T}$ by the vector x^* :*

$$\mathbf{T}(x^*) = x^* + \mathbf{M} = \{x^* + h \in \mathbb{R}^n : g'(x^*).h = 0\}$$

$$= \{x \in \mathbb{R}^n : g'(x^*).(x - x^*) = 0\}.$$

$\mathbf{T}(x^*)$ *is the tangent plane to the surface $[g(x) = c]$ passing through x^*.*

FIGURE 3.4: Horizontal tangent line at local extreme points

Remark 3.1.2 *Tangent plane at a point of a surface $z = f(x)$*

* *Suppose x^* is an interior point of a surface $z = f(x)$ where f is a C^1 function. Then, the tangent plane at $(x^*, f(x^*))$ is given by*

$$z = f(x^*) + f'(x^*).(x - x^*)$$

Indeed, setting $g(x, z) = z - f(x) = 0,$ *then*

$$g'(x, z) = \langle -f'(x), 1 \rangle \neq \mathbf{0}   \text{and}   rank(g'(x^*, f(x^*))) = 1.$$

The tangent plane at $(x^*, f(x^*))$ is characterized by

$$g'(x^*, f(x^*)).\langle x - x^*, z - f(x^*) \rangle = \langle -f'(x^*), 1 \rangle.\langle x - x^*, z - f(x^*) \rangle = 0.$$

** *If x^* is an interior stationary point, then $\nabla f(x^*) = \mathbf{0}$, and the tangent plane $T(x^*, f(x^*))$ is the horizontal plane $z = f(x^*)$.*

*** *The graph of the tangent plane is the graph of the linear approximation $L(x) = f(x^*) + f'(x^*).(x - x^*)$. Thus, we have*

$$f(x) \approx_{x^*} L(x)   \text{for x close to x^*.}$$

Example 1. **The tangent plane to a curve $y = f(x)$ at a point $(x_0, f(x_0))$** corresponds to the tangent line to the curve at that point described by the equation

$$y = f(x_0) + f'(x_0)(x - x_0).$$

The following examples, in Table 3.1, show that the tangent line is horizontal at local extreme points and separates the graph into two parts at an inflection point; see Figure 3.4 and Figure 3.5.

f(x)	point x_0	$f'(x_0)$	tangent line
$(x-1)^2 - 1$	1 : global minimum	$2(x-1)\Big]_{x=1} = 0$	$y = -1$
$1 - x^2$	0 : global maximum	$-2x\Big]_{x=0} = 0$	$y = 1$
$(x-1)^3 + 1$	0 : inflection point	$3(x-1)^2\Big]_{x=1} = 0$	$y = 1$
$\ln x$	e	$\dfrac{1}{x}\Big]_{x=e} = \dfrac{1}{e}$	$y - 1 = \dfrac{1}{e}(x - e)$

TABLE 3.1: Tangent planes in one dimension

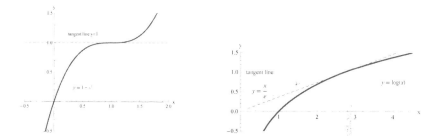

FIGURE 3.5: Tangent lines at an inflection and at an ordinary points

Example 2. *The tangent plane to a surface* $z = f(x, y)$ at a point $(x_0, y_0, f(x_0, y_0))$ corresponds to the usual tangent plane to the surface at that point described by the equation

$$z = f(x_0, y_0) + f_x(x_0, y_0)(x - x_0) + f_y(x_0, y_0)(y - y_0).$$

A normal vector to this plane is

$$\mathbf{n} = \langle f_x(x_0, y_0), f_y(x_0, y_0), -1 \rangle = f_x(x_0, y_0)\mathbf{i} + f_y(x_0, y_0)\mathbf{j} - \mathbf{k}.$$

A normal line to the surface $z = f(x, y)$ at $(x_0, y_0, f(x_0, y_0))$ is the line parallel to the vector $\mathbf{n}$.

The examples, given in Table 3.2 and graphed in Figures 3.6 and 3.7, show that the tangent plane is horizontal at local extreme points and separates the graph into two parts at a saddle point.

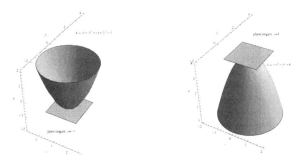

FIGURE 3.6: Horizontal tangent planes at local extreme points

The corresponding tangent planes at (x_0, y_0) are respectively

$$a) \quad z = -1, \qquad b) \quad z = 4, \qquad c) \quad z = 0, \qquad d) \quad z = 2x + 2y - 3$$

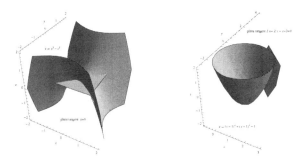

FIGURE 3.7: Tangent planes at a saddle and ordinary points

Example 3. Find the tangent plane at the point $(0,1,0)$ to the set $g = (g_1, g_2) = \langle 1, 1 \rangle$ with

$$g_1(x,y,z) = x + y + z, \qquad \text{and} \qquad g_2(x,y,z) = x^2 + y^2 + z^2.$$

Solution: The surface $g(x,y,z) = \langle 1, 1 \rangle$ is the intersection of the two surfaces $g_1(x,y,z) = 1$ and $g_2(x,y,z) = 1$. So, it is a curve in the space $\mathbb{R}^3$. We have

$$g'(x,y,z) = \begin{bmatrix} \frac{\partial g_1}{\partial x} & \frac{\partial g_1}{\partial y} & \frac{\partial g_1}{\partial z} \\ \\ \frac{\partial g_2}{\partial x} & \frac{\partial g_2}{\partial y} & \frac{\partial g_2}{\partial z} \end{bmatrix} = \begin{bmatrix} 1 & 1 & 1 \\ 2x & 2y & 2z \end{bmatrix}.$$

$\mathbf{z = f(x, y)}$	$\mathbf{f'(x_0, y_0)}$
a) $(x-1)^2 + (y+1)^2 - 1$	$\langle 2(x-1), 2(y+1) \rangle \Big]_{(x,y)=(1,-1)} = \langle 0, 0 \rangle$
b) $4 - x^2 - y^2$	$\langle -2x, -2y \rangle \Big]_{(x,y)=(0,0)} = \langle 0, 0 \rangle$
c) $y^2 - x^2$	$\langle -2x, 2y \rangle \Big]_{(x,y)=(0,0)} = \langle 0, 0 \rangle$
d) $(x-1)^2 + (y+1)^2 - 1$	$\langle 2(x-1), 2(y+1) \rangle \Big]_{(x,y)=(2,0)} = \langle 2, 2 \rangle$

TABLE 3.2: Examples in two dimension

$$g'(0,1,0) = \begin{bmatrix} 1 & 1 & 1 \\ 0 & 2 & 0 \end{bmatrix} \qquad \text{has rank 2}$$

The tangent plane is the set of points (x, y, z) such that

$$g'(0,1,0).\langle x-0, y-1, z-0 \rangle = \begin{bmatrix} 1 & 1 & 1 \\ 0 & 2 & 0 \end{bmatrix} \cdot \begin{bmatrix} x \\ y-1 \\ z \end{bmatrix} = \begin{bmatrix} 0 \\ 0 \end{bmatrix}$$

$$\Longleftrightarrow \qquad \begin{cases} x + y - 1 + z = 0 \\ 2(y-1) = 0. \end{cases}$$

A parametrization of the tangent plane to the two surfaces at $(0,1,0)$ is the line (see Figure 3.8)

$$x = t \qquad\qquad y = 1 \qquad\qquad z = -t, \qquad\qquad t \in \mathbb{R}.$$

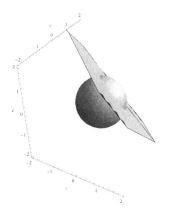

FIGURE 3.8: Tangent plane at $(0, 1, 0)$ to $[g = \langle 1, 1 \rangle]$

Remark 3.1.3 *Note that the representation of the tangent plane obtained in the theorem has used the fact that the point was regular. When this hypothesis is omitted, the representation is not necessary true. Indeed, if S is the set defined by*

$$g(x, y) = 0 \qquad \text{with} \qquad g(x, y) = x^2,$$

then S is the y axis. No point of S is regular since we have

$$g'(x, y) = \langle 2x, 0 \rangle \qquad \text{and} \qquad g'(0, y) = \langle 0, 0 \rangle \qquad \text{on the y-axis.}$$

We deduce that at each point $(0, y_0) \in S$, we have

$$M = \{ h = (h_1, h_2) : \quad g'(0, y_0).h = 0 \} = \mathbb{R}^2.$$

However, the line

$$x(t) = 0 \qquad\qquad y(t) = y_0 + t$$

passes through the point $(0, y_0)$ at $t = 0$ with direction $\langle x'(0), y'(0) \rangle = \langle 0, 1 \rangle$ and remains included in S. Hence, the tangent plane is equal to S.

Solved Problems

1. – Find an equation of the tangent plane to the ellipsoid $x^2 + 4y^2 + z^2 = 18$ at the point $(1, 2, 1)$.

Solution: Set $g(x, y, z) = x^2 + 4y^2 + z^2 = 18$. Then, $g'(x, y, z) = 2xi + 8yj + 2zk$,

$$g'(1, 2, 1) = 2i + 16j + 2k \neq \mathbf{0} \implies rank(g'(1, 2, 1)) = 1.$$

The tangent plane (see Figure 3.9) is the set of points (x, y, z) such that

$$g'(1, 2, 1).\langle x - 1, y - 2, z - 1 \rangle = \begin{bmatrix} 2 & 16 & 2 \end{bmatrix} . \begin{bmatrix} x - 1 \\ y - 2 \\ z - 1 \end{bmatrix} = 0$$

$$\iff \quad 2(x - 1) + 16(y - 2) + 2(z - 1) = 0 \quad \iff \quad x + 8y + z - 18 = 0.$$

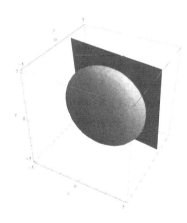

FIGURE 3.9: Tangent plane at $(1, 2, 1)$ to the ellipsoid

2. – Find all points on the surface

$$2x^2 + 3y^2 + 4z^2 = 9$$

at which the tangent plane is parallel to the plane $x - 2y + 3z = 5$.

Solution: Set $g(x, y, z) = 2x^2 + 3y^2 + 4z^2 = 9$. We have

$$g'(x, y, z) = \begin{bmatrix} 4x & 6y & 8z \end{bmatrix} \neq \mathbf{0} \quad \text{on } [g = 9] \quad \Longrightarrow \quad rank(g'(x, y, z)) = 1$$

since

$$g'(x, y, z) = \mathbf{0} \quad \Longleftrightarrow \quad (x, y, z) = \mathbf{0} \quad \text{and} \quad g(\mathbf{0}) \neq 9.$$

The tangent plane to the surface $g(x, y, z) = 9$ at a point (x_0, y_0, z_0) is the set of points (x, y, z) such that

$$g'(x_0, y_0, z_0).\langle x - x_0, y - y_0, z - z_0 \rangle = \begin{bmatrix} 4x_0 & 6y_0 & 8z_0 \end{bmatrix}. \begin{bmatrix} x - x_0 \\ y - y_0 \\ z - z_0 \end{bmatrix} = 0$$

$$\Longleftrightarrow \quad 4x_0(x - x_0) + 6y_0(y - y_0) + 8z_0(z - z_0) = 0.$$

This tangent plane will be parallel to the plane $x - 2y + 3z = 5$ if the two planes have their respective normals $g'(x_0, y_0, z_0)$ and $\langle 1, -2, 3 \rangle$ parallel. So, we have to solve the following system

$$\begin{cases} \text{find } t \in \mathbb{R} : \\ g'(x_0, y_0, z_0) = t\langle 1, -2, 3 \rangle \\ g(x_0, y_0, z_0) = 9 \end{cases} \Longleftrightarrow \begin{cases} 4x_0 = t \\ 6y_0 = -2t \\ 8z_0 = 3t \\ 2x_0^2 + 3y_0^2 + 4z_0^2 = 9 \end{cases}$$

$$\Longrightarrow \quad 2\left(\frac{t}{4}\right)^2 + 3\left(-\frac{t}{3}\right)^2 + 4\left(\frac{3t}{8}\right)^2 = 9 \quad \Longrightarrow \quad t = \pm\frac{12}{7}\sqrt{3}.$$

The needed points on the surface are

$$\left(\frac{3}{7}\sqrt{3}, -\frac{4}{7}\sqrt{3}, \frac{9}{14}\sqrt{3}\right), \qquad \left(-\frac{3}{7}\sqrt{3}, \frac{4}{7}\sqrt{3}, -\frac{9}{14}\sqrt{3}\right).$$

The equations of the tangent planes to the surface (see Figure 3.10) at these points are

$$\left(x - \frac{3}{7}\sqrt{3}\right) - 2\left(y + \frac{4}{7}\sqrt{3}\right) + 3\left(z - \frac{9}{14}\sqrt{3}\right) = 0,$$

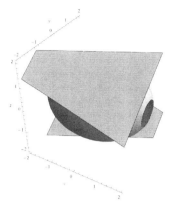

FIGURE 3.10: Parallel tangent planes to an ellipsoid

$$\left(x + \frac{3}{7}\sqrt{3}\right) - 2\left(y - \frac{4}{7}\sqrt{3}\right) + 3\left(z + \frac{9}{14}\sqrt{3}\right) = 0.$$

3. – Show that the surfaces

$$z = \sqrt{x^2 + y^2} \qquad \text{and} \qquad z = \frac{1}{10}(x^2 + y^2) + \frac{5}{2}$$

intersect at $(3, 4, 5)$ and have a common tangent plane at that point.

Solution: Set

$$g_1(x, y, z) = z - \sqrt{x^2 + y^2} \qquad g_2(x, y, z) = z - \frac{1}{10}(x^2 + y^2) - \frac{5}{2}.$$

Since $g_1(3, 4, 5) = 0$ and $g_2(3, 4, 5) = 0$, then the point $(3, 4, 5)$ is a common point to the surfaces $g_1(x, y, z) = 0$ and $g_2(x, y, z) = 0$. We have

$$g_1'(x, y, z) = -\frac{x}{\sqrt{x^2 + y^2}} i - \frac{y}{\sqrt{x^2 + y^2}} j + k$$

$$g_2'(x, y, z) = -\frac{x}{5} i - \frac{y}{5} j + k$$

$$g_1'(3, 4, 5) = -\frac{3}{5} i - \frac{4}{5} j + k \neq \mathbf{0}, \qquad rank(g_1'(3, 4, 5)) = 1$$

$$g_2'(3, 4, 5) = -\frac{3}{5} i - \frac{4}{5} j + k \neq \mathbf{0}, \qquad rank(g_2'(3, 4, 5)) = 1.$$

Note that the normal vectors $g_1'(3,4,5)$ and $g_2'(3,4,5)$ of the tangent planes to the surfaces $g_1(x,y,z) = 0$ and $g_2(x,y,z) = 0$ respectively are the same. Hence, the two surfaces have a common tangent plane at this point with the equation

$$-\frac{3}{5}(x-3) - \frac{4}{5}(y-4) + (z-5) = 0.$$

4. – Find two unit vectors that are normal to the surface

$$\sin(xz) - 4\cos(yz) = 4$$

at the point $P(\pi, \pi, 1)$.

Solution: A vector that is normal to the surface $g(x,y,z) = \sin(xz) - 4\cos(yz) = 4$ is normal to the tangent plane to this surface at this point and we have

$$g'(x,y,z) = z\cos(xz)i + 4z\sin(yz)j + (x\cos(xz) + 4y\sin(yz))k$$
$$g'(\pi,\pi,1) = -i - \pi k \neq \mathbf{0} \qquad \Longrightarrow \qquad rank(g'(\pi,\pi,1)) = 1.$$

A normal vector to the tangent plane is $g'(\pi,\pi,1) = -i - \pi k$ and two unit vectors that are normal to the surface $\sin(xz) - 4\cos(yz) = 4$ at the point $P(\pi,\pi,1)$ are

$$\pm\frac{g'(\pi,\pi,1)}{\|g'(\pi,\pi,1)\|} = \pm\langle -\frac{1}{\sqrt{1+\pi^2}}, 0, -\frac{\pi}{\sqrt{1+\pi^2}} \rangle.$$

3.2   Necessary Condition for Local Extreme Points-Equality Constraints

Before setting the results rigorously, we will try to give an intuitive approach of the comparison of the values of f close to a local maximum value $f(x^*)$ under the constraints $g(x) = c$. We will follow the unconstrained case in parallel.

• **Unconstrained case:** *We compare values of f taken in a neighborhood of x^* in all directions*

$$f(x^* + th) \leqslant f(x^*) \qquad for \ h \in \mathbb{R}^n \qquad |t| < \delta$$

or equivalently, for each $i = 1, \dots, n$

$$f(x^* + te_i) \leqslant f(x^*) \qquad |t| < \delta$$

then for $|t| < \delta$, we have

$$\frac{f(x^* + te_i) - f(x^*)}{t} \leqslant 0 \qquad if \quad t > 0 \qquad and$$

$$\frac{f(x^* + te_i) - f(x^*)}{t} \geqslant 0 \qquad if \quad t > 0.$$

Since f is differentiable, we obtain as $t \longrightarrow 0^+$ and $t \longrightarrow 0^-$ respectively

$$f_{x_i}(x^*) \leqslant 0 \qquad and \qquad f_{x_i}(x^*) \geqslant 0$$

So

$$f_{x_i}(x^*) - 0 \qquad for \ each \quad i - 1, \dots, n.$$

• **Constrained case:** *We cannot choose points around x^* in any direction because we need to remain on the set $[g = c]$. A way to do that, is to consider curves $t \longmapsto x(t)$ satisfying $x(t) \in [g = c]$ for $t \in (-a, a)$ and $x(0) = x^*$. Then, we have*

$$f(x(t)) \leqslant f(x^*) \qquad \forall t \in (-a, a)$$

and 0 is local maximum point for the function $t \longmapsto f(x(t))$. Hence, for regular functions, we have

$$\frac{d}{dt}\Big[f(x(t))\Big]\Big|_{t=0} = f'(x(t)).x'(t)\Big|_{t=0} = 0 \qquad \Longrightarrow \qquad f'(x^*).x'(0) = 0.$$

$x'(0)$ is a tangent vector to the curve $x(t)$ at the point $x(0) = x^*$. This equality musn't depend on a particular curve. Thus, it must be satisfied for any $y = x'(0) \in M$, which is summarized below:

Lemma 3.2.1 *Let f and $g = (g_1, \ldots, g_m)$ be C^1 functions in a neighborhood of $x^* \in [g = c]$. If x^* is a regular point and a local extreme point of f subject to these constraints, then we have*

$$\forall y \in \mathbb{R}^n : \qquad g'(x^*)y = 0 \qquad \Longrightarrow \qquad f'(x^*)y = 0.$$

The lemma says that $f'(x^*)$ is orthogonal to the plane tangent at x^* to the surface $g(x) = c$. As a consequence, we will see that $f'(x^*)$ is a linear combination of $g'_1(x^*), \ldots, g'_m(x^*)$.

Theorem 3.2.1 *Let f and $g = (g_1, \ldots, g_m)$ be C^1 functions in a neighborhood of $x^* \in [g = c]$. If x^* is a regular point and a local extreme point of f subject to these constraints, then there exists unique numbers $\lambda_1^*, \ldots, \lambda_m^*$ such that*

$$\frac{\partial f}{\partial x_i}(x^*) - \sum_{j=1}^m \lambda_j^* \frac{\partial g_j}{\partial x_i}(x^*) = 0 \qquad i = 1, \ldots, n.$$

Proof. The proof uses a simple argument of linear algebra. Indeed,

$$A = g'(x^*) = \begin{bmatrix} \frac{\partial g_1}{\partial x_1} & \frac{\partial g_1}{\partial x_2} & \cdots & \frac{\partial g_1}{\partial x_n} \\ \frac{\partial g_2}{\partial x_1} & \frac{\partial g_2}{\partial x_2} & \cdots & \frac{\partial g_2}{\partial x_n} \\ \vdots & \vdots & \vdots & \vdots \\ \frac{\partial g_m}{\partial x_1} & \frac{\partial g_m}{\partial x_2} & \cdots & \frac{\partial g_m}{\partial x_n} \end{bmatrix} \qquad rank(A) = m$$

$$b = f'(x^*) = \left[\frac{\partial f}{\partial x_1}, \ldots, \frac{\partial f}{\partial x_n} \right] \qquad b \in \mathbb{R}^n.$$

From the previous lemma, we have

$$\forall y \in \mathbb{R}^n : \quad Ay = 0 \quad \Longrightarrow \quad b.y = 0.$$

In other words, we have

$$KerA = Ker \begin{bmatrix} A \\ b \end{bmatrix}$$

where $KerN$ denotes the Kernel [10] of the linear transformation induced by the matrix N. Since we have [10]

$$dim\mathbb{R}^n = dim(kerA) + rank(A) = dim(ker \begin{bmatrix} A \\ b \end{bmatrix}) + rank(\begin{bmatrix} A \\ b \end{bmatrix})$$

then

$$rank(A) = rank(\begin{bmatrix} A \\ b \end{bmatrix})$$

which means that the vector b is linearly dependent on the line vectors of A, so there exists a unique vector $\lambda^* = (\lambda_1^*, \ldots, \lambda_m^*) \in \mathbb{R}^m$ such that

$$^t b =^t A\lambda^* \qquad \Longleftrightarrow \qquad \frac{\partial f}{\partial x_i}(x^*) = \sum_{j=1}^{m} \lambda_j^* \frac{\partial g_j}{\partial x_i}(x^*), \qquad i = 1, \ldots, n.$$

Finally, to look for extreme points of f subject to the constraint $g(x) = c$, we are lead to solve the system

$$\frac{\partial f}{\partial x_i}(x) - \sum_{j=1}^{m} \lambda_j \frac{\partial g_j}{\partial x_i}(x) = 0 \qquad i = 1, \cdots, n$$

$$g_j(x) - c_j = 0 \qquad j = 1, \cdots, m.$$

These equations suggest to introduce the function

$$\mathcal{L}(x, \lambda) = f(x) - \lambda_1(g_1(x) - c_1) - \cdots - \lambda_m(g_m(x) - c_m)$$

called **Lagrange function** or **Lagrangian** and $\lambda_1, \ldots, \lambda_m$ the **Lagrange multipliers**.

The necessary conditions can be, then, expressed in the form

$$\begin{cases} \dfrac{\partial \mathcal{L}}{\partial x_i}(x, \lambda) = \dfrac{\partial f}{\partial x_i}(x) - \sum_{j=1}^{m} \lambda_j \dfrac{\partial g_j}{\partial x_i}(x) = 0 \qquad i = 1, \ldots, n \\[4mm] \dfrac{\partial \mathcal{L}}{\partial \lambda_j}(x, \lambda) = -(g_j(x) - c_j) = 0 \qquad j = 1, \ldots, m \end{cases}$$

or simply $\qquad \nabla\mathcal{L}(x, \lambda) = 0.$

We may reformulate the previous theorem as follow:

Theorem 3.2.2 *Let f and $g = (g_1, \ldots, g_m)$ be C^1 functions in a neighborhood of $x^* \in [g = c]$.*

$$x^* \text{ is a regular point and a local extreme point of } f$$

$$\implies \qquad \exists! \, \lambda^* \in \mathbb{R}^m \text{ such that } \nabla \mathcal{L}(x^*, \lambda^*) = 0.$$

Remark 3.2.1 *When $m = 1$, the necessary condition is reduced to*

$$\exists! \lambda^* \in \mathbb{R}: \qquad \nabla f = \lambda^* \nabla g \qquad \implies \qquad \nabla f \, // \, \nabla g.$$

The vectors $g'(x^*)$ and $f'(x^*)$ are respectively normal to the level curves $g(x) = c$ and $f(x) = f(x^*)$. When the extreme point is attained then the two vectors $g'(x^*)$ and $f'(x^*)$ are parallel. Thus the two level curves have a common tangent plane at x^*. When, using a graphic utility, it is where the level curves are tangent, the constrained extreme points may locate.

Example 1. At what points on the circle $x^2 + y^2 = 1$ does $f(x, y) = xy$ have its maximum and minimum?

Solution: Set

$$g(x, y) = x^2 + y^2 \qquad\qquad S = \{(x, y): \quad g(x, y) = x^2 + y^2 = 1\}$$

By the extreme-value theorem, f attains its maximum and minimum values on S since f is continuous on the closed and bounded unit circle S; see Figure 3.11.

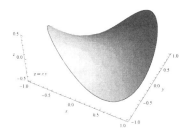

FIGURE 3.11: Graph of $f(x, y) = xy$ on the unit disk $[x^2 + y^2 \leqslant 1]$

Next, the functions f and g are C^1 around each point $(x,y) \in \mathbb{R}^2$ and in particular each point of S is relatively interior to S and is a regular point since we have

$$g'(x,y) = 2xi + 2yj \neq \mathbf{0} \qquad \Longrightarrow \qquad rank(g'(x,y)) = 1.$$

Thus, introducing the Lagrangian

$$\mathcal{L}(x,y,\lambda) = xy - \lambda(x^2 + y^2 - 1),$$

we can apply Lagrange multipliers method to look for the interior extreme points as solutions of the system

$$\begin{cases} \mathcal{L}_x = f_x(x,y) - \lambda g_x(x,y) = 0 \\ \mathcal{L}_y = f_y(x,y) - \lambda g_y(x,y) = 0 \\ \mathcal{L}_\lambda = -(g(x,y) - 1) = 0 \end{cases} \Longleftrightarrow \begin{cases} y - 2x\lambda = 0 \\ x - 2y\lambda = 0 \\ x^2 + y^2 - 1 = 0 \end{cases}$$

$$\Longleftrightarrow \begin{cases} y - 2x\lambda = 0 \\ x(1 - 4\lambda^2) = 0 \\ x^2 + y^2 - 1 = 0 \end{cases} \Longleftrightarrow \begin{cases} y - 2x\lambda = 0 \\ x = 0 \quad \text{or} \quad \lambda = \pm\frac{1}{2} \\ x^2 + y^2 - 1 = 0. \end{cases}$$

* $\quad x = 0$ leads to $y = 0$ and $(0,0)$ is not a point on the constrained curve.

** $\quad \lambda = \frac{1}{2}$ leads to $y = x$ and from the constraint equation, we deduce that $x = \pm 1/\sqrt{2}$.

*** $\quad \lambda = -\frac{1}{2}$ leads to $y = -x$ and from the constraint equation, we deduce that $x = \pm 1/\sqrt{2}$.

So, the stationary points, for the Lagrangian, are the four points

$$(\frac{1}{\sqrt{2}}, \frac{1}{\sqrt{2}}), \qquad (-\frac{1}{\sqrt{2}}, -\frac{1}{\sqrt{2}}), \qquad (\frac{1}{\sqrt{2}}, -\frac{1}{\sqrt{2}}), \qquad (-\frac{1}{\sqrt{2}}, \frac{1}{\sqrt{2}})$$

at which f takes its maximum and minimum values respectively

$$f(\frac{1}{\sqrt{2}}, \frac{1}{\sqrt{2}}) = f(-\frac{1}{\sqrt{2}}, -\frac{1}{\sqrt{2}}) = \frac{1}{2}, \qquad f(\frac{1}{\sqrt{2}}, -\frac{1}{\sqrt{2}}) = f(-\frac{1}{\sqrt{2}}, \frac{1}{\sqrt{2}}) = -\frac{1}{2}.$$

The problem can be solved graphically, as illustrated in Figure 3.12.

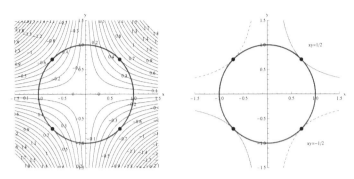

FIGURE 3.12: The constraint $[x^2 + y^2 = 1]$ and the level curves $xy = -\dfrac{1}{2}, \dfrac{1}{2}$ are tangent

Remark 3.2.2 *Note that the Lagrange's method doesn't transform a constrained optimization problem into one finding an unconstrained extreme point of the Lagrangian.*

Example 2. Consider the problem

$$\max x\,y \qquad\qquad \text{subject to} \qquad x + y = 2, \qquad x \geqslant 0, \qquad y \geqslant 0.$$

Using the Lagrange multiplier method, prove that $(x, y) = (1, 1)$ solves the problem with $\lambda = 1$. Prove also that $(1, 1, 1)$ does not maximize the Lagrangian $\mathcal{L}$.

Solution: Since x and y must be positive and satisfy the sum $x + y = 2$, we may look for the extreme points in the set $[0, 2] \times [0, 2]$. Let us denote

$$f(x, y) = xy \qquad\qquad g(x, y) = x + y \qquad\qquad \Omega = [0, 2] \times [0, 2].$$

First, the optimization problem has a solution by the extreme-value theorem. Indeed, f is continuous on the line segment (see Figure 3.13)

$$S = \{(x, y) : \quad g(x, y) = 2, \quad x \geqslant 0, \quad y \geqslant 0\}$$

which is a closed and bounded subset of $\mathbb{R}^2$.
Next, the functions f and g are C^1 around each point $(x, y) \in (0, 2) \times (0, 2)$ which is a regular point since we have

$$g'(x, y) = i + j \neq 0 \qquad\Longrightarrow\qquad rank(g'(x, y)) = 1.$$

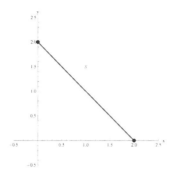

FIGURE 3.13: Set of the constraints

So, by applying the method of Lagrange Multipliers, we introduce the Lagrangian

$$\mathcal{L}(x, y, \lambda) = f(x, y) - \lambda(g(x, y) - 2) = xy - \lambda(x + y - 2)$$

and look for the interior extreme points as solutions of the system

$$\begin{cases} \mathcal{L}_x = f_x - \lambda g_x = 0 \\ \mathcal{L}_y = f_y - \lambda g_y = 0 \\ \mathcal{L}_\lambda = -(g - 2) = 0 \end{cases} \iff \begin{cases} y - \lambda = 0 \\ x - \lambda = 0 \\ x + y - 2 = 0 \end{cases} \iff x = y = \lambda = 1.$$

So, the point $(1, 1, 1)$ is a stationary point for the Lagrangian $\mathcal{L}$. But, it is not an extreme point for $\mathcal{L}$. Indeed, the second test derivative gives

$$H_{\mathcal{L}}(x, y, \lambda) = \begin{bmatrix} 0 & 1 & -1 \\ 1 & 0 & -1 \\ 1 & 1 & 0 \end{bmatrix} \qquad \text{the Hessian matrix of } \mathcal{L}.$$

The leading principal minors of $H_{\mathcal{L}}$ at $(1, 1, 1)$ are

$$D_1 = \begin{vmatrix} 0 \end{vmatrix}, \qquad D_2 = \begin{vmatrix} 0 & 1 \\ 1 & 0 \end{vmatrix} = -1, \qquad D_3 = \begin{vmatrix} 0 & 1 & -1 \\ 1 & 0 & -1 \\ 1 & 1 & 0 \end{vmatrix} = -2 \neq 0.$$

Hence, $(1, 1, 1)$ is a saddle point.

It remains to show that the point $(1, 1)$ is the maximum point for the problem; see Figure 3.14 for a graphical solution using level curves. Indeed, since it is the only interior point to the segment, it suffices to compare the value of f at $(1, 1)$ with its value at the end points of the segment. We have

$$f(1, 1) = 1 \qquad\qquad f(2, 0) = 0 \qquad\qquad f(0, 2) = 0.$$

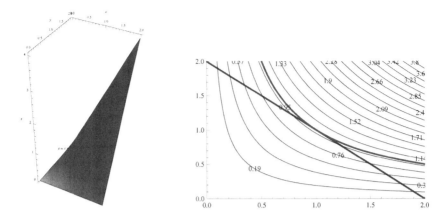

FIGURE 3.14: The constraint $x + y = 2$ and the level curve $f = xy = 1$ are tangent

So f attains its maximum value at $(1, 1)$ under the constraint $g(x, y) = 2$.

Remark 3.2.3 *A function subject to a constraint needs not to have a local extremum at every stationary point of the associated Lagrangian. The Lagrangian multiplier method transforms a constrained optimization problem into one of finding the appropriate stationary points of the Lagrangian.*

Example 3. Consider the problem

$$\min\, x\,y \qquad \text{subject to} \qquad x + y = 2, \qquad x \geqslant 0, \qquad y \geqslant 0.$$

Using the Lagrange multiplier method, prove that $(x, y) = (1, 1)$ doesn't solve the problem with $\lambda = 1$.

Solution: Arguing as in the Example 2, the problem has a solution by the extreme-value theorem. But, by applying the method of Lagrange multiplier, we found the only point candidate $(1, 1)$ and it realizes the maximum of f. So the minimum point of f is not necessary a stationary point of $\mathcal{L}$. In fact, f attains its minimum value 0, under the constraint $g(x, y) = 2$ at $(2, 0)$ and $(0, 2)$.

Solved Problems

1. –

   i) Show that the Lagrange equations for

$$\max\,(\min)\ f(x,y) = x+y+3 \qquad \text{subject to} \qquad g(x,y) = x-y = 0$$

   have no solution.

  ii) Show that any point of the constraints' set is a regular point.

 iii) What can you conclude about the minimum and maximum values of f subject to $g = 0$? Show this directly.

Solution: i) Set

$$\mathcal{L}(x,y,\lambda) = f(x,y) - \lambda(g(x,y) - 0) = x + y + 3 - \lambda(x - y).$$

By applying Lagrange's multipliers method, we look for the interior extreme points as a solution of the system

$$\begin{cases} \mathcal{L}_x = f_x - \lambda g_x = 0 \\[2mm] \mathcal{L}_y = f_y - \lambda g_y = 0 \\[2mm] \mathcal{L}_\lambda = -(g - 0) = 0 \end{cases} \qquad \Longleftrightarrow \qquad \begin{cases} 1 - \lambda = 0 \\[2mm] 1 + \lambda = 0 \\[2mm] x - y = 0 \end{cases}$$

which leads to a contradiction with $\lambda = 1$ and $\lambda = -1$. So the system has no solution.

ii) Any point of the constraint is a regular point since we have

$$g'(x,y) = i - j \neq 0 \qquad \Longrightarrow \qquad rank(g'(x,y)) = 1.$$

iii) We can conclude that f has no maximum nor minimum on the set of the constraints since if they existed, they would be solution of the above

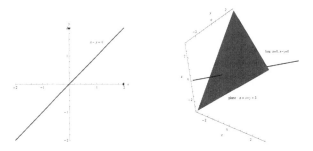

FIGURE 3.15: No solution for the constrained optimization problem

system. Indeed, all conditions of the theorem on the necessary conditions for a constrained candidate point are satisfied.

The problem is equivalent to optimize

$$F(x) = f(x, x) = 2x + 3 \qquad \text{for } x \in \mathbb{R}.$$

We can see that

$$\lim_{x \to -\infty} F(x) = -\infty \qquad\qquad \lim_{x \to +\infty} F(y) = +\infty.$$

Therefore, f cannot reach a finite lower and upper bound on the set of the constraints.

The graph of f is a plane; see Figure 3.15. The level curves $y+1 = k$ are parallel lines that intersects the constraint line $x - y = 0$ at the points $(k-1, k-1)$. This shows that f takes large values (see Figure 3.16).

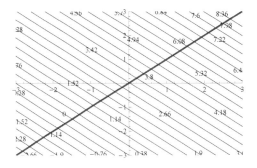

FIGURE 3.16: $\nabla f = \langle 1, 1 \rangle \nparallel \langle 1, -1 \rangle = \nabla g$

2. – Consider the problem of minimizing

$$f(x, y) = y + 1 \qquad \text{subject to} \qquad g(x, y) = x^4 - (y - 2)^5 = 0.$$

i) Show, without using calculus, that the minimum occurs at $(0, 2)$. Is it a regular point?

ii) Show that the Lagrange condition $\nabla f = \lambda \nabla g$ is not satisfied for any value of λ.

iii) Does this contradicts the theorem on the necessary conditions for a constrained candidate point?

Solution: i) Note that we have,

$$g(x, y) = x^4 - (y - 2)^5 = 0 \qquad \Longleftrightarrow \qquad (y - 2)^5 = x^4 \geqslant 0 \qquad \Longrightarrow \qquad y \geqslant 2.$$

So, on the set of the constraint (see Figure 3.17), we have

$$f(x, y) = y + 1 \geqslant 3 = f(x, 2) \qquad \forall (x, y) \in [g = 0].$$

Since $g(0, 2) = 0$, then $(0, 2) \in [g = 0]$. Thus

$$f(x, y) \geqslant f(0, 2) \qquad \forall (x, y) \in [g = 0]$$

and $(0, 2)$ is a global minimum point.

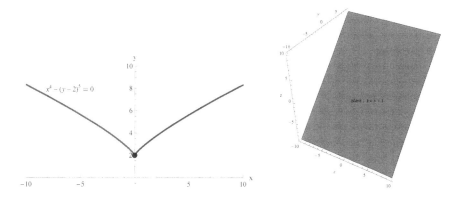

FIGURE 3.17: Minimal value of f on the constraint set $g = 0$

ii) Let

$$\mathcal{L}(x, y, \lambda) = f(x, y) - \lambda(g(x, y) - 0) = y + 1 - \lambda(x^4 - (y - 2)^5).$$

An interior extreme point, if it exists, is a solution of the system

$$
\begin{cases}
\mathcal{L}_x = f_x - \lambda g_x = 0 \\[2mm]
\mathcal{L}_y = f_y - \lambda g_y = 0 \\[2mm]
\mathcal{L}_\lambda = -(g - 0) = 0
\end{cases}
\quad\Longleftrightarrow\quad
\begin{cases}
0 - 4\lambda x^3 = 0 \\[2mm]
1 - 5\lambda(y - 2)^4 = 0 \\[2mm]
x^4 - (y - 2)^5 = 0
\end{cases}
$$

Note that $\lambda = 0$ is not possible by the second equation. So, we deduce that $x = 0$, from the first equation, and then $y = 2$ from the third equation. But, this leads to a contradiction by the second equation. So the system has no solution. No level curve is tangent to the constraint set in Figure 3.18.

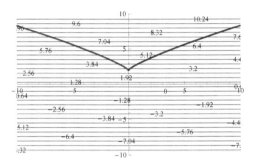

FIGURE 3.18: No solution with Lagrange method

iii) This does not contradict the theorem on the necessary conditions for a constrained candidate point since the theorem is true if all assumptions are satisfied which is not the case for the regularity of the point $(0, 2)$. Indeed, we have

$$g'(x, y) = 4x^3 i - 5(y-2)^4 j \qquad g'(0, 2) = \langle 0, 0\rangle \quad\Longrightarrow\quad rank(g'(0, 2)) = 0 \neq 1.$$

3. – At what points on the curve $g(x, y) = x^4 + y^4 = 1$ does $f(x, y) = x^2 + y^2$ have its maximum and minimum values?

Give a geometric interpretation of the problem.

Solution: Note that, the optimization problem has a solution by the extreme-value theorem since f is continuous on the closed and bounded subset $[g = 1] = g^{-1}(\{1\})$ of $\mathbb{R}^2$.

Next, the functions f and g are C^1 around each point $(x,y) \in \mathbb{R}^2$. In particular each point of $[g = 1]$ is relatively interior to $[g = 1]$. Indeed, if $(x_0, y_0) \in [g = 1]$, then the point (x_0^2, y_0^2) is on the unit circle. Thus, (x_0^2, y_0^2) is an interior point and we conclude, by using the preimage of an open ball by the continuous function $(x,y) \longmapsto (x^2, y^2)$ is an open set.

Moreover, each point of $[g = 1]$ is a regular point since we have

$$g'(x,y) = 4x^3 i + 4y^3 j \neq 0 \quad \text{on} \quad [g = 1] \quad \Longrightarrow \quad rank(g'(x,y)) = 1.$$

So, by setting

$$\mathcal{L}(x, y, \lambda) = x^2 + y^2 - \lambda (x^4 + y^4 - 1)$$

we are led to solve the system

$$
\begin{cases}
\mathcal{L}_x = 2x - 4\lambda x^3 = 0 \\
\mathcal{L}_y = 2y - 4\lambda y^3 = 0 \\
\mathcal{L}_\lambda = -(x^4 + y^4 - 1) = 0
\end{cases}
\Longleftrightarrow
\begin{cases}
2x(1 - 2\lambda x^2) = 0 \\
2y(1 - 2\lambda y^2) = 0 \\
x^4 + y^4 = 1
\end{cases}
$$

$$
\Longleftrightarrow
\begin{cases}
x = 0 \quad \text{or} \quad 2\lambda x^2 = 1 \\
y = 0 \quad \text{or} \quad 2\lambda y^2 = 1 \\
x^4 + y^4 = 1
\end{cases}
\Longleftrightarrow
$$

$$
\begin{cases}
x = 0 \\
y = \pm 1 \\
\lambda = 1/2
\end{cases}
\text{or}
\begin{cases}
y = 0 \\
x = \pm 1 \\
\lambda = 1/2
\end{cases}
\text{or}
\begin{cases}
x^2 = y^2 \\
x^4 = 1/2 \\
\lambda = 1/(2x^2).
\end{cases}
$$

So, the stationary points for the Lagrangian are

$$(0, \pm 1), \quad (\pm 1, 0), \quad (\frac{1}{2^{1/4}}, \pm \frac{1}{2^{1/4}}), \quad (-\frac{1}{2^{1/4}}, \pm \frac{1}{2^{1/4}})$$

at which f takes its maximum and minimum values respectively

$$\max_{g=1} f = f(\frac{1}{2^{1/4}}, \pm \frac{1}{2^{1/4}}) = f(-\frac{1}{2^{1/4}}, \pm \frac{1}{2^{1/4}}) = \sqrt{2}$$

$$\min_{g=1} f \;=\; f(\pm 1, 0) = f(0, \pm 1) = 1.$$

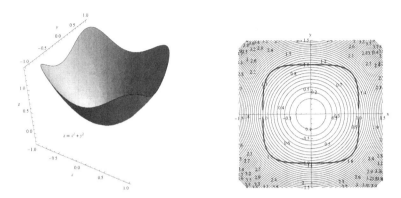

FIGURE 3.19: The constraint $[x^4 + y^4 = 1]$ and the level curves $f = 1$, $\sqrt{2}$ are tangent

Since $f(x, y) = \|(x, y) - (0, 0)\|^2$, then the problem looks for points (x, y) on the curve $x^4 + y^4 = 1$ that are closest and farthest from the origin; see Figure 3.19.

4. – Figures A an B (see Figure 3.20) show the level curves of f and the constraint curve $g(x, y) = 0$ graphed thickly. Estimate the maximum and minimum values of f subject to the constraint. Locate the point(s), if any, where an extreme value occurs.

Solution: Figure A. Two level curves of f are tangent to the constraint curve $g = 0$. Comparing the values of f taken at these level curves, we deduce that

$$local \max_{g=0} f \approx 15 \approx f(-1.5, 1.5) \qquad\qquad local \min_{g=0} f \approx 3.64 \approx f(-1.5, 1.5).$$

Figure B. One level curve of f is tangent to the constraint curve $g = 0$. Comparing the values of f taken at different level curves, we remark that f keeps taking large values on the constraint set. Therefore, we deduce that

$$local \max_{g=0} f \quad \text{doesn't exist} \qquad\qquad local \min_{g=0} f \approx 19.2 \approx f(3, -2).$$

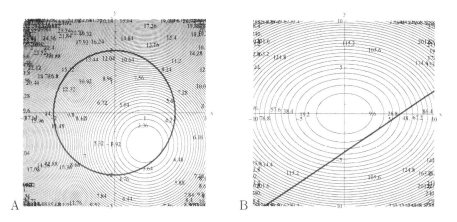

FIGURE 3.20: Level curves of f and the constraint curve $g = 0$

5. – Find the points on the sphere $x^2 + y^2 + z^2 = 1$ that are closest to and farthest from the point $(1, 2, 2)$.

Solution: The distance of a point (x, y, z) to the point $(1, 2, 2)$ is given by

$$D(x, y, z) = \sqrt{(x-1)^2 + (y-2)^2 + (z-2)^2}.$$

To look for the shortest and the farthest distance when (x, y, z) remains into the unit sphere is equivalent to optimize $D^2(x, y, z)$ under the constraint $x^2 + y^2 + z^2 = 1$. So, let us denote

$$f(x, y, z) = (x-1)^2 + (y-2)^2 + (z-2)^2$$
$$g(x, y, z) = x^2 + y^2 + z^2 \qquad S = [g = 1].$$

First, the optimization problem has a solution by the extreme value theorem since f is continuous on the unit sphere S, which is a closed and bounded subset of $\mathbb{R}^3$.

Next, f and g are C^∞ around each point $(x, y, z) \in \mathbb{R}^3$. In particular, each point of S a relatively interior point and is a regular point since we have

$$g'(x, y, z) = 2xi + 2yj + 2zk \neq \mathbf{0} \qquad \Longrightarrow \qquad rank(g'(x, y, z)) = 1.$$

So, consider the Lagrangian

$$\mathcal{L}(x, y, z, \lambda) = (x-1)^2 + (y-2)^2 + (z-2)^2 - \lambda(x^2 + y^2 + z^2 - 1)$$

and apply Lagrange multipliers method to look for the interior extreme points
by solving the system

$$
\begin{cases}
\mathcal{L}_x = f_x(x,y,z) - \lambda g_x(x,y,z) = 0 \\
\mathcal{L}_y = f_y(x,y,z) - \lambda g_y(x,y,z) = 0 \\
\mathcal{L}_z = f_z(x,y,z) - \lambda g_z(x,y,z) = 0 \\
\mathcal{L}_\lambda = -(g(x,y,z) - 1) = 0
\end{cases}
\iff
\begin{cases}
2(x-1) - 2x\lambda = 0 \\
2(y-2) - 2y\lambda = 0 \\
2(z-2) - 2z\lambda = 0 \\
x^2 + y^2 + z^2 - 1 = 0.
\end{cases}
$$

If $x = 0$, then the first equation leads to $x = 1$ which is impossible. We cannot
have also $y = 0$ and $z = 0$. So, we deduce from the system that

$$
\lambda = 1 - \frac{1}{x} = 1 - \frac{2}{y} = 1 - \frac{2}{z}
$$

from which we deduce

$$
\begin{cases}
y = z = 2x \\
\lambda = 1 - 1/x \\
x^2 + y^2 + z^2 = 1
\end{cases}
\iff
\begin{cases}
y = z = 2x \\
\lambda = 1 - 1/x \\
x^2 + 4x^2 + 4x^2 - 1 = 0 \iff x = \pm\dfrac{1}{3}.
\end{cases}
$$

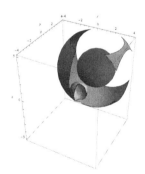

FIGURE 3.21: The constraint $[g = 1]$ and the level curves $f = 4,\ 16$ are tangent

So, the stationary points for the Lagrangian are the two points

$$
(\frac{1}{3}, \frac{2}{3}, \frac{2}{3}, -2), \qquad\qquad (-\frac{1}{3}, -\frac{2}{3}, -\frac{2}{3}, 4)
$$

and f takes its maximum and minimum values respectively

$$
f(-\frac{1}{3}, -\frac{2}{3}, -\frac{2}{3}) = 16 \qquad \text{and} \qquad f(\frac{1}{3}, \frac{2}{3}, \frac{2}{3}) = 4.
$$

The level curves passing by these points are spheres tangents to the constraint
as shown in Figure 3.21.

3.3   Classification of Local Extreme Points-Equality Constraints

To classify a local extreme point x^* in the case of an unconstrained optimization problem, we compared values $f(x^* + h)$ with $f(x^*)$ using Taylor's formula and the fact that $\nabla f(x^*) = 0$. In this constrained case, we also need to make this comparison, but, we have to take into account the presence of the constraints. The Lagrangian function links the values of f to those of g. Therefore, we will apply Taylor's formula to compare values $\mathcal{L}(x^* + h)$ with $\mathcal{L}(x^*)$ using the fact that $\nabla \mathcal{L}(x^*, \lambda^*) = 0$. More precisely, we establish a second derivative test under specific assumptions.

Consider the optimization problem with equality constraints,

$$local \max(\min) f(x) \qquad \text{subject to} \qquad g(x) = c$$

where

$$g(x) = \langle g_1(x), \ldots, g_m(x) \rangle, \qquad c = \langle c_1, \ldots, c_m \rangle \qquad (m < n).$$

The associated Lagrangian is

$$\mathcal{L}(x, \lambda) = f(x) - \lambda_1(g_1(x) - c_1) - \lambda_2(g_2(x) - c_2) - \ldots - \lambda_m(g_m(x) - c_m).$$

Theorem 3.3.1 *Sufficient conditions for a strict local constrained extreme point*

Let f and $g = (g_1, \ldots, g_m)$ be C^2 functions in a neighborhood of x^* in $\mathbb{R}^n$ such that:

$$g(x^*) = c \qquad\qquad rank(g'(x^*)) = m,$$

$$\nabla \mathcal{L}(x^*, \lambda^*) = 0 \qquad \text{for a unique vector} \qquad \lambda^* = \langle \lambda_1^*, \ldots, \lambda_m^* \rangle.$$

Then

$$(i) \qquad (-1)^m \mathbb{B}_r(x^*) > 0 \quad \forall r = m+1, \ldots, n$$
$$\implies \qquad x^* \text{ is a strict local minimum point}$$

$$(ii) \qquad (-1)^r \mathbb{B}_r(x^*) > 0 \quad \forall r = m+1, \ldots, n$$
$$\implies \qquad x^* \text{ is a strict local maximum point}.$$

For $r = m+1, \ldots, n$, $\mathbb{B}_r(x^*)$ is the bordered Hessian determinant defined by

$$\mathbb{B}_r(x^*) = \begin{vmatrix} 0 & \cdots & 0 & \frac{\partial g_1}{\partial x_1}(x^*) & \cdots & \frac{\partial g_1}{\partial x_r}(x^*) \\ \vdots & \ddots & \vdots & \vdots & \ddots & \vdots \\ 0 & \cdots & 0 & \frac{\partial g_m}{\partial x_1}(x^*) & \cdots & \frac{\partial g_m}{\partial x_r}(x^*) \\ \frac{\partial g_1}{\partial x_1}(x^*) & \cdots & \frac{\partial g_m}{\partial x_1}(x^*) & \mathcal{L}_{x_1 x_1}(x^*, \lambda^*) & \cdots & \mathcal{L}_{x_1 x_r}(x^*, \lambda^*) \\ \vdots & \ddots & \vdots & \vdots & \ddots & \vdots \\ \frac{\partial g_1}{\partial x_r}(x^*) & \cdots & \frac{\partial g_m}{\partial x_r}(x^*) & \mathcal{L}_{x_r x_1}(x^*, \lambda^*) & \cdots & \mathcal{L}_{x_r x_r}(x^*, \lambda^*) \end{vmatrix}$$

The variables are **renumbered** in order to make the first m columns in the matrix $g'(x^*)$ linearly independent.

Remark 3.3.1 *If we introduce the notations:*

$$Q(h) = Q(h_1, \ldots, h_n) = \sum_{i=1}^{n} \sum_{j=1}^{n} \mathcal{L}_{x_i x_j}(x^*, \lambda^*) h_i h_j$$

the $(m+n) \times (m+n)$ bordered matrix $\begin{bmatrix} 0_{m \times m} & g'(x^*) \\ {}^t g'(x^*) & [\mathcal{L}_{x_i x_j}(x^*, \lambda^*)]_{n \times n} \end{bmatrix}$

$$\boldsymbol{M} = \{h \in \mathbb{R}^n : g'(x^*).h = 0\}$$

the theorem says that

$$Q(h) > 0 \quad \forall h \in \boldsymbol{M}, \; h \neq 0 \quad \Longrightarrow \quad x^* \text{ is a strict local minimum}$$

$$Q(h) < 0 \quad \forall h \in \boldsymbol{M}, \; h \neq 0 \quad \Longrightarrow \quad x^* \text{ is a strict local maximum.}$$

It suffices then to study the definite positivity (negativity) of the quadratic form on the tangent plan M to the constraint $g = c$ at the point x^* (see the reminder at the end of this section).

Before proving the theorem, we will see its application through some examples.

Example 1. Consider the problem

$$local \max \ f(x,y) = xy \quad \text{subject to} \quad g(x,y) = x+y = 2, \ x \geqslant 0, \ y \geqslant 0.$$

Lagrange multiplier method shows that $(1,1)$ is a regular candidate point. Prove that it is a local maximum to the constrained optimization problem.

Solution: Considering the Lagrangian

$$\mathcal{L}(x,y,\lambda) = f(x,y) - \lambda(g(x,y) - 2) = xy - \lambda(x+y-2),$$

we can study the nature of the point $(1,1)$ using the second derivatives test. Here, we have $n = 2$ and $m = 1$. The first column vector of $g'(1,1) = \langle 1,1 \rangle$ is linearly independent. So, we keep the matrix $g'(1,1)$ without renumbering the variables. Then, we have to consider the sign of the bordered Hessian determinant $(r = m+1 = 2 = n)$

$$(-1)^2 \mathbb{B}_2(1,1) = \begin{vmatrix} 0 & g_x(1,1) & g_y(1,1) \\ g_x(1,1) & \mathcal{L}_{xx}(1,1,1) & \mathcal{L}_{xy}(1,1,1) \\ g_y(1,1) & \mathcal{L}_{xy}(1,1,1) & \mathcal{L}_{yy}(1,1,1) \end{vmatrix} = \begin{vmatrix} 0 & 1 & 1 \\ 1 & 0 & 1 \\ 1 & 1 & 0 \end{vmatrix} = 2 > 0.$$

We conclude that the point $(1,1)$ is a local maximum to the problem.

Example 2. Solve the problem

$$local \max f(x,y,z) = xy + yz + xz \quad \text{subject to} \quad g(x,y,z) = x+y+z = 3.$$

Solution: Note that f and g are C^1 in $\mathbb{R}^3$ and

$$g'(x,y,z) = i + j + k \neq \mathbf{0} \quad \Longrightarrow \quad rank(g'(x,y,z)) = 1.$$

Thus, any point, interior to the constraint set $[g = 3]$ (see Figure 3.22), is a regular point.
Consider the Lagrangian

$$\mathcal{L}(x,y,z,\lambda) = f(x,y,z) - \lambda(g(x,y,z) - 3)$$
$$= xy + yz + xz - \lambda(x+y+z-3)$$

and let us look for its stationary points solutions of the system

$$\nabla\mathcal{L}(x,y,z,\lambda) = \langle 0,0,0,0 \rangle \Longleftrightarrow \begin{cases} \mathcal{L}_x = y+z-\lambda = 0 \\ \mathcal{L}_y = x+z-\lambda = 0 \\ \mathcal{L}_z = y+x-\lambda = 0 \\ \mathcal{L}_\lambda = -(x+y+z-3) = 0. \end{cases}$$

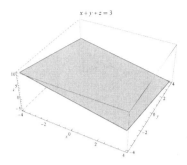

FIGURE 3.22: The constraint set $[g = 3]$

From the first three equations, we deduce that $\dfrac{\lambda}{2} = x = y = z$, which inserted into the last equation gives

$$x = y = z = 1 \qquad\qquad \lambda = 2.$$

Now, let us study the nature of the point $(1, 1, 1)$. For this we use the second derivative test since f and g are C^2 around this point. The first column vector of $g'(1, 1, 1)$ is linearly independent. So, we keep the matrix $g'(1, 1, 1)$ without renumbering the variables. As $n = 3$ and $m = 1$, we have to consider the signs of the following bordered Hessian determinants:

$$(-1)^2 \mathbb{B}_2(1,1,1) = \begin{vmatrix} 0 & g_x(1,1,1) & g_y(1,1,1) \\ g_x(1,1,1) & \mathcal{L}_{xx}(1,1,1,2) & \mathcal{L}_{xy}(1,1,1,2) \\ g_y(1,1,1) & \mathcal{L}_{xy}(1,1,1,2) & \mathcal{L}_{yy}(1,1,1,2) \end{vmatrix} = \begin{vmatrix} 0 & 1 & 1 \\ 1 & 0 & 1 \\ 1 & 1 & 0 \end{vmatrix} = 2 > 0.$$

$$(-1)^3 \mathbb{B}_3(1,1,1) = - \begin{vmatrix} 0 & g_x(1,1,1) & g_y(1,1,1) & g_z(1,1,1) \\ g_x(1,1,1) & \mathcal{L}_{xx}(1,1,1,2) & \mathcal{L}_{xy}(1,1,1,2) & \mathcal{L}_{xz}(1,1,1,2) \\ g_y(1,1,1) & \mathcal{L}_{yx}(1,1,1,2) & \mathcal{L}_{yy}(1,1,1,2) & \mathcal{L}_{yz}(1,1,1,2) \\ g_z(1,1,1) & \mathcal{L}_{zx}(1,1,1,2) & \mathcal{L}_{zy}(1,1,1,2) & \mathcal{L}_{zz}(1,1,1,2) \end{vmatrix}$$

$$- - \begin{vmatrix} 0 & 1 & 1 & 1 \\ 1 & 0 & 1 & 1 \\ 1 & 1 & 0 & 1 \\ 1 & 1 & 1 & 0 \end{vmatrix} = 3 > 0.$$

We conclude that the point $(1, 1, 1)$ is a local maximum to the constrained maximization problem.

Proof. We will prove assertion i). Assertion ii) can be established similarly. We follow for this the proof in [25] with more details in the steps involved.

Step 1 : Let Ω be a neighborhood of x^*. For $h \in \mathbb{R}^n$ such that $x^* + h \in \overset{\circ}{\Omega}$, we have from Taylor's formula, for some $\tau \in (0,1)$,

$$\mathcal{L}(x^*+h, \lambda^*) = \mathcal{L}(x^*, \lambda^*) + \sum_{i=1}^{n} \mathcal{L}_{x_i}(x^*, \lambda^*)h_i + \frac{1}{2}\sum_{i=1}^{n}\sum_{j=1}^{n}\mathcal{L}_{x_i x_j}(x^*+\tau h, \lambda^*)h_i h_j.$$

Since $x^* \in \overset{\circ}{\Omega}$ and (x^*, λ^*) is a local stationary point of $\mathcal{L}$ then, in particular,

$$\mathcal{L}_{x_i}(x^*, \lambda^*) = 0 \qquad i = 1, \cdots, n.$$

Moreover, we have

$$g_1(x^*) - c_1 = g_2(x^*) - c_2 = \ldots = g_m(x^*) - c_m = 0$$

$$\mathcal{L}(x^*, \lambda^*) = f(x^*) - \lambda_1^*(g_1(x^*) - c_1) - \ldots - \lambda_m^*(g_m(x^*) - c_m) = f(x^*)$$

$$\mathcal{L}(x^* + h, \lambda^*) = f(x^* + h) - \lambda_1^*(g_1(x^* + h) - c_1) - \ldots - \lambda_m^*(g_m(x^* + h) - c_m)$$

from which we deduce

$$f(x^*+h) - f(x^*) = \sum_{k=1}^{n}\lambda_k^*[g_k(x^*+h) - c_k] + \frac{1}{2}\sum_{i=1}^{n}\sum_{j=1}^{n}\mathcal{L}_{x_i x_j}(x^*+\tau h, \lambda^*)h_i h_j.$$

Using Taylor's formula for each g_k, $k = 1, \ldots, m$, we obtain

$$g_k(x^* + h) - c_k = g_k(x^* + h) - g_k(x^*) = \sum_{j=1}^{n}\frac{\partial g_k}{\partial x_j}(x^* + \tau_k h)h_j \qquad \tau_k \in (0,1).$$

Step 2 : Now consider the $(m + n) \times (m + n)$ bordered Hessian matrix

$$\mathbf{B}(\mathbf{x}^0, \mathbf{x}^1, \ldots, \mathbf{x}^m) = \begin{bmatrix} \mathbf{0} & \mathbf{G}(\mathbf{x}^1, \ldots, \mathbf{x}^m) \\ {}^t\mathbf{G}(\mathbf{x}^1, \ldots, \mathbf{x}^m) & \mathcal{H}_{\mathcal{L}(.,\lambda^*)}(\mathbf{x}^0) \end{bmatrix}$$

where

$$\mathbf{G}(\mathbf{x}^1, \ldots, \mathbf{x}^m) = \left(\frac{\partial g_i}{\partial x_j}(\mathbf{x}^i) \right)_{m \times n} = \begin{bmatrix} \frac{\partial g_1}{\partial x_1}(\mathbf{x}^1) & \cdots & \frac{\partial g_1}{\partial x_n}(\mathbf{x}^1) \\ \vdots & \ddots & \vdots \\ \frac{\partial g_m}{\partial x_1}(\mathbf{x}^m) & \cdots & \frac{\partial g_m}{\partial x_n}(\mathbf{x}^m) \end{bmatrix}$$

$\mathbf{x}^1, \ldots, \mathbf{x}^m$ are arbitrary vectors in some open ball around x^*

$\mathcal{H}_{\mathcal{L}(.,\lambda^*)}(\mathbf{x}^0) :$ is the Hessian matrix of $\mathcal{L}$ with respect to x evaluated at $\mathbf{x}^0$.

For $r = m+1, \ldots, n$, let $det\mathbf{B}^r(\mathbf{x}^0, \mathbf{x}^1, \ldots, \mathbf{x}^m)$ be the $(m+r) \times (m+r)$ leading principal minor of the matrix $\mathbf{B}(\mathbf{x}^0, \mathbf{x}^1, \ldots, \mathbf{x}^m)$.

Suppose that $(-1)^m \mathbb{B}_r(x^*) > 0$ for all $r = m+1, \ldots, n$, then by continuity of the second-order partial derivatives of f and g, and since

$$det\mathbf{B}^r(x^*, x^*, \ldots, x^*) = \mathbb{B}_r(x^*)$$

there exists $\rho > 0$ such that, $\forall r = m+1, \ldots, n$,

$$(-1)^m det\mathbf{B}^r(\mathbf{x}^0, \mathbf{x}^1, \ldots, \mathbf{x}^m) > 0 \qquad \forall \mathbf{x}^0, \mathbf{x}^1, \ldots, \mathbf{x}^m \in B_\rho(x^*).$$

As a consequence, for $\mathbf{x}^0, \mathbf{x}^1, \ldots, \mathbf{x}^m \in B_\rho(x^*)$, the quadratic form

$$Q(t) = Q(t_1, \ldots, t_n) = \sum_{i=1}^n \sum_{j=1}^n \mathcal{L}_{x_i x_j}(\mathbf{x}^0, \lambda^*) t_i t_j,$$

with the associated symmetric matrix $\left[\mathcal{L}_{x_i x_j}(\mathbf{x}^0) \right]_{n \times n}$, is definite positive subject to the constraints

$$\mathbf{G}(\mathbf{x}^1, \ldots, \mathbf{x}^m).t = 0 \qquad \Longleftrightarrow \qquad \sum_{j=1}^n \frac{\partial g_k}{\partial x_j}(\mathbf{x}^k) t_j = 0 \qquad k = 1, \ldots, m.$$

Step 3 : Because $\tau, \tau_k \in (0, 1)$, we have, for $x^* + h \in B_\rho(x^*)$,

$$\mathbf{x}^0 = x^* + \tau h, \quad \mathbf{x}^1 = x^* + \tau_1 h, \ldots, \mathbf{x}^m = x^* + \tau_m h \in B_\rho(x^*).$$

Then

$$\sum_{i=1}^n \sum_{j=1}^n \mathcal{L}_{x_i x_j}(x^* + \tau h, \lambda^*) t_i t_j > 0 \quad \forall t \neq 0 \quad \text{such that}$$

$$\sum_{j=1}^{n} \frac{\partial g_k}{\partial x_j}(x^* + \tau_k h)t_j = 0 \quad k = 1, \dots, m.$$

In particular, for $t = h$ such that

$$\sum_{j=1}^{n} \frac{\partial g_k}{\partial x_j}(x^* + \tau_k h)h_j = 0 \qquad k = 1, \dots, m, \tag{1}$$

we have

$$f(x^* + h) - f(x^*) = \frac{1}{2}\sum_{i=1}^{n}\sum_{j=1}^{n} \mathcal{L}_{x_i x_j}(x^* + \tau h, \lambda^*)h_i h_j > 0. \tag{2}$$

This shows that the stationary point x^* is a strict local minimum point for f subject to the constraint $g(x) = c$ in particular directions.

Step 4 : Suppose that x^* is not a strict relative minimum point. Then, there exists a sequence of points y_l satisfying

$$y_l \longrightarrow x^* \qquad g(y_l) = c \qquad f(y_l) \leqslant f(x^*).$$

Write each y_l in the form

$$y_l = x^* + \delta_l s_l \neq 0 \qquad s_l \in \mathbb{R}^n \qquad \|s_l\| = 1 \qquad \delta_l > 0 \qquad \forall l.$$

Note that we have

$$\delta_l = \|\delta_l s_l\| = \|y_l - x^*\| \longrightarrow 0.$$

Hence, there exists $l_0 > 1$ such that for all $l \geqslant l_0$, $y_l \in B_\rho(x^*)$. Choose in steps 1 and 3, $h = \delta_l s_l = y_l - x^*$. Then

$$g(x^* + h) - g(x^*) = g(y_{l_k}) - g(x^*) = c - c = 0$$

$$g_k(x^* + h) - g_k(x^*) = \sum_{j=1}^{n} \frac{\partial g_k}{\partial x_j}(x^* + \tau_k h)h_j = 0 \quad \tau_k \in (0,1), \quad k = 1, \dots, m$$

and we should have from (1) and (2)

$$0 \geqslant f(y_l) - f(x^*) = f(x^* + h) - f(x^*) = \frac{1}{2}\sum_{i=1}^{n}\sum_{j=1}^{n} \mathcal{L}_{x_i x_j}(x^* + \tau h)h_i h_j > 0$$

which is a contradiction.

Theorem 3.3.2 *Necessary conditions for local extreme points*

Let f and $g = (g_1, \ldots, g_m)$ be C^2 functions in a neighborhood of x^* in $\mathbb{R}^n$ such that:

$$g(x^*) = c \qquad rank(g'(x^*)) = m,$$
$$\nabla \mathcal{L}(x^*, \lambda^*) = 0 \qquad \text{for a unique vector} \quad \lambda^* = \langle \lambda_1^*, \ldots, \lambda_m^* \rangle.$$

Then,

(i) x^* is a local minimum point $\implies H_{\mathcal{L}} = (\mathcal{L}_{x_i x_j}(x^*, \lambda^*))_{n \times n}$
is positive semi definite on $\boldsymbol{M}$: $\,^t y H_{\mathcal{L}} y \geqslant 0 \quad \forall y \in \boldsymbol{M}$

(ii) x^* is a local maximum point $\implies H_{\mathcal{L}} = (\mathcal{L}_{x_i x_j}(x^*, \lambda^*))_{n \times n}$
is negative semi definite on $\boldsymbol{M}$: $\,^t y H_{\mathcal{L}} y \leqslant 0 \quad \forall y \in \boldsymbol{M}$

where $\boldsymbol{M} = \{h \in \mathbb{R}^n : g'(x^*).h = 0\}$ is the tangent plane to the surface $g(x) = c$ at the point x^*.

Proof. We prove $i)$, then $ii)$ can be established similarly.

Let $x(t)$ be a two differentiable curve on the constraint surface $g(x) = c$ with $x(0) = x^*$. Suppose that x^* is a local minimum point for f subject to the constraint $g(x) = c$. Then there exists $r > 0$ such that

$$f(x^*) \leqslant f(x(t)) \qquad \qquad \forall t \in (-r, r).$$

Then

$$\widetilde{f}(0) = f(x^*) \leqslant f(x(t)) = \widetilde{f}(t) \qquad \qquad \forall t \in (-r, r).$$

So $\widetilde{f}$ is a one variable function that has an interior minimum at $t = 0$. Consequently, it satisfies $\widetilde{f}'(0) = 0$ and $\widetilde{f}''(0) \geqslant 0$ or equivalently

$$\nabla f(x^*).x'(0) = 0 \qquad \text{and} \qquad \frac{d^2}{dt^2} f(x(t))\Big|_{t=0} \geqslant 0.$$

We have

$$\frac{d^2}{dt^2} f(x(t)) = \,^t x'(t) H_f(x(t)) x'(t) + \nabla f(x(t)) x''(t)$$

$$\frac{d^2}{dt^2} f(x(t))\Big|_{t=0} = \,^t x'(0) H_f(x^*) x'(0) + \nabla f(x^*).x''(0).$$

Moreover, differentiating the relation $g(x(t)) = c$ twice, we obtain

$$\,^t x'(t) H_g(x(t)) x'(t) + \nabla g(x(t)) x''(t) = 0 \qquad \implies$$

$$^t x'(0) H_g(x^*) x'(0) \; + \; \nabla g(x^*) x''(0) = 0.$$

Hence

$$0 \leqslant \frac{d^2}{dt^2} f(x(t)) \Big|_{t=0} \; = \; [^t x'(0) H_f(x^*) x'(0) \; + \; \nabla f(x^*) x''(0)]$$

$$- \; ^t \lambda^* [^t x'(0) H_g(x^*) x'(0) \; + \; \nabla g(x^*) x''(0)]$$

$$= \; ^t x'(0) [H_f(x^*) - ^t \lambda^* H_g(x^*)] x'(0) \; + \; [\nabla f(x^*) \; + \; ^t \lambda^* \nabla g(x^*)] x''(0)$$

$$= \; ^t x'(0) [H_{\mathcal{L}}(x^*)] x'(0) \qquad \text{since} \qquad \nabla f(x^*) \; + \; ^t \lambda^* \nabla g(x^*) = 0$$

and the result follows since $x'(0)$ is an arbitrary element of $\mathbf{M}$.

Quadratic Forms with Linear Constraints

Consider the symmetric quadratic form in n variables

$$Q(h) = \sum_{i=1}^{n} \sum_{j=1}^{n} a_{ij} h_i h_j \qquad (a_{ij} = a_{ji})$$

subject to m linear homogeneous constraints

$$b_{11} h_1 + \ldots + b_{1n} h_n = 0$$
$$\vdots \qquad \qquad \vdots$$
$$b_{m1} h_1 + \ldots + b_{mn} h_n = 0$$

Set

$$A = \begin{bmatrix} a_{11} & \cdots & a_{1n} \\ \vdots & \ddots & \vdots \\ a_{n1} & \cdots & a_{nn} \end{bmatrix} \qquad B = \begin{bmatrix} b_{11} & \cdots & b_{1n} \\ \vdots & \ddots & \vdots \\ b_{m1} & \cdots & b_{mn} \end{bmatrix} \qquad h = \begin{bmatrix} h_1 \\ \vdots \\ h_n \end{bmatrix}$$

Definition.

   $Q(h) \; = \; ^t h A h$ is positive (resp. negative) definite subject to the linear
   constraints $Bh = 0$ if $Q(h) > 0$ (resp. < 0) for all $h \neq 0$ that satisfy
   $Bh = 0$.

We have the following necessary and sufficient condition for a quadratic form
Q to be positive (resp. negative) definite subject to linear constraints.

Theorem: Assume the first m columns in the matrix $B = (b_{ij})$ are linearly independent. Then

$$Q \text{ is positive definite subject to the constraints } Bh = 0$$

$$\Longleftrightarrow \quad (-1)^m \mathbb{B}_r > 0 \quad r = m+1, \ldots, n$$

$$Q \text{ is negative definite subject to the constraints } Bh = 0$$

$$\Longleftrightarrow \quad (-1)^r \mathbb{B}_r > 0 \quad r = m+1, \ldots, n$$

where $\mathbb{B}_r$ are the symmetric determinants

$$\mathbb{B}_r = \begin{vmatrix} 0 & \cdots & 0 & b_{11} & \cdots & b_{1r} \\ \vdots & \ddots & \vdots & \vdots & & \vdots \\ 0 & \cdots & 0 & b_{m1} & \cdots & b_{mr} \\ b_{11} & \cdots & b_{m1} & a_{11} & \cdots & a_{1r} \\ \vdots & & \vdots & \vdots & \ddots & \vdots \\ b_{1r} & \cdots & b_{mr} & a_{r1} & \cdots & a_{rr} \end{vmatrix} \qquad \text{for} \qquad r = m+1, \ldots, n.$$

Solved Problems

1. – Consider the problem

$$\max(\min) f(x,y) = x^2 + 2y^2 \qquad \text{subject to} \qquad g(x,y) = x^2 + y^2 = 1.$$

i) Find the four points that satisfy the first-order conditions.

ii) Classify them by using the second derivatives test.

iii) Graph some level curves of f and the graph of $g = 1$. Explain, where the extreme points occur.

Solution: i) First, each of the optimization problems has a solution by the extreme-value theorem; see Figure 3.23. Indeed, f is continuous on the unit circle

$$S = \{(x,y) : \ g(x,y) = 1\}$$

which is a closed and bounded subset of $\mathbb{R}^2$.

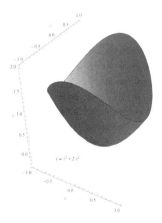

FIGURE 3.23: Graph of f on the set $[x^2 + y^2 \leqslant 1]$

Next, the functions f and g are C^1 in $\mathbb{R}^2$ and any point on the unit circle is regular since, for each $(x, y) \in S$, we have

$$g'(x, y) = (2x, 2y) \neq (0, 0) \quad \Longrightarrow \quad rank(g'(x, y)) = 1.$$

Thus, if we introduce the Lagrangian

$$\mathcal{L}(x, y, \lambda) = f(x, y) - \lambda(g(x, y) - 1) = x^2 + 2y^2 - \lambda(x^2 + y^2 - 1),$$

then, by applying Lagrange multipliers method, the interior extreme points candidates are solutions of the system $\quad \nabla\mathcal{L}(x, y, \lambda) = \langle 0, 0, 0 \rangle$

$$\Longleftrightarrow \begin{cases} \mathcal{L}_x = 2x - \lambda(2x) = 0 \\ \mathcal{L}_y = 4y - \lambda(2y) = 0 \\ \mathcal{L}_\lambda = -(x^2 + y^2 - 1) = 0 \end{cases} \Longleftrightarrow \begin{cases} x = 0 \quad \text{or} \quad \lambda = 1 \\ y = 0 \quad \text{or} \quad \lambda = 2 \\ x^2 + y^2 - 1 = 0. \end{cases}$$

We cannot have $x = y = 0$ since the constraint is not satisfied. If $x = 0$ and $\lambda = 2$, we deduce from the third equation $y = \pm 1$. Then, if $y = 0$ and $\lambda = 1$, we get $x = \pm 1$. So the four points that satisfy the necessary conditions are

$$(1, 0) \qquad (-1, 0) \qquad (0, 1) \qquad (0, -1).$$

ii) Now, because f and g are C^2, we may study the nature of the four points by using the second derivatives test. Here, we have $n = 2$ and $m = 1$. Then, we have to consider the sign of the bordered Hessian determinant $\mathbb{B}_2$ at each point.

Nature of the points $(\pm 1, 0)$ ***where*** $\lambda = 1$: First, we have

$$g'(x, y) = (2x, 2y), \qquad g'(\pm 1, 0) = (\pm 2, 0), \qquad rank(g'(\pm 1, 0)) = 1,$$

and the first column vector of $g'(\pm 1, 0)$ is linearly independent. We have

$$\mathbb{B}_2(x, y) = \begin{vmatrix} 0 & g_x(x, y) & g_y(x, y) \\ g_x(x, y) & \mathcal{L}_{xx}(x, y, \lambda) & \mathcal{L}_{xy}(x, y, \lambda) \\ g_y(x, y) & \mathcal{L}_{xy}(x, y, \lambda) & \mathcal{L}_{yy}(x, y, \lambda) \end{vmatrix} = \begin{vmatrix} 0 & 2x & 2y \\ 2x & 2-2\lambda & 0 \\ 2y & 0 & 4-2\lambda \end{vmatrix}$$

$$\mathbb{B}_2(1, 0) = \begin{vmatrix} 0 & 2 & 0 \\ 2 & 0 & 0 \\ 0 & 0 & 2 \end{vmatrix} = -8 \qquad \mathbb{B}_2(-1, 0) = \begin{vmatrix} 0 & -2 & 0 \\ -2 & 0 & 0 \\ 0 & 0 & 2 \end{vmatrix} = -8.$$

For $m = 1$, we have

$$(-1)^1 \mathbb{B}_2(1, 0) = 2 > 0 \qquad (-1)^1 \mathbb{B}_2(-1, 0) = 2 > 0$$

and the points $(\pm 1, 0)$ are local minima.

Nature of the points $(0, \pm 1)$ ***where*** $\lambda = 2$ **:** We have

$$g'(x, y) = (2x, 2y), \qquad g'(0, \pm 1) = (0, \pm 2) \implies rank(g'(0, \pm 1)) = 1.$$

Note that the first column vector of $g'(0, \pm 1)$ is linearly dependent and the second column vector is linearly independent. So, we renumber the variables so that the second column vector of $g'(0, \pm 1)$ is in the first position. Hence $\mathbb{B}_2$ will be written as

$$\mathbb{B}_2(x, y) = \begin{vmatrix} 0 & g_y(x, y) & g_x(x, y) \\ g_y(x, y) & \mathcal{L}_{yy}(x, y, \lambda) & \mathcal{L}_{yx}(x, y, \lambda) \\ g_x(x, y) & \mathcal{L}_{xy}(x, y, \lambda) & \mathcal{L}_{xx}(x, y, \lambda) \end{vmatrix} = \begin{vmatrix} 0 & 2y & 2x \\ 2y & 4 - 2\lambda & 0 \\ 2x & 0 & 2 - 2\lambda \end{vmatrix}$$

$$\mathbb{B}_2(0, 1) = \begin{vmatrix} 0 & 2 & 0 \\ 2 & 0 & 0 \\ 0 & 0 & -2 \end{vmatrix} = 8 \qquad \mathbb{B}_2(0, -1) = \begin{vmatrix} 0 & -2 & 0 \\ -2 & 0 & 0 \\ 0 & 0 & -2 \end{vmatrix} = 8.$$

For $r = m + 1 = 2 = n$, we have

$$(-1)^2 \mathbb{B}_2(0, 1) = 8 > 0 \qquad\qquad (-1)^2 \mathbb{B}_2(0, -1) = 8 > 0$$

and the points $(0, \pm 1)$ are local maxima.

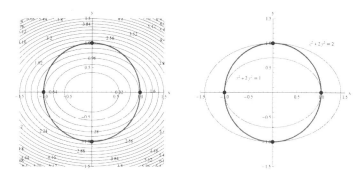

FIGURE 3.24: Level curves $f = 1$ and $f = 2$ are tangent to the constraint $g = 1$

iii) ***Conclusion:*** We have

$$f(\pm 1, 0) = 1 \qquad\qquad f((0, \pm 1) = 2.$$

Subject to the constraint $g(x, y) = 1$, f attains its maximum value 2 at the points $(0, \pm 1)$ and its minimum value 1 at the points $(\pm 1, 0)$. At these points, the level curves $x^2 + 2y^2 = 1$, $x^2 + 2y^2 = 2$ and the constraint $x^2 + y^2 = 1$, sketched in Figure 3.24, are tangent.

2. – Consider the problem

$$\min f(x, y, z) = (x - x_0)^2 + (y - y_0)^2 + (z - z_0)^2$$

$$\text{subject to} \qquad g(x, y, z) = ax + by + cz + d = 0$$

for $(x_0, y_0, z_0) \in \mathbb{R}^3$, $d \in \mathbb{R}$ and $(a, b, c) \neq (0, 0, 0)$.

i) Find the points that satisfy the first-order conditions.

ii) Show that the second-order conditions for a local minimum are satisfied.

iii) Give a geometric argument for the existence of a minimum solution.

iv) Does the maximization problem have any solution?

v) Solve

$$\min x^2 + y^2 + z^2 \qquad \text{subject to} \qquad x + y + z = 1.$$

Solution: i) Note that f and g are C^1 in $\mathbb{R}^3$. In particular, each point of $[g = 0]$ is a relative interior and regular point since we have

$$g'(x, y, z) = ai + bj + ck \neq 0 \qquad \Longrightarrow \qquad rank(g'(x, y, z)) = 1.$$

So, by applying Lagrange multipliers method, we will look for the candidate extreme points as stationary points for the Lagrangian

$$\mathcal{L}(x, y, z, \lambda) = (x - x_0)^2 + (y - y_0)^2 + (z - z_0)^2 - \lambda(ax + by + cz + d).$$

These points are solution of the system

$$\nabla \mathcal{L}(x, y, z, \lambda) = \langle 0, 0, 0, 0 \rangle \quad \Longleftrightarrow \quad \begin{cases} \mathcal{L}_x = 2(x - x_0) - \lambda a = 0 \\[2mm] \mathcal{L}_y = 2(y - y_0) \quad \lambda b = 0 \\[2mm] \mathcal{L}_z = 2(z - z_0) - \lambda c = 0 \\[2mm] \mathcal{L}_\lambda = -(ax + by + cz + d) = 0 \end{cases}$$

from which we deduce

$$\begin{cases} x = \dfrac{\lambda}{2}a + x_0 \qquad\qquad y = \dfrac{\lambda}{2}b + y_0 \qquad\qquad z = \dfrac{\lambda}{2}c + z_0 \\[3mm] a\left(\dfrac{\lambda}{2}a + x_0\right) + b\left(\dfrac{\lambda}{2}b + y_0\right) + c\left(\dfrac{\lambda}{2}c + z_0\right) + d = 0 \end{cases}$$

and that

$$\frac{\lambda}{2} = -\frac{ax_0 + by_0 + cz_0 + d}{a^2 + b^2 + c^2} = \frac{\lambda^*}{2}.$$

Thus, we have only one critical point denoted (x^*, y^*, z^*) with $\lambda = \lambda^*$.

ii) First, note that

$$g'(x^*, y^*, z^*) = (a, b, c) \neq (0, 0, 0)$$

and discuss:

Case $a \neq 0$.

The first column vector of $g'(x^*, y^*, z^*)$ is linearly independent, and because $n = 3$ and $m = 1$, we have to consider the signs of the following bordered Hessian determinants:

$$\mathbb{B}_2(x^*, y^*, z^*) = \begin{vmatrix} 0 & g_x & g_y \\ g_x & \mathcal{L}_{xx} & \mathcal{L}_{xy} \\ g_y & \mathcal{L}_{xy} & \mathcal{L}_{yy} \end{vmatrix} = \begin{vmatrix} 0 & a & b \\ a & 2 & 0 \\ b & 0 & 2 \end{vmatrix} = -2(a^2 + b^2) < 0.$$

The partial derivatives of g are taken at (x^*, y^*, z^*) and those of $\mathcal{L}$ at $(x^*, y^*, z^*, \lambda^*)$.

$$\mathbb{B}_3 = \begin{vmatrix} 0 & g_x & g_y & g_z \\ g_x & \mathcal{L}_{xx} & \mathcal{L}_{xy} & \mathcal{L}_{xz} \\ g_y & \mathcal{L}_{yx} & \mathcal{L}_{yy} & \mathcal{L}_{yz} \\ g_z & \mathcal{L}_{zx} & \mathcal{L}_{zy} & \mathcal{L}_{zz} \end{vmatrix} = \begin{vmatrix} 0 & a & b & c \\ a & 2 & 0 & 0 \\ b & 0 & 2 & 0 \\ c & 0 & 0 & 2 \end{vmatrix} = -4(a^2 + b^2 + c^2) < 0.$$

Case $a = 0$ & $b \neq 0$.

The first column vector of $g'(x^*, y^*, z^*)$ is linearly dependent and the second is linearly independent. We renumber the variables in the order y, x, z and obtain

$$\mathbb{B}_2 = \begin{vmatrix} 0 & b & a \\ b & 2 & 0 \\ a & 0 & 2 \end{vmatrix} = -2(a^2 + b^2) \qquad \mathbb{B}_3 = \begin{vmatrix} 0 & b & a & c \\ b & 2 & 0 & 0 \\ a & 0 & 2 & 0 \\ c & 0 & 0 & 2 \end{vmatrix} = -4(a^2 + b^2 + c^2).$$

Case $a = 0, \ b = 0, \& \ c \neq 0.$

The first and second column vector of $g'(x^*, y^*, z^*)$ are linearly dependent and the third is linearly independent. We renumber the variables in the order z, x, y and obtain

$$
\mathbb{B}_2 = \begin{vmatrix} 0 & c & a \\ c & 2 & 0 \\ a & 0 & 2 \end{vmatrix} = -2(a^2 + c^2) \qquad \mathbb{B}_3 = \begin{vmatrix} 0 & c & a & b \\ c & 2 & 0 & 0 \\ a & 0 & 2 & 0 \\ b & 0 & 0 & 2 \end{vmatrix} = -4(a^2 + b^2 + c^2).
$$

Conclusion. In each case, we have, with $m = 1$,

$$(-1)^m \mathbb{B}_2(x^*, y^*, z^*) > 0 \qquad\qquad (-1)^m \mathbb{B}_3(x^*, y^*, z^*) = 4(a^2 + b^2 + c^2) > 0.$$

We conclude that the point (x^*, y^*, z^*) is a local minimum to the constrained minimization problem.

iii) *Geometric interpretation of the minimization problem:*

If $M(x, y, z), \ M_0(x_0, y_0, z_0) \in \mathbb{R}^3$, then

$$f(x, y, z) = (x - x_0)^2 + (y - y_0)^2 + (z - z_0)^2 = M_0 M^2$$

is the square of the distance of the point M to the point M_0. The constraint surface

$$g(x, y, z) = ax + by + cz + d = 0 \quad \text{is the plane with normal} \quad \langle a, b, c \rangle.$$

The minimization problem consists in finding a point M in the plane that is located at a shortest distance from M_0. Such a point exists and is obtained by considering the intersection of the line passing through the point M_0 and perpendicular to the plane. A direction of this line is given by the normal to the plane $\langle a, b, c \rangle$. Therefore, parametric equations of the line are

$$x = x_0 + ta \qquad\qquad y = y_0 + tb \qquad\qquad z = z_0 + tc \qquad\qquad t \in \mathbb{R}.$$

Clearly the intersection of the line with the plane gives

$$a\Big(x_0 + ta\Big) + b\Big(y_0 + tb\Big) + c\Big(z_0 + tc\Big) + d = 0 \quad \Longleftrightarrow$$

$$t = \frac{\lambda^*}{2} = -\frac{ax_0 + by_0 + cz_0 + d}{a^2 + b^2 + c^2}.$$

f takes its minimum value

$$f\Big(\frac{\lambda^*}{2}a + x_0, \frac{\lambda^*}{2}b + y_0, \frac{\lambda^*}{2}c + z_0\Big) = \Big(\frac{\lambda^*}{2}a\Big)^2 + \Big(\frac{\lambda^*}{2}b\Big)^2 + \Big(\frac{\lambda^*}{2}c\Big)^2 = \frac{\lambda^{*2}}{4}(a^2 + b^2 + c^2).$$

The shortest distance of M_0 to the plan $g = 0$ is

$$D = \sqrt{\frac{\lambda^{*2}}{4}(a^2 + b^2 + c^2)} = \frac{|\lambda^*|}{2}\sqrt{a^2 + b^2 + c^2} = \frac{|ax_0 + by_0 + cz_0 + d|}{\sqrt{a^2 + b^2 + c^2}}.$$

iv) *The maximization problem doesn't have a solution:*

Suppose that there exists (x_m, y_m, z_m) a solution to the maximization problem. Then the points $(x_m + t, y_m - t, z_m)$ for $t \in \mathbb{R}$ are located in the plane and satisfy

$$f(x_m + t, y_m - t, z_m) = (x_m + t)^2 + (y_m - t)^2 + z^2 \longrightarrow +\infty \qquad \text{as} \qquad t \longrightarrow +\infty.$$

v) From, the previous study, choose $(a, b, c) = (1, 1, 1), \ d = -1, \ (x_0, y_0, z_0) = (0, 0, 0)$. Then

$$\frac{\lambda}{2} = \frac{1}{3} \qquad \text{and} \qquad (x^*, y^*, z^*) = \left(\frac{1}{3}, \frac{1}{3}, \frac{1}{3}\right).$$

We conclude that the point $\left(\frac{1}{3}, \frac{1}{3}, \frac{1}{3}\right)$ is a local minimum to the constrained minimization problem. At this point, the two level surfaces

$$x^2 + y^2 + z^2 = \frac{1}{3} = f\left(\frac{1}{3}, \frac{1}{3}, \frac{1}{3}\right), \qquad \text{and} \qquad x + y + z = 1$$

are tangent, as it is described in Figure 3.25.

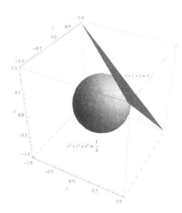

FIGURE 3.25: The level surface and the plane are tangent

3. – The planes $x + y + z = 3$ and $x - y = 2$ intersect in a straight line. Find the point on that line that is closest to the origin.

Solution: i) We formulate the problem as follows

$$\min \ f(x, y, z) = x^2 + y^2 + z^2 \quad \text{subject to} \quad \begin{cases} g_1(x, y, z) = x + y + z = 3 \\ \\ g_2(x, y, z) = x - y = 2. \end{cases}$$

Note that f, g_1 and g_2 are C^1 in $\mathbb{R}^3$ and any point of the set of the constraints, sketched in Figure 3.26 and defined by $g = (g_1, g_2) = (3, 2)$, is an interior point and regular since we have

$$g'(x, y, z) = \begin{bmatrix} 1 & 1 & 1 \\ 1 & -1 & 0 \end{bmatrix} \qquad rank(g'(x, y, z)) = 2.$$

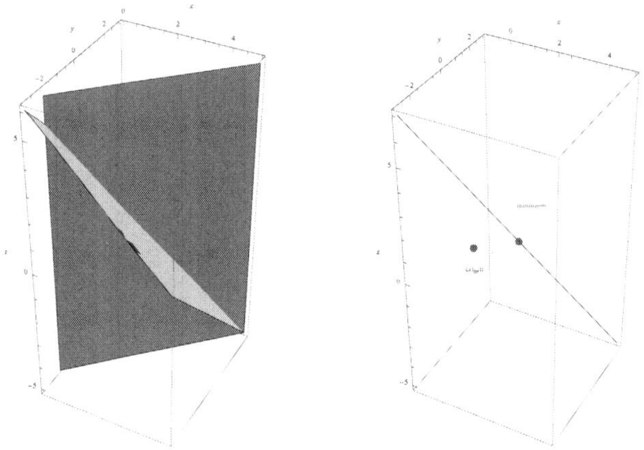

FIGURE 3.26: The constraints, the origin and the minimum point

Consider the Lagrangian

$$\mathcal{L}(x, y, z, \lambda_1, \lambda_2) = f(x, y, z) - \lambda_1(g_1(x, y, z) - 3) - \lambda_2(g_2(x, y, z) - 2)$$

$$= x^2 + y^2 + z^2 - \lambda_1(x + y + z - 3) - \lambda_2(x - y - 2)$$

and look for the stationary points solutions of $\nabla \mathcal{L}(x, y, z, \lambda_1, \lambda_2) = 0_{\mathbb{R}^5}$

$$\Longleftrightarrow \begin{cases} (1) & \mathcal{L}_x = 2x - \lambda_1 - \lambda_2 = 0 \\[2mm] (2) & \mathcal{L}_y = 2y - \lambda_1 + \lambda_2 = 0 \\[2mm] (3) & \mathcal{L}_z = 2z - \lambda_1 = 0 \\[2mm] (4) & \mathcal{L}_{\lambda_1} = -(x + y + z - 3) = 0 \\[2mm] (5) & \mathcal{L}_{\lambda_2} = -(x - y - 2) = 0. \end{cases}$$

From equations (1), (2) and (3), we deduce that

$$\begin{cases} x = \frac{1}{2}(\lambda_1 + \lambda_2) \\[2mm] y = \frac{1}{2}(\lambda_1 - \lambda_2) \\[2mm] z = \frac{1}{2}\lambda_1 \end{cases}$$

then substituting these values into equations (4) and (5), we obtain

$$\begin{cases} \frac{1}{2}(\lambda_1 + \lambda_2) + \frac{1}{2}(\lambda_1 - \lambda_2) + \frac{1}{2}\lambda_1 = 3 \\[2mm] \frac{1}{2}(\lambda_1 + \lambda_2) - \frac{1}{2}(\lambda_1 - \lambda_2) = 2 \end{cases} \Longrightarrow (\lambda_1, \lambda_2) = (2, 2).$$

The only critical point for $\mathcal{L}$ is $(x^*, y^*, z^*, \lambda_1^*, \lambda_2^*) = (2, 0, 1, 2, 2)$.

ii) Note that the first two column vectors of $g'(x, y, z)$ are linearly independent. We can, therefore, keep the matrix without renumbering the variables, and consider the sign of the following bordered Hessian determinant ($n = 3$, $m = 2$, $r = m + 1 = 3$):

$$\mathbb{B}_3(2,0,1) = \begin{vmatrix} 0 & 0 & \frac{\partial g_1}{\partial x} & \frac{\partial g_1}{\partial y} & \frac{\partial g_1}{\partial z} \\[2mm] 0 & 0 & \frac{\partial g_2}{\partial x} & \frac{\partial g_2}{\partial y} & \frac{\partial g_2}{\partial z} \\[2mm] \frac{\partial g_1}{\partial x} & \frac{\partial g_2}{\partial x} & \mathcal{L}_{xx} & \mathcal{L}_{xy} & \mathcal{L}_{xz} \\[2mm] \frac{\partial g_1}{\partial y} & \frac{\partial g_2}{\partial y} & \mathcal{L}_{yx} & \mathcal{L}_{yy} & \mathcal{L}_{yz} \\[2mm] \frac{\partial g_1}{\partial z} & \frac{\partial g_2}{\partial z} & \mathcal{L}_{zx} & \mathcal{L}_{zy} & \mathcal{L}_{zz} \end{vmatrix} = \begin{vmatrix} 0 & 0 & 1 & 1 & 1 \\ 0 & 0 & 1 & -1 & 0 \\ 1 & 1 & 2 & 0 & 0 \\ 1 & -1 & 0 & 2 & 0 \\ 1 & 0 & 0 & 0 & 2 \end{vmatrix} = 12$$

We have

$$(-1)^m \mathbb{B}_3(2,0,1) = (-1)^2 \mathbb{B}_3(2,0,1) = 12 > 0.$$

We conclude that the point $(2, 0, 1)$ is a local minimum to the constrained optimization problem.

iii) *To show that the point is the global minimum point, we use the following parametrization of the set of the constraints; see Figure 3.26:*

$$x = t + 2, \qquad\qquad y = t, \qquad\qquad z = 1 - 2t \qquad\qquad t \in \mathbb{R}.$$

So the optimization problem is reduced to

$$\min_{t \in \mathbb{R}} \; F(t) = f(t + 2, t, 1 - 2t) = (t + 2)^2 + t^2 + (2t - 1)^2.$$

We have

$$F'(t) = 2(t + 2) + 2t + 2(2t - 1)(2) = 12t = 0 \quad\Longleftrightarrow\quad t = 0$$

and

$$F''(t) = 12 > 0 \qquad \forall t \in \mathbb{R}.$$

Hence 0 is a global minimum for F. That is, the point $(2, 0, 1)$ is the solution to the minimization problem.

In Section 3.4, we will see that using the convexity of the Lagrangian in (x, y, z), when $(\lambda_1, \lambda_2) = (2, 2)$, we can conclude that the local minimum point $(2, 0, 1)$ is the global minimum point. Therefore, it solves the problem. The advantage, in arguing in this way, prevents us from exploring the geometry of the constraint set.

3.4   Global Extreme Points-Equality Constraints

The following theorem gives sufficient conditions for a critical point of the Lagrangian to be a global extreme point for the associated constrained optimization problem.

Theorem 3.4.1 *Let* $\Omega \subset \mathbb{R}^n$, Ω *be an open set and* $f, g_1, \ldots, g_m : \Omega \longrightarrow \mathbb{R}$ *be* C^1 *functions. Let* $S \subset \Omega$ *be convex,* $x^* \in \overset{\circ}{S}$ *and* $\mathcal{L}$ *be the Lagrangian*

$$\mathcal{L}(x, \lambda) = f(x) - \lambda_1(g_1(x) - c_1) - \ldots - \lambda_m(g_m(x) - c_m).$$

Then, we have

$$\left. \begin{array}{c} \exists \lambda^* = \langle \lambda_1^*, \ldots, \lambda_m^* \rangle \ : \ \nabla_{x,\lambda}\mathcal{L}(x^*, \lambda^*) = 0 \\[2mm] \mathcal{L}(., \lambda^*) \ \text{is concave (resp. convex) in } x \in S \end{array} \right\}$$

$$\implies \quad f(x^*) = \max_{\{x \in S: \, g(x) = c\}} f(x) \quad (\text{ resp. min})$$

Proof. Suppose that the Lagrangian $\mathcal{L}(., \lambda^*)$ is concave in x and that

$$\frac{\partial \mathcal{L}}{\partial x_i}(x^*, \lambda^*) = \frac{\partial f}{\partial x_i}(x^*) - \sum_{j=1}^{m} \lambda_j^* \frac{\partial g_j}{\partial x_i}(x^*) = 0 \qquad i = 1, \ldots, n,$$

then x^* is a stationary point for $\mathcal{L}(., \lambda^*)$. Therefore, x^* is a global maximum for $\mathcal{L}(., \lambda^*)$ in S (by Theorem 2.3.4) and we have

$$\mathcal{L}(x^*, \lambda^*) = f(x^*) - \lambda_1^*(g_1(x^*) - c_1) - \ldots - \lambda_m^*(g_m(x^*) - c_m)$$

$$\geqslant f(x) - \lambda_1^*(g_1(x) - c_1) - \ldots - \lambda_m^*(g_m(x) - c_m) = \mathcal{L}(x, \lambda^*) \quad \forall x \in S.$$

Since, we have

$$\frac{\partial \mathcal{L}}{\partial \lambda_j}(x^*, \lambda^*) = -(g_j(x^*) - c_j) = 0 \qquad j = 1, \ldots, m$$

then

$$g_1(x^*) - c_1 = g_2(x^*) - c_2 = \ldots = g_m(x^*) - c_m = 0.$$

So, the previous inequality reduces to

$$f(x^*) \geqslant f(x) - \lambda_1^*(g_1(x) - c_1) - \ldots - \lambda_m^*(g_m(x) - c_m).$$

In particular, we have

$$f(x^*) \geqslant f(x) \qquad \forall x \in \{x \in S: \quad g(x) = c\}.$$

Thus x^* solves the constrained maximization problem.

The minimization case can be established similarly.

Remark 3.4.1 * *Note that there is no regularity assumption on the point x^* in the theorem. The proof uses the characterization of a C^1 convex function on a convex set.*

** *The concavity/convexity hypothesis is a sufficient condition. We may have a global extreme point with a Lagrangian that is neither concave nor convex (see Example 3).*

Example 1. Economy. If the cost of capital K and labor L is r and w dollars per unit respectively, find the values of K and L that minimize the cost to produce the output $Q = c\,K^a\,L^b$, where c, a and b are positive parameters satisfying $a + b < 1$.

Solution: The inputs K and L minimizing the cost must solve the problem

$$\min \ rK + wL \qquad \text{subject to} \qquad cK^aL^b = Q.$$

We look for the extreme points in the set $\Omega = (0, +\infty) \times (0, +\infty)$ since K and L must satisfy $cK^aL^b = Q$. Denote

$$f(K,L) = rK + wL \qquad g(K,L) = cK^aL^b \qquad S = \Omega.$$

Note that f and g are C^1 in the open convex set Ω.
Consider the Lagrangian

$$\mathcal{L}(K,L,\lambda) = f(K,L) - \lambda(g(K,L) - Q) = rK + wL - \lambda(cK^aL^b - Q)$$

and Lagrange's necessary conditions

$$\nabla \mathcal{L}(K,L,\lambda) = \langle 0,0,0 \rangle \qquad \Longleftrightarrow \qquad
\begin{cases}
\mathcal{L}_K = r - \lambda ca K^{a-1}L^b - 0 \\[2mm]
\mathcal{L}_L = w - \lambda cb K^a L^{b-1} = 0 \\[2mm]
\mathcal{L}_\lambda = -(cK^aL^b - Q) = 0.
\end{cases}$$

Multiplying each side of the first equality by K, each side of the second equality by L, we obtain

$$rK = \lambda ca K^a L^b = \lambda a Q \qquad\qquad wL = \lambda cb K^a L^b = \lambda b Q$$

then using the third equality, we deduce the unique solution of the system

$$K^* = \lambda^* \frac{aQ}{r} \qquad L^* = \lambda^* \frac{bQ}{w} \qquad \lambda^* = \left(\frac{Q}{c}\right)^{\frac{1}{a+b}} \left(\frac{r}{aQ}\right)^{\frac{a}{a+b}} \left(\frac{w}{bQ}\right)^{\frac{b}{a+b}}.$$

Convexity of $\mathcal{L}$ in (K, L). The Hessian matrix of $\mathcal{L}$ is

$$H_{\mathcal{L}(.,.,\lambda^*)} = \begin{bmatrix} -\lambda^* ca(a-1)K^{a-2}L^b & -\lambda^* cab K^{a-1}L^{b-1} \\ -\lambda^* cab K^{a-1}L^{b-1} & -\lambda^* cb(b-1)K^a L^{b-2} \end{bmatrix}.$$

The leading principal minors are

$$D_1(K, L) = -\lambda^* ca(a-1)K^{a-2}L^b > 0 \quad \text{since } 0 < a < a + b < 1$$

$$D_2(K, L) = \begin{vmatrix} -\lambda^* ca(a-1)K^{a-2}L^b & -\lambda^* cab K^{a-1}L^{b-1} \\ -\lambda^* cab K^{a-1}L^{b-1} & -\lambda^* cb(b-1)K^a L^{b-2} \end{vmatrix}$$
$$= (\lambda^*)^2 c^2 ab K^{2a-2}L^{2b-2}(1 - (a+b)) > 0.$$

Hence, $\mathcal{L}(.,.,\lambda^*)$ is strictly convex in (K, L) in Ω, and we conclude that the point (K^*, L^*) is the solution to the constrained minimization problem.

Example 2. Two-constraint problem. Solve the problem

$$\min \text{ (max) } f(x, y, z) = x - z \qquad \text{subject to} \qquad \begin{cases} g_1(x, y, z) = x^2 + y^2 = 1 \\ \\ g_2(x, y, z) = x^2 + z^2 = 1. \end{cases}$$

Solution: i) Consider the Lagrangian

$$\mathcal{L}(x, y, z, \lambda_1, \lambda_2) = f(x, y, z) - \lambda_1(g_1(x, y, z) - 1) - \lambda_2(g_2(x, y, z) - 1)$$
$$= x - z - \lambda_1(x^2 + y^2 - 1) - \lambda_2(x^2 + z^2 - 1)$$

and look for its stationary points, solution of the system

$$\nabla \mathcal{L}(x, y, z, \lambda_1, \lambda_2) = 0_{\mathbb{R}^5} \iff \begin{cases} (1) & \mathcal{L}_x = 1 - 2x\lambda_1 - 2x\lambda_2 = 0 \\\\ (2) & \mathcal{L}_y = 0 - 2y\lambda_1 = 0 \\\\ (3) & \mathcal{L}_z = -1 - 2z\lambda_2 = 0 \\\\ (4) & \mathcal{L}_{\lambda_1} = -(x^2 + y^2 - 1) = 0 \\\\ (5) & \mathcal{L}_{\lambda_2} = -(x^2 + z^2 - 1) = 0. \end{cases}$$

From equation (2), we deduce that

$$\lambda_1 = 0 \qquad \text{or} \qquad y = 0.$$

* If $y = 0$, then from (4) and (5) we deduce that

$$x = \pm 1 \qquad \text{and} \qquad z = 0.$$

But (3) is not possible.

* If $\lambda_1 = 0$, then (1) and (3) reduce to

$$1 - 2x\lambda_2 = 0 \qquad \text{or} \qquad -1 - 2z\lambda_2 = 0.$$

Since λ_2 cannot be equal to zero, we deduce that

$$x = -z = \frac{1}{2\lambda_2}.$$

Inserting $x = -z$ in (5), we obtain

$$2x^2 = 1 \qquad \Longleftrightarrow \qquad x = \pm 1/\sqrt{2}.$$

Then, from (4), we get

$$\frac{1}{2} + y^2 = 1 \qquad \Longleftrightarrow \qquad y = \pm 1/\sqrt{2}.$$

So, the critical points of $\mathcal{L}$ are

$$\left(\frac{1}{\sqrt{2}}, \pm\frac{1}{\sqrt{2}}, -\frac{1}{\sqrt{2}}, \lambda_1^*, \lambda_2^*\right) \qquad \text{with} \qquad (\lambda_1^*, \lambda_2^*) = \left(0, \frac{1}{\sqrt{2}}\right),$$

$$\left(-\frac{1}{\sqrt{2}}, \pm\frac{1}{\sqrt{2}}, \frac{1}{\sqrt{2}}, \lambda_1^*, \lambda_2^*\right) \qquad \text{with} \qquad (\lambda_1^*, \lambda_2^*) = \left(0, -\frac{1}{\sqrt{2}}\right).$$

The values taken by f at these points are

$$f(\frac{1}{\sqrt{2}}, \pm\frac{1}{\sqrt{2}}, -\frac{1}{\sqrt{2}}) = \sqrt{2} \qquad f(-\frac{1}{\sqrt{2}}, \pm\frac{1}{\sqrt{2}}, \frac{1}{\sqrt{2}}) = -\sqrt{2}.$$

ii) To study the convexity of $\mathcal{L}$ in (x, y, z), consider the Hessian matrix

$$H_{\mathcal{L}(x,y,z,\lambda_1,\lambda_2)} = \begin{bmatrix} \mathcal{L}_{xx} & \mathcal{L}_{xy} & \mathcal{L}_{xz} \\ \mathcal{L}_{yx} & \mathcal{L}_{yy} & \mathcal{L}_{yz} \\ \mathcal{L}_{zx} & \mathcal{L}_{zy} & \mathcal{L}_{zz} \end{bmatrix} = \begin{bmatrix} -2(\lambda_1 + \lambda_2) & 0 & 0 \\ 0 & -2\lambda_1 & 0 \\ 0 & 0 & -2\lambda_2 \end{bmatrix}$$

* With $(\lambda_1^*, \lambda_2^*) = (0, \frac{1}{\sqrt{2}})$, the Hessian is

$$H_{\mathcal{L}(x,y,z,0,\frac{1}{\sqrt{2}})} = \begin{bmatrix} -\sqrt{2} & 0 & 0 \\ 0 & 0 & 0 \\ 0 & 0 & -\sqrt{2} \end{bmatrix}$$

and

$$\Delta_1^{12} = \left| -\sqrt{2} \right| = -\sqrt{2} \qquad \Delta_1^{13} = \left| 0 \right| = 0 \qquad \Delta_1^{23} = \left| -\sqrt{2} \right| = -\sqrt{2}$$

$$\Delta_2^1 = \begin{vmatrix} 0 & 0 \\ 0 & -\sqrt{2} \end{vmatrix} = 0 \qquad \Delta_2^2 = \begin{vmatrix} -\sqrt{2} & 0 \\ 0 & -\sqrt{2} \end{vmatrix} = 2 \qquad \Delta_2^3 = \begin{vmatrix} -\sqrt{2} & 0 \\ 0 & 0 \end{vmatrix} = 0$$

$$\Delta_3 = \begin{vmatrix} -\sqrt{2} & 0 & 0 \\ 0 & 0 & 0 \\ 0 & 0 & -\sqrt{2} \end{vmatrix} = 0 \qquad (-1)^k \Delta_k \geqslant 0 \quad k = 1, 2, 3.$$

Thus $\mathcal{L}(., 0, \frac{1}{\sqrt{2}})$ is concave in $\mathbb{R}^3$ and the points $(\frac{1}{\sqrt{2}}, \pm\frac{1}{\sqrt{2}}, -\frac{1}{\sqrt{2}})$ are maxima points.

** Similarly, we show that $\mathcal{L}(., 0, -\frac{1}{\sqrt{2}})$ is convex and the points $(-\frac{1}{\sqrt{2}}, \pm\frac{1}{\sqrt{2}}, \frac{1}{\sqrt{2}})$ are minima points.

iii) **Comments.** The constraint set, illustrated in Figure 3.27, is the intersection of two cylinders. A parametrization of this set is described by the equations

$$x(t) = \pm\sqrt{1 - t^2}, \quad y(t) = t, \quad z(t) = t \quad \text{or} \quad -t \quad t \in [-1, 1].$$

The set is closed since g_1 and g_2 are continuous on $\mathbb{R}^3$ and

$$[(g_1, g_2) = (1, 1)] = g_1^{-1}(\{1\}) \cap g_2^{-1}(\{1\}).$$

It is bounded since, for any $(x, y, z) \in [(g_1, g_2) = (1, 1)]$, we have

$$\|(x, y, z)\| = x^2 + y^2 + z^2 \leqslant (x^2 + y^2) + (x^2 + z^2) \leqslant 1 + 1 = 2.$$

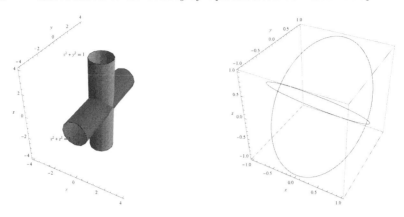

FIGURE 3.27: The constraint set

As f is continuous on the closed bounded constraint set $[(g_1, g_2) = (1, 1)]$, it attains its maximum and minimum values on this set by the extreme value theorem. Thus, the solution of the problem is found by comparing the values of f taken at the candidate points obtained in i).

Example 3. No concavity nor convexity. Consider the problem

$$\max f(x, y, z) = xy + yz + xz \qquad \text{subject to} \qquad g(x, y, z) = x + y + z = 3$$

and the associated Lagrangian

$$\mathcal{L}(x, y, z, \lambda) = f(x, y, z) - \lambda(g(x, y, z) - 3) = xy + yz + xz - \lambda(x + y + z - 3).$$

Show that the local maximum point $(1, 1, 1)$ of the constrained optimization problem, with $\lambda = 2$, is a global maximum, but $\mathcal{L}(., 2)$ is not concave.

Solution: We have

$$\mathcal{L}_x = y + z - \lambda \qquad\qquad \mathcal{L}_y = x + z - \lambda$$

$$\mathcal{L}_z = y + x - \lambda \qquad\qquad \mathcal{L}_\lambda = -(x + y + z - 3).$$

To study the concavity of $\mathcal{L}$ in (x, y, z) when $\lambda = 2$, consider the Hessian matrix

$$H_{\mathcal{L}(x,y,z,2)} = \begin{bmatrix} \mathcal{L}_{xx} & \mathcal{L}_{xy} & \mathcal{L}_{xz} \\ \mathcal{L}_{yx} & \mathcal{L}_{yy} & \mathcal{L}_{yz} \\ \mathcal{L}_{zx} & \mathcal{L}_{zy} & \mathcal{L}_{zz} \end{bmatrix} = \begin{bmatrix} 0 & 1 & 1 \\ 1 & 0 & 1 \\ 1 & 1 & 0 \end{bmatrix}.$$

The principal minors are

$$\Delta_1^{12} = \mid 0 \mid = 0 \qquad \Delta_1^{13} = \mid 0 \mid = 0 \qquad \Delta_1^{23} = \mid 0 \mid = 0$$

$$\Delta_2^1 = \begin{vmatrix} 0 & 1 \\ 1 & 0 \end{vmatrix} = -1 \qquad \Delta_2^2 = \begin{vmatrix} 0 & 1 \\ 1 & 0 \end{vmatrix} = -1 \qquad \Delta_2^3 = \begin{vmatrix} 0 & 1 \\ 1 & 0 \end{vmatrix} = -1$$

$$\Delta_3 = \begin{vmatrix} 0 & 1 & 1 \\ 1 & 0 & 1 \\ 1 & 1 & 0 \end{vmatrix} = 2.$$

So $\mathcal{L}(.,2)$ is neither concave nor convex in (x, y, z). Thus, we cannot conclude, by using the theorem, whether the point $(1, 1, 1)$ is a global maximum or not.

Now, to show that $(1, 1, 1)$ is a global maximum point, we can proceed as follows.

Consider the values of f taken on the plane $g(x, y, z) = 3$:

$$f(x, y, 3 - (x+y)) = xy + (y+x)[3 - (x+y)] = xy + 3(x+y) - (x+y)^2 = \theta(x, y).$$

The maximization problem is equivalent to solve the following unconstrained problem

$$\max_{(x,y)\in\mathbb{R}^2} \theta(x, y).$$

Since θ is C^1, the critical points are solutions of

$$\nabla\theta(x, y) = \langle y + 3 - 2(x+y), x + 3 - 2(x+y) \rangle = \langle 3 - 2x - y, 3 - x - 2y \rangle = \langle 0, 0 \rangle$$

$$\Longleftrightarrow \quad \begin{cases} 2x + y = 3 \\ x + 2y = 3 \end{cases} \quad \Longleftrightarrow \quad (x, y) = (1, 1).$$

$(1, 1)$ is the only critical point for θ. Moreover, we have

$$H_\theta(x, y) \quad = \quad \begin{bmatrix} \theta_{xx} & \theta_{xy} \\ \theta_{yx} & \theta_{yy} \end{bmatrix} \quad = \quad \begin{bmatrix} -2 & -1 \\ -1 & -2 \end{bmatrix}$$

$$D_1(x, y) = -2 \quad \Longrightarrow \quad (-1)^1 D_1(x, y) = 2 > 0$$

$$D_2(x, y) = \begin{vmatrix} -2 & -1 \\ -1 & -2 \end{vmatrix} = 3 \quad \Longrightarrow \quad (-1)^2 D_2(x, y) = 3 > 0.$$

θ is strictly concave on $\mathbb{R}^2$. Thus $(1, 1)$ is a global maximum of θ on $\mathbb{R}^2$. Therefore, $(1, 1, 1)$ is a global maximum of f on $[g = 3]$.

Solved Problems

Part 1. – *A constrained optimization problem.* [29] i) Solve the following constrained minimization problem

$$\min \ x_1^2 + x_2^2 + \ldots + x_n^2 \qquad \text{subject to} \qquad x_1 + x_2 + \ldots + x_n = c$$

where $c \in \mathbb{R}$.

ii) Use part (i) to show that if $x_1, x_2, \ldots, x_n$ are given numbers, then

$$n \sum_{i=1}^{i=n} x_i^2 \geqslant \left(\sum_{i=1}^{i=n} x_i \right)^2.$$

When does the equality hold ?

Solution: i) Denote by f and g the C^∞ functions in $\mathbb{R}^n$:

$$f(x_1, \ldots, x_n) = x_1^2 + x_2^2 + \ldots + x_n^2 \qquad g(x_1, \ldots, x_n) = x_1 + x_2 + \ldots + x_n.$$

Consider the Lagrangian

$$\begin{aligned} \mathcal{L}(x_1, \ldots, x_n, \lambda) &= f(x_1, \ldots, x_n) - \lambda(g(x_1, \ldots, x_n) - c) \\ &= x_1^2 + x_2^2 + \ldots + x_n^2 - \lambda(x_1 + x_2 + \ldots + x_n - c). \end{aligned}$$

Note that any point of the hyperplane $g = c$ is a regular point since we have

$$g'(x_1, \ldots, x_n) = \langle 1, \ldots, 1 \rangle \qquad \Longrightarrow \qquad rank(g'(x_1, \ldots, x_n)) = 1.$$

The stationary points of the Lagrangian are solutions of the system

$$\nabla \mathcal{L}(x_1, \ldots, x_n, \lambda) = \langle 0, \ldots, 0, 0 \rangle \Longleftrightarrow \begin{cases} \mathcal{L}_{x_1} = 2x_1 - \lambda = 0 \\ \vdots \\ \mathcal{L}_{x_i} = 2x_i - \lambda = 0 \\ \vdots \\ \mathcal{L}_{x_n} = 2x_n - \lambda = 0 \\ \mathcal{L}_\lambda = -(x_1 + x_2 + \ldots + x_n - c) = 0. \end{cases}$$

We deduce, from the n first equations, that

$$\lambda = 2x_1 = \ldots = 2x_i = \ldots = 2x_n \qquad \Longrightarrow \qquad x_i = \frac{\lambda}{2} \qquad i = 1, \ldots, n,$$

which inserted into the last equation gives $\dfrac{\lambda}{2} n = c$. Hence the unique solution to the system is

$$\lambda = \frac{2c}{n} \qquad\qquad x_i = \frac{c}{n} \qquad i - 1, \ldots, n.$$

Now, let us study the convexity of $\mathcal{L}$ in $(x_1, \ldots, x_n)$ when $\lambda = \dfrac{2c}{n}$.

The corresponding Hessian matrix is

$$\begin{bmatrix} 2 & \cdots & 0 \\ \vdots & \ddots & \vdots \\ 0 & \cdots & 2 \end{bmatrix}$$

The leading principal minors are

$$D_1 = 2 > 0, \quad D_2 = 2^2 > 0, \ldots D_i = 2^i, \ldots D_n = 2^n > 0.$$

Hence, $\mathcal{L}$ is strictly convex in $(x_1, \ldots, x_n)$, and we conclude that the point

$$\left(\frac{c}{n}, \ldots, \frac{c}{n} \right)$$

is the solution to the constrained minimization problem.

ii) Let $x_1, x_2, \ldots, x_n$ be given numbers. Denote by c their sum. From part i), we have

$$f\left(\frac{c}{n}, \ldots, \frac{c}{n} \right) \leqslant f(t_1, \ldots, t_n) \qquad \forall (t_1, \ldots, t_n) \in [t_1 + \ldots + t_n = c].$$

In particular, for the given x_i, we can write

$$\left(\frac{c}{n} \right)^2 + \left(\frac{c}{n} \right)^2 + \ldots + \left(\frac{c}{n} \right)^2 \leqslant x_1^2 + x_2^2 + \ldots + x_n^2$$

$$\Longleftrightarrow \qquad n\frac{c^2}{n^2} = \frac{c^2}{n} \leqslant x_1^2 + x_2^2 + \ldots + x_n^2$$

$$\Longleftrightarrow \qquad c^2 = (x_1 + x_2 + \ldots + x_n)^2 \leqslant n(x_1^2 + x_2^2 + \ldots + x_n^2).$$

The equality holds only at the minimum point whose coordinates are equal to $(x_1 + x_2 + \ldots + x_n)/n$.

Part 2. – *Method of least squares.* [1]

Consider n points $(x_1, y_1), \ldots, (x_n, y_n)$ such that $x_1, \ldots, x_n$ are not all equal. Find the slope m and the y-intercept b of the line $y = mx + b$, that minimize the quantity

$$D(m,b) = \sum_{i=1}^{n} (mx_i + b - y_i)^2 = (mx_1 + b - y_1)^2 + \ldots + (mx_n + b - y_n)^2$$

which represents the sum of the squares of the vertical distances $d_i = [y_i - (mx_i + b)]$ from these points to the line. This line is called the regression line or the least squares' line of best fit.

(Hint: find the point candidate and check its global optimality by using Part 1 (ii))

Solution: consider the following unconstrained minimization problem:

$$\min_{(m,b)} D(m,b) = [y_1 - (mx_1 + b)]^2 + \ldots + [y_n - (mx_n + b)]^2$$

Since D is regular, then the local extreme points are stationary points of the gradient of D, i.e, solution of $\nabla D(m,b) = \langle 0,0 \rangle$

$$\Longleftrightarrow \begin{cases} \dfrac{\partial D}{\partial m} = -2[y_1 - (mx_1 + b)]x_1 - \ldots - 2[y_n - (mx_n + b)]x_n = 0 \\ \dfrac{\partial D}{\partial b} = -2[y_1 - (mx_1 + b)] - \ldots - 2[y_n - (mx_n + b)] = 0 \end{cases}$$

$$\Longleftrightarrow \begin{cases} \sum_{i=1}^{n} y_i x_i = m[\sum_{i=1}^{n} x_i^2] + b[\sum_{i=1}^{n} x_i] \\ \sum_{i=1}^{n} y_i = m[\sum_{i=1}^{n} x_i] + b[n]. \end{cases}$$

The determinant of this 2×2 linear system is

$$\begin{vmatrix} \sum_{i=1}^{n} x_i^2 & \sum_{i=1}^{n} x_i \\ \sum_{i=1}^{n} x_i & n \end{vmatrix} = n\sum_{i=1}^{n} x_i^2 - \left(\sum_{i=1}^{n} x_i\right)^2 \neq 0$$

since $x_1, \ldots, x_n$ are different (see Part 1). Therefore, there exists a unique solution to the system. It remains to show that it is the minimum point. For this, we study the convexity of D where its Hessian matrix is given by

$$
\mathcal{H}_D(m,b) = \begin{bmatrix} 2\sum_{i=1}^{n} x_i^2 & 2\sum_{i=1}^{n} x_i \\[2ex] 2\sum_{i=1}^{n} x_i & 2n \end{bmatrix}
$$

The leading principal minors values are

$$
D_1(m,b) = 2\sum_{i=1}^{n} x_i^2 > 0 \qquad D_2(m,b) = 4\left(n\sum_{i=1}^{n} x_i^2 - \left(\sum_{i=1}^{n} x_i\right)^2 \right) > 0.
$$

So D is convex and the unique critical point (m^*, b^*) is the global minimum. The regression line equation is $y = m^*x + b^*$ with

$$
m^* = \frac{\begin{vmatrix} \sum_{i=1}^{n} y_i x_i & \sum_{i=1}^{n} x_i \\[2ex] \sum_{i=1}^{n} y_i & n \end{vmatrix}}{\begin{vmatrix} \sum_{i=1}^{n} x_i^2 & \sum_{i=1}^{n} x_i \\[2ex] \sum_{i=1}^{n} x_i & n \end{vmatrix}} \qquad b^* = \frac{\begin{vmatrix} \sum_{i=1}^{n} x_i^2 & \sum_{i=1}^{n} y_i x_i \\[2ex] \sum_{i=1}^{n} x_i & \sum_{i=1}^{n} y_i \end{vmatrix}}{\begin{vmatrix} \sum_{i=1}^{n} x_i^2 & \sum_{i=1}^{n} x_i \\[2ex] \sum_{i=1}^{n} x_i & n \end{vmatrix}}
$$

Part 3. – Students' scores. In a math course, Table 3.3 lists the scores x_i of 14 students on the midterm exam and their scores y_i on the final exam.

i) Plot the data. Do the data appear to lie along a straight line?

ii) Find the least squares' line of best fit of y as a function of x.

iii) Plot the points and the regression line on the same graph.

iv) Use your answer from ii) to predict the final exam score of a student whose midterm score was 41 and who dropped the course.

x_i	100	95	81	71	83	48	92	100	85	63	78	58	73	60
y_i	95	88	53	58	80	31	91	78	85	52	78	74	60	60

TABLE 3.3: Students' scores

Solution: i) The plot, in Figure 3.28, shows that 10 points are close to a line. The plot is obtained using the Mathematica coding below:

$$fp = \{\{100, 95\}, \{95, 88\}, \{81, 53\}, \{71, 58\}, \{83, 80\}, \{48, 31\}, \{92, 91\},$$
$$\{100, 78\}, \{85, 85\}, \{63, 52\}, \{78, 78\}, \{58, 74\}, \{73, 60\}, \{60, 60\}\};$$
$$gp = ListPlot[fp]$$

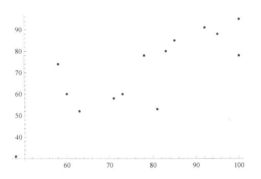

FIGURE 3.28: The data shows an alignment

ii) Using the results from Part 2, we have

$$\sum_{i=1}^{14} x_i = 1087 \quad \sum_{i=1}^{14} x_i^2 = 87855 \quad \sum_{i=1}^{14} y_i = 983 \quad \sum_{i=1}^{14} x_i y_i = 7942815178$$

$$m^* = \frac{1499}{1669} \approx 0.8981426 \qquad b^* = \frac{801}{1669} \approx 0.4799281$$

and the regression line will be

$$y = 0.8981426\,x + 0.4799281.$$

iii) To check the equation of the line of best fit, we use the instruction
$$line = Fit[fp, \{1, x\}, x]$$
$$0.479928 + 0.898143\,x$$

To sketch the line (see Figure 3.29) with the data, we add the following Mathematica coding:

$$gl = Plot[line, \{x, 25, 110\}];$$
$$Show[gl, gp]$$

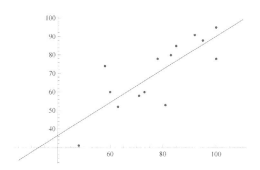

FIGURE 3.29: Data and line $y = 0.479928 + 0.898143\,x$

iv) The student who dropped the course would have at the final exam the approximate mark of $y(41) \approx 0.8981426(41) + 0.4799281 = 36.4056$. The student would have failed if he didn't improve his understanding of the material studied. However, this is only a relative prediction that doesn't take into account other factors involving the learning experience of the student.

Part 4. – University tuition. [12] The following, in Table 3.4, are the tuition fees that were charged at Vanderbilt University from 1982 to 1991.

i) Plot the data.

ii) To fit these data with a model of the form $y = \beta_0 e^{\beta_1 x}$, find the least squares' line of best fit of $\ln y$ as a function of $\ln x$. Deduce approximate values of β_0 and β_1.

iii) Sketch the curve in ii) with the data plot in i).

iv) Suppose the exponential model is accurate for a period of time. In which year would the tuition attain a rate of $40000?

Solution: i) The data, of points (x, y), appear to lie along a straight line. The plot, shown in Figure 3.30, is obtained using the Mathematica coding below:

year	year after 1981, x	tuition (in thousands \$), y
1982	1	6.1
1983	2	6.8
1984	3	7.5
1985	4	8.5
1986	5	9.3
1987	6	10.5
1988	7	11.5
1989	8	12.625
1990	9	13.975
1991	10	14.975

TABLE 3.4: University tuition

$fp1 = \{\{1, 6.1\}, \{2, 6.8\}, \{3, 7.5\}, \{4, 8.5\}, \{5, 9.3\}, \{6, 10.5\}, \{7, 11.5\},$
$\qquad \{8, 12.625\}, \{9, 13.975\}, \{10, 14.975\}\};$
$gp1 = ListPlot[fp1]$

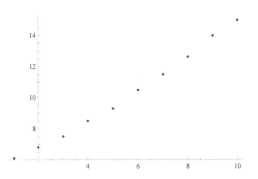

FIGURE 3.30: The data (x_i, y_i) lie along a straight line

The plot of the data, of points $(x_i, \ln y_i)$, appears also to lie along a straight line (see Figure 3.31).

$fp2 = \{\{1, Log[6.1]\}, \{2, Log[6.8]\}, \{3, Log[7.5]\}, \{4, Log[8.5]\}, \{5, Log[9.3]\},$
$\qquad \{6, Log[10.5]\}, \{7, Log[11.5])\}, \{8, Log[12.625]\}, \{9, Log[13.975]\}, \{10, Log[14.975]\}\};$
$gp2 = ListPlot[fp2]$

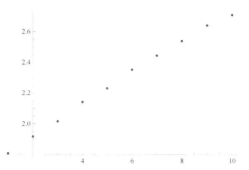

FIGURE 3.31: The data $(x_i, \ln y_i)$ are positioned along a straight line

ii) Using the results from Part 2, the least squares' line of best fit is given by $\ln(y) = \ln(\beta_0) + \beta_1 x$; where $b^* = \ln(\beta_0)$ and $m^* = \beta_1$, are the solution of the linear system

$$p = A.m^* + B.b^*, \qquad\qquad q = B.m^* + 10.b^*$$

where

$$B = \sum_{i=1}^{10} x_i = 55 \qquad\qquad A = \sum_{i=1}^{10} x_i^2 = 385$$

$$p = \sum_{i=1}^{10} \ln(y_i) = 22.7832 \qquad\qquad q = \sum_{i=1}^{10} x_i \ln(y_i) = 22.7832$$

$$m^* = \frac{p.10 - qB}{10A - B^2} \approx 0.10156 \qquad\qquad b^* = \frac{qA - Bp}{10A - B^2} \approx 1.71975$$

and the regression line will be

$$\ln y = 0.10156x + 1.71975.$$

Thus
$$\beta_0 = e^{b^*} \approx 5.583112000 \qquad\qquad \beta_1 = m^* \approx 0.10156.$$

iii) Using Matematica, we find the equation of the line of best fit
$$line = Fit[fp2, \{1, x\}, x]$$
$$1.71975 + 0.10156\, x$$

We sketch the line with the data $(x_i, \ln(y_i))$, in Figure 3.32, using the coding:

$$gl = Plot[line, \{x, 1/2, 11\}];$$
$$Show[gl, gp2]$$

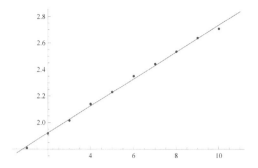

FIGURE 3.32: Data $(x_i, \ln(y_i))$ and line $y = 1.71975 + 0.10156\,x$

Finally, we sketch, in Figure 3.33, the curve $y = f(x) = \beta_0 e^{\beta_1 x}$, with the original data (x_i, y_i):

$$curve = Plot[5.583112Exp[0.10156x], \{x, 1/2, 11\}];$$
$$Show[curve, gp1]$$

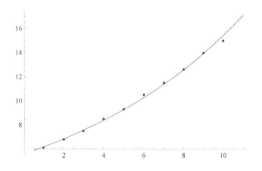

FIGURE 3.33: Data (x_i, y_i) and curve model $f(x) = 5.583112e^{0.10156x}$

vi) *Using the formula for prediction.* We need to solve the equation

$$1000f(x) = 40000 \qquad \Longleftrightarrow \qquad x = \frac{1}{0.10156}\ln(\frac{40}{5.583112}) \approx 19.38886498$$

Thus in the year $1981 + 19 = 2000$, the tuition fees reached the rate of \$40000.

Chapter 4

Constrained Optimization-Inequality Constraints

In this chapter, we are interested in optimizing functions $f : \Omega \subset \mathbb{R}^n \longrightarrow \mathbb{R}$ over subsets described by inequalities

$$g(\mathbf{x}) = (g_1(\mathbf{x}), g_2(\mathbf{x}), \dots, g_m(\mathbf{x})) \leqslant \mathbf{b}_{\mathbb{R}^m} \quad \Longleftrightarrow \quad \begin{cases} g_1(\mathbf{x}) \leqslant b_1 \\ \vdots \quad \vdots \\ g_m(\mathbf{x}) \leqslant b_m \end{cases} \quad \mathbf{x} \in \mathbb{R}^n.$$

Denote the set of the constraints

$$S = [g(x) \leqslant \mathbf{b}] = [g_1(\mathbf{x}) \leqslant b_1] \cap [g_2(\mathbf{x}) \leqslant b_2] \cap \dots \cap [g_m(\mathbf{x}) \leqslant b_m].$$

Example.

* $\quad S = [g_1(x,y) = x^2 + y^2 \leqslant 1] \cap [g_2(x,y) = x - y \leqslant 0]$ is the plane region inside the unit disk and above the line $y = x$. Here $\quad (n = 2, \quad m = 2)$.

** $\quad S = [g_1(x,y,z) = 9 - (x^2 + y^2 + z^2) \leqslant 0] = [x^2 + y^2 + z^2 \geqslant 9]$ is the domain outside the sphere centered at the origin with radius 3. Here $\quad (n = 3, \quad m = 1)$.

*** $\quad S = [g(x,y) = x^2 \leqslant 0] = \{(0,y) : \quad y \in \mathbb{R}\}$ is the y-axis. Here $\quad (n = 2, m = 1)$.

Note that sets defined by inequalities contain interior points and boundary points. So, for comparing the values of a function f taken around an extreme point x^*, it will be suitable to consider curves $x(t)$ passing through x^* and included in the constraint set $[g \leqslant \mathbf{b}]$. We will consider, this time, curves $t \longmapsto x(t)$ such that the set $\{x(t) : t \in [0,a], x(0) = x^*\}$, for some $a > 0$, is included in $[g \leqslant b]$. Then, if x^* is a local maximum of f, then we have

$$f(x(t)) \leqslant f(x^*) \qquad \forall t \in [0,a].$$

Thus, 0 is local maximum point for the function $t \longmapsto f(x(t))$. Hence

$$\frac{d}{dt}\Big[f(x(t))\Big]\Big|_{t=0} = f'(x(t)).x'(t)\Big|_{t=0} \leqslant 0 \qquad \Longrightarrow \qquad f'(x^*).x'(0) \leqslant 0.$$

$x'(0)$ is a tangent vector to the curve $x(t)$ at the point $x(0) = x^*$. This inequality musn't depend on a particular curve $x(t)$. So, we should have

$$f'(x^*).x'(0) \leqslant 0 \qquad \text{for any curve } x(t) \text{ such that } g(x(t)) \leqslant b.$$

In this chapter, we will first characterize, in Section 4.1, the set of tangent vectors to such curves, then establish, in Section 4.2, the equations satisfied by a local extreme point x^*. In Section 4.3, we identify the candidates points for optimality, and in Section 4.4, we explore the global optimality of a constrained local candidate point. Finally, we establish, in Section 4.5, the dependence of the optimal value of the objective function with respect to certain parameters involved in the problem.

4.1   Cone of Feasible Directions

Let
$$x^* \in S = [g(x) \leqslant \mathbf{b}]$$

Definition 4.1.1 *The set defined by*

$$T = \{\, x'(0): \quad t \longmapsto x(t) \in S, \quad x \in C^1[0,a], \quad a > 0, \quad x(0) = x^* \,\}$$

of all tangent vectors at x^* to differentiable curves included in S, is called cone of feasible directions at x^* to the set $[g \leqslant \mathbf{b}]$.

We have the following characterization of the cone T at an interior point x^* of S.

Remark 4.1.1 *We have*

$$g \text{ continuous on } \Omega \quad and \quad x^* \in [g(x) < b] \implies T = \mathbb{R}^n.$$

That is, when x^* is an interior point of S, then the cone at x^* coincides with the whole space.

Indeed, we have $T \subset \mathbb{R}^n$. Let us prove that $\mathbb{R}^n \subset T$. Let $y \in \mathbb{R}^n$.

$*$ If $y = 0$, then the constant curve $x(t) = x^*$ with $t \in [0,1]$ satisfies:

$$x \in C^1[0,1], \quad x(0) = x^*, \quad x'(t) = 0, \quad x'(0) = 0 = y, \quad x(t) = x^* \in S \ \forall t \in [0,1].$$

So $y = 0 \in T$.

$**$ Suppose $y \neq 0$. We have $x^* \in \bigcap_{j=1}^{m}[g_j(x) < b_j]$ which is an open subset of $\mathbb{R}^n$. So there exists $\delta > 0$ such that

$$B_\delta(x^*) \subset \bigcap_{j=1}^{m}[g_j(x) < b_j].$$

Now

$$x(t) = x^* + ty \in B_\delta(x^*) \qquad \forall t \in [-\frac{\delta}{2|y|}, \frac{\delta}{2|y|}]$$

since

$$|x(t) - x^*| = |t||y| \leqslant \frac{\delta}{2|y|}|y| = \frac{\delta}{2} < \delta.$$

We deduce that $y \in T$ since the curve satisfies: $x \in C^1[0, \frac{\delta}{2|y|}]$,

$$x(0) = x^*, \quad x'(t) = y, \quad x'(0) = y, \quad x(t) = x^* + ty \in S \quad \forall t \in [0, \frac{\delta}{2|y|}].$$

Example 1. Find and sketch the cone of feasible directions at the point $(-1/2, 1/2)$ belonging to the set

$$S = \{(x,y) \in \mathbb{R}^2 : g_1(x,y) = x^2 + y^2 - 1 \leqslant 0 \quad and \quad g_2(x,y) = x - y \leqslant 0\}.$$

Solution: The set S is the part of the unit disk located above the line $y = x$. The point $(-1/2, 1/2)$ is an interior point of S; see Figure 4.1. Thus $T = \mathbb{R}^2$.

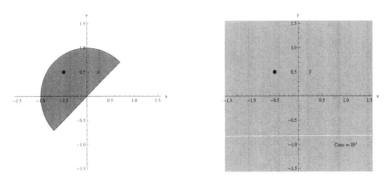

FIGURE 4.1: S and the cone at $(-1/2, 1/2)$

We know a representation of the cone T when x^* is a regular point of S.

Definition 4.1.2 *A point $x^* \in S = [g \leqslant b]$ is said to be **a regular point** of the constraints if the gradient vectors $\nabla g_i(x^*)$, $i \in I(x^*)$ are linearly independent, and where*

$$I(x^*) = \Big\{ i, i \in \{1, \ldots, m\} : \quad g_i(x) = b_i \Big\}.$$

Theorem 4.1.1

At a regular point $x^* \in S = [g \leqslant b]$, where g is C^1 in a neighborhood of x^*, the cone of feasible directions T is equal to the convex cone

$$C = \{ y \in \mathbb{R}^n : \quad g_i'(x^*) y \leqslant 0, \quad i \in I(x^*) \}.$$

Before giving the proof, we give some remarks and identify some cones.

Remark 4.1.2 *The cone of feasible directions at a point $x^* \in S$ with vertex x^* is the translation of C by the vector x^* given by*

$$C(x^*) = x^* + C = x^* + \{ h \in \mathbb{R}^n : \quad g_i'(x^*).h \leqslant 0, \quad i \in I(x^*) \}$$

$$= \{x^* + h \in \mathbb{R}^n : \quad g_i'(x^*).h \leqslant 0, \qquad i \in I(x^*)\}$$

$$= \{x \in \mathbb{R}^n : \quad g_i'(x^*).(x - x^*) \leqslant 0, \qquad i \in I(x^*)\}$$

$C(x^*)$ is the cone of feasible directions to the constraint set $[g(x) \leqslant \mathbf{b}]$ passing through x^*.

Example 2. Find and sketch the cone of feasible directions $C(x, y)$ with vertex $(x, y) = (-1/2, -1/2)$, $(0, 1)$ and $(1/\sqrt{2}, 1/\sqrt{2})$. The points belong to the set

$$S = \{(x, y) \in \mathbb{R}^2 : \quad g_1(x, y) = x^2 + y^2 - 1 \leqslant 0 \quad \text{and} \quad g_2(x, y) = x - y \leqslant 0\}.$$

Solution: Note that the three points belong to ∂S; see Figure 4.2.

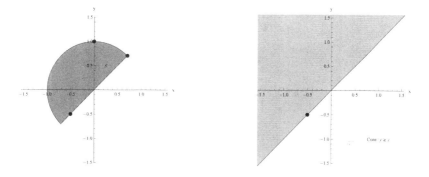

FIGURE 4.2: Location of the points on S and $C(-1/2, -1/2)$

To determine the cone of feasible directions at each point (see Figures 4.2 and 4.3), we need to discuss the regularity of each point. First, we will need:

$$g'(x, y) = \begin{bmatrix} \frac{\partial g_1}{\partial x} & \frac{\partial g_1}{\partial y} \\ \frac{\partial g_2}{\partial x} & \frac{\partial g_2}{\partial y} \end{bmatrix} = \begin{bmatrix} 2x & 2y \\ 1 & -1 \end{bmatrix}.$$

$*$ At $(-1/2, -1/2)$, the equality constraints $g_2 = 0$ is satisfied and the point is regular. We have

$$g_2'(x, y) = \begin{bmatrix} 1 & -1 \end{bmatrix} \qquad \text{and} \qquad rank(g_2'(-1/2, -1/2)) = 1$$

$$C(-1/2, -1/2) = \left\{ (x,y) \in \mathbb{R}^2 : \quad \begin{bmatrix} 1 & -1 \end{bmatrix} \cdot \begin{bmatrix} x + \frac{1}{2} \\ y + \frac{1}{2} \end{bmatrix} \leqslant 0 \right\}$$

$$= \left\{ (x,y) \in \mathbb{R}^2 : \quad x - y \leqslant 0 \right\}.$$

** At $(0,1)$, only the equality-constraint $g_1 = 0$ is satisfied and the point is regular. We have

$$g_1'(0,1) = \begin{bmatrix} 0 & 2 \end{bmatrix} \qquad \text{and} \qquad rank(g_1'(0,1)) = 1$$

$$C(0,1) = \left\{ (x,y) \in \mathbb{R}^2 : \quad \begin{bmatrix} 0 & 2 \end{bmatrix} \cdot \begin{bmatrix} x - 0 \\ y - 1 \end{bmatrix} \leqslant 0 \right\}$$

$$= \left\{ (x,y) \in \mathbb{R}^2 : \quad y \leqslant 1 \right\}.$$

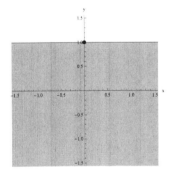 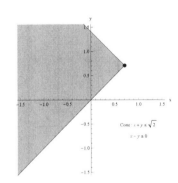

FIGURE 4.3: $C(0,1) = [y \leqslant 1]$ and $C(1/\sqrt{2}, 1/\sqrt{2}) = [x + y \leqslant \sqrt{2}] \cap [x \leqslant y]$

*** At $(1/\sqrt{2}, 1/\sqrt{2})$, the two equality constraints $g_1 = g_2 = 0$ are satisfied and the point is regular. We have

$$g'(\frac{1}{\sqrt{2}}, \frac{1}{\sqrt{2}}) = \begin{bmatrix} \sqrt{2} & \sqrt{2} \\ 1 & -1 \end{bmatrix} \qquad \text{and} \qquad rank(g'(\frac{1}{\sqrt{2}}, \frac{1}{\sqrt{2}})) = 2$$

$$C(\frac{1}{\sqrt{2}}, \frac{1}{\sqrt{2}}) = \left\{ (x,y) \in \mathbb{R}^2 : \quad \begin{bmatrix} \sqrt{2} & \sqrt{2} \\ 1 & -1 \end{bmatrix} \cdot \begin{bmatrix} x - \frac{1}{\sqrt{2}} \\ y - \frac{1}{\sqrt{2}} \end{bmatrix} \leqslant 0 \right\}$$

$$= \left\{ (x,y) \in \mathbb{R}^2 : \quad x + y - \sqrt{2} \leqslant 0 \quad \text{and} \quad x - y \leqslant 0 \right\}.$$

Remark 4.1.3 *The conclusion of the theorem is also true when the point* x^* *satisfies any one of the following regularity conditions [5] :*

i) Each constraint $g_j(x)$ *is affine for* $j \in I(x^*)$.

ii) There exists $\bar{x}$ *such that*

$$g_j(\bar{x}) \leqslant b_j$$
$$\forall j \in I(x^*).$$
$$g_j(\bar{x}) < b_j \quad if \quad g_j \quad is \ not \ affine$$

Example 3. Suppose that all the constraints are affine and that the set S is described by

$$S = \{x \in \mathbb{R}^n : \ \sum_{j=1}^n a_{ij} x_j \leqslant b_i, \quad i = 1, \ldots, m\} \ = \ \{x \in \mathbb{R}^n : \quad Ax \leqslant b\}$$

where $A = (a_{ij})$ is an $m \times n$ matrix and $b \in \mathbb{R}^m$.

Here $g(x) = Ax$, $g'(x) = A$ and $g_i'(x) = \begin{bmatrix} a_{i1} & a_{i2} & \cdots & a_{in} \end{bmatrix}$. Thus, from the previous remark, any point of S is a regular point and the cone of feasible directions at a point $x^* \in S$ with vertex x^* is given by the polyhedra

$$C(x^*) = \{x \in \mathbb{R}^n : \ g_i'(x).(x - x^*) = \sum_{j=1}^n a_{ij}(x_j - x_j^*) \leqslant 0, \quad i \in I(x^*)\}$$

$$= \{x \in \mathbb{R}^n : \ \sum_{j=1}^n a_{ij} x_j \leqslant \sum_{j=1}^n a_{ij} x_j^* = b_i, \quad i \in I(x^*)\}.$$

Example 4. Suppose f is a C^1 function and

$$S = [z \leqslant f(x)] = \{(x, z) \in \Omega \times \mathbb{R} : \quad z \leqslant f(x)\} \qquad \Omega \subset \mathbb{R}^n.$$

Let x^* be a relative interior point of the surface $z = f(x)$. Find the cone at x^*.

Solution: If we set $\quad g(x, z) = z - f(x), \quad$ then the set S can be described by

$$S = [g(x, z) \leqslant 0] = \{(x, z) \in \Omega \times \mathbb{R} : \quad g(x, z) \leqslant 0\}$$

and the point $(x^*, f(x^*))$ is a regular point since we have

$$g'(x^*, f(x^*)) = [\ -f'(x^*) \quad 1 \ ] \neq \mathbf{0} \qquad rank(g'(x^*, f(x^*))) = 1.$$

The cone of feasible directions at the point $(x^*, f(x^*))$ with vertex $(x^*, f(x^*))$ is given by

$$C(x^*, f(x^*)) = \left\{ (x, z) \in \mathbb{R}^n \times \mathbb{R} : \quad g'(x^*, f(x^*)).\begin{bmatrix} x - x^* \\ z - f(x^*) \end{bmatrix} \leqslant 0 \right\}.$$

We have

$$g'(x^*, f(x^*)).\begin{bmatrix} x - x^* \\ z - f(x^*) \end{bmatrix} = [\ -f'(x^*) \quad 1 \ ].\begin{bmatrix} x - x^* \\ z - f(x^*) \end{bmatrix}$$

$$= -f'(x^*).(x - x^*) + z - f(x^*) \leqslant 0 \quad \Longleftrightarrow \quad z \leqslant f(x^*) + f'(x^*).(x - x^*).$$

Hence

$$C(x^*, f(x^*)) = \left\{ (x, z) \in \mathbb{R}^n \times \mathbb{R} : \quad z \leqslant f(x^*) + f'(x^*).(x - x^*) \right\}.$$

The cone is the region below the hyperplane $z = f(x^*) + f'(x^*).(x - x^*)$, which is also the tangent plane to the surface $z = f(x)$ at x^*.

In particular, when x^* is a stationary point, i.e $f'(x^*) = \mathbf{0}$, the cone of feasible directions at x^* is the region below the horizontal tangent plane $z = f(x^*)$.

Remark 4.1.4 *Note that the representation of the cone of feasible directions obtained in the theorem used the fact that the point was regular. When, this hypothesis is omitted the representation is not necessary valid.*

Indeed, if we consider the set S defined by

$$g(x, y) \leqslant 0 \qquad \text{with} \qquad g(x, y) = x^2,$$

then S is reduced to the y axis. No point of S is regular since we have

$$g'(x, y) = [\ 2x \quad 0 \ ] \qquad \text{and} \qquad g'(0, y) = [\ 0 \quad 0 \ ] \qquad \text{on the y-axis.}$$

We deduce that at each point $(0, y_0)$, we have

$$C(0, y_0) = \left\{ (x, y) : \quad g'(0, y_0).\begin{bmatrix} x - 0 \\ y - y_0 \end{bmatrix} \leqslant 0 \right\}$$

$$= \left\{ (x, y) : \quad [\ 0 \quad 0 \ ].\begin{bmatrix} x - 0 \\ y - y_0 \end{bmatrix} = 0 \right\} = \mathbb{R}^2.$$

However, the line
$$x(t) = 0 \qquad\qquad y(t) = y_0 + t$$
remains included in S, passes through the point $(0, y_0)$ at $t = 0$, and has the direction $\begin{bmatrix} x'(0) \\ y'(0) \end{bmatrix} = \begin{bmatrix} 0 \\ 1 \end{bmatrix}$. Hence, the cone of feasible directions at each point of S is equal to S. Note that, it also coincides with the tangent plane at each point, since

$$g(x, y) = x^2 \leqslant 0 \qquad\qquad \Longleftrightarrow \qquad\qquad g(x, y) = x^2 = 0.$$

Proof. We have:

T $\subset$ C : Indeed, let $y \in T$, $y \neq 0$, then

$$\exists\, x(t) \text{ differentiable such that } g(x(t)) \leqslant \mathbf{b} \quad \forall t \in [0, a] \text{ for some } a > 0,$$
$$x(0) = x^*, \qquad\qquad x'(0) = y.$$

So 0 is a minimum for the function $\phi_i(t) = g_i(x(t)) - b_i$, $(i \in I(x^*))$, over the interval $[0, a]$ since we have

$$\phi_i(t) = g_i(x(t)) - b_i \leqslant 0 = \phi_i(0)$$

$$\phi_i(0) = g_i(x^*) - b_i = 0 \quad \text{because } i \in I(x^*).$$

Since g_i and $x(.)$ are C^1, then ϕ_i is C^1 and Taylor's formula gives

$$\phi_i(t) - \phi_i(0) = \phi_i'(0)t + t\alpha(t) = t\left(\phi_i'(0) + \alpha(t)\right) \qquad \text{with} \qquad \lim_{t \to 0^+} \alpha(t) = 0.$$

If $\phi_i'(0) > 0$ then there exists $a_0 \in (0, a)$ such that

$$\|\alpha(t)\| < \frac{\phi_i'(0)}{2} \qquad \forall t \in (0, a_0) \qquad \Longrightarrow \qquad \alpha(t) > -\frac{\phi_i'(0)}{2} \qquad \forall t \in (0, a_0).$$

We deduce that

$$\phi_i(t) - \phi_i(0) > t\left(\phi_i'(0) - \frac{\phi_i'(0)}{2}\right) = t\frac{\phi_i'(0)}{2} > 0 \qquad \forall t \in (0, a_0)$$

which contradicts that 0 is a maximum for ϕ_i on $[0, a]$. So $y \in C$ since we have

$$\phi_i'(0) = \frac{d}{dt}(g_i(x(t)))\Big]_{t=0} = \nabla g_i(x(t)).x'(t)\Big]_{t=0} = g_i'(x^*).y \leqslant 0.$$

$\mathbf{C} \subset \mathbf{T}$: Let $y \in C \setminus \{0\}$. We distinguish between two situations:

First   case   Suppose that

$$g_i'(x^*).y < 0 \qquad \forall i \in I(x^*).$$

Since $x^* \in [g_j(x) < b_j]$ for $j \notin I(x^*)$ and g continuous, there exists $\delta > 0$ such that

$$B_\delta(x^*) \ \subset \ \bigcap_{j \notin I(x^*)} [g_j(x) < b_j].$$

Consider the curve

$$x(t) = x^* + ty \qquad t > 0 \qquad \text{where} \qquad x(0) = x^* \quad \text{and} \quad x'(0) = y.$$

We claim that

$$\exists \delta_0 \in \left(0, \min(\delta, \frac{\delta}{|y|})\right) \quad \text{such that} \quad x(t) \in S = [g(x) \leqslant b] \quad \forall t \in [0, \delta_0].$$

Indeed, for $j \in I(x^*)$, we have

$$g_j(x(t)) = g_j(x^* + ty) = g_j(x^*) + tg_j'(x^*).y + t\varepsilon_j(t) \quad \text{with} \quad \lim_{t \to 0} \varepsilon_j(t) = 0.$$

Since $g_j(x^*) \ = \ b_j$ and $g_j'(x^*).y \ < \ 0$, we deduce the existence of $\delta_0^j \in (0, \min(\delta, \frac{\delta}{|y|}))$ such that

$$|\varepsilon_j(t)| < -\frac{1}{2}g_j'(x^*).y.$$

Consequently, for $\delta_0 = \min_{j \in I(x^*)} \delta_0^j$, we have $\quad \forall j \in I(x^*)$,

$$g_j(x(t)) < b_j + tg_j'(x^*).y - \frac{t}{2}g_j'(x^*).y = b_j + \frac{t}{2}g_j'(x^*).y < b_j \quad \forall t \in (0, \delta_0).$$

Second   case   Suppose that

$$g_i'(x^*).y = 0 \qquad \forall i \in \{i_1, i_2, \dots, i_p\} \subset I(x^*) \qquad \text{and}$$

$$g_i'(x^*).y < 0 \qquad \forall i \in I(x^*) \setminus \{i_1, i_2, \dots, i_p\} \qquad p < n.$$

Consider the system of equations

$$F(t,u) = G\Big(x^* + ty + {}^t\, G'(x^*)u\Big) - B = 0$$

where, for t fixed, $u \in \mathbb{R}^p$ is the unknown, and where

$$G = (g_{i_1}, g_{i_2}, \ldots, g_{i_p}), \qquad B = (b_{i_1}, b_{i_2}, \ldots, b_{i_p}), \qquad rank(G'(x^*)) = p.$$

Note that F is well defined on an open subset of $\mathbb{R} \times \mathbb{R}^p$. Indeed, if g is C^1 on

$$B_\delta(x^*) \subset \{x \in \mathbb{R}^n : g_j(x) < b_j, \quad j \notin I(x^*)\},$$

then $\forall (t,u) \in (-\delta_0, \delta_0) \times B_{\delta_0}(\mathbf{0})$ with $\delta_0 = \min\left(\dfrac{\delta}{2\|y\|}, \dfrac{\delta}{2\|G'(x^*)\|}\right)$, we have

$$\|(x^* + ty + {}^t\, G'(x^*)u) - x^*\| \leqslant |t|\|y\| + \|u\|\|G'(x^*)\|$$

$$< \dfrac{\delta}{2\|y\|}\|y\| + \dfrac{\delta}{2\|G'(x^*)\|}\|G'(x^*)\| = \dfrac{\delta}{2} + \dfrac{\delta}{2} = \delta$$

$$\implies \quad (x^* + ty + {}^t\, G'(x^*)u) \in B_\delta(x^*).$$

We have

$$F(t,u) = G(X(t,u)) - B \qquad\qquad X(t,u) = x^* + ty + {}^t\, G'(x^*)u$$

$$X_j(t,u) = x_j^* + ty_j + \sum_{l=1}^{p} \dfrac{\partial G_{i_l}}{\partial x_j}(x^*)u_l \qquad\qquad \dfrac{\partial X_j}{\partial u_p} = \dfrac{\partial G_{i_p}}{\partial x_j}(x^*)$$

$$\dfrac{\partial F_k}{\partial u_p}(t,u) = \sum_{j=1}^{n} \dfrac{\partial G_{i_k}}{\partial X_j}\dfrac{\partial X_j}{\partial u_p} = \sum_{j=1}^{n} \dfrac{\partial G_{i_k}}{\partial x_j}(X(t,u))\dfrac{\partial G_{i_p}}{\partial x_j}(x^*)$$

$$\left[\dfrac{\partial F_k}{\partial u_i}(t,u)\right]_{k,i=1,\cdots,m} = G'(X(t,u))\Big({}^t\, G'(x^*)\Big).$$

By hypotheses, we have

– F is a C^1 function in the open set $A = (-\delta_0, \delta_0) \times B_{\delta_0}(\mathbf{0})$

– $F(0, \mathbf{0}) = G(x^*) - B = 0$

– $(0, \mathbf{0}) \in (-\delta_0, \delta_0) \times B_{\delta_0}(\mathbf{0})$, so $(0, \mathbf{0})$ is an interior point

– $det(\nabla_u F(0, \mathbf{0})) = \dfrac{\partial(F_1, \ldots, F_p)}{\partial(u_1, \ldots, u_p)} = det\left[G'(x^*)\Big({}^t\, G'(x^*)\Big)\right] \neq 0$ since $G'(x^*)$ has rank p.

Then, by the implicit function theorem, there exists open balls $B_\epsilon(0) \subset (-\delta_0, \delta_0)$, $B_\eta(\mathbf{0}) \subset B_{\delta_0}(\mathbf{0})$, $\epsilon, \eta > 0$ with $B_\epsilon(0) \times B_\eta(\mathbf{0}) \subseteq A$, and such that

$$det(\nabla_u F(t, u)) \neq 0 \qquad \text{in} \qquad B_\epsilon(0) \times B_\eta(\mathbf{0})$$

$$\forall t \in B_\epsilon(0), \quad \exists! u \in B_\eta(\mathbf{0}): \qquad F(t, u) = 0$$

$$u: (-\epsilon, \epsilon) \longrightarrow B_\eta(\mathbf{0}); \qquad t \longmapsto u(t) \quad \text{is a } C^1 \text{ function.}$$

Thus, the curve

$$x(t) = X(t, u(t)) = x^* + ty +^t G'(x^*)u(t)$$

is, by construction, a curve in S since we have for each $t \in (-\epsilon, \epsilon)$

$$G(x(t)) - B = 0 \quad \Longleftrightarrow \quad g_j(x(t)) - b_j = 0 \quad \forall j \in \{i_1, i_2, \ldots, i_p\} \subset I(x^*)$$

$$x(t) \in B_\delta(x^*) \subset \{x \in \mathbb{R}^n : g_j(x) < b_j, \quad j \notin I(x^*)\}$$

$$\Longleftrightarrow \quad g_j(x(t)) - b_j < 0 \quad \forall j \notin I(x^*).$$

By differentiating both sides of

$$F(t, u(t)) = G(x(t)) - B = G(X(t, u(t))) - B = 0$$

with respect to t, we get

$$0 = \frac{d}{dt} G(x(t)) = \sum_{j=1}^n \frac{\partial G}{\partial X_j} \frac{\partial X_j}{\partial t}$$

$$X_j(t, u) = x_j^* + ty_j + \sum_{l=1}^m \frac{\partial G_{i_l}}{\partial x_j}(x^*)u_l \qquad \frac{\partial X_j}{\partial t} = y_j + \sum_{l=1}^m \frac{\partial G_{i_l}}{\partial x_j}(x^*)\frac{\partial u_l}{\partial t}$$

$$0 = \frac{d}{dt} G(x(t))\Big]_{t=0} = \sum_{j=1}^n \frac{\partial G}{\partial x_j}(X(t, u))\Big[y_j + \sum_{l=1}^m \frac{\partial G_l}{\partial x_j}(x^*)\frac{\partial u_l}{\partial t}\Big]_{t=0}$$

$$= G'(x^*)y + G'(x^*)^t G'(x^*)u'(0).$$

Since we have $G'(x^*)y = \mathbf{0}$ and that $G'(x^*)^t G'(x^*)$ is nonsingular and definite positive, we conclude that

$$G'(x^*)^t G'(x^*)u'(0) = G'(x^*)y = \mathbf{0} \qquad \Longrightarrow \qquad u'(0) = \mathbf{0}.$$

Hence

$$x'(0) = y +^t G'(x^*)u'(0) = y.$$

Now, for $j \in I(x^*) \setminus \{i_1, i_2, \ldots, i_p\}$, we have

$$g_j(x(t)) = g_j(x(0)) + tg_j'(x^*).x'(0) + t\eta(t) = b_j + tg_j'(x^*).y + t\eta(t)$$

with $\lim_{t \to 0} \eta(t) = 0$. Then, from the first case, there exists $\epsilon_0 \in (0, \epsilon)$ such that

$$g_j(x(t)) < b_j \qquad\qquad \forall t \in (0, \epsilon_0)$$

thus

$$x(t) \in [g_j(x) \leqslant b_j] \qquad \text{for all} \qquad j \in I(x^*) \setminus \{i_1, i_2, \ldots, i_p\}.$$

Finally, y is a tangent vector to the curve $x(t)$ included in S for $t \in [0, \epsilon_0/2]$, so $y \in T$.

*C **is a cone of** $\mathbb{R}^n$ since for $y \in C$ and $\kappa \in \mathbb{R}^+$, we have

$$g_i'(x^*)(\kappa y) = \kappa g_i'(x^*)y \leqslant 0$$

for $i \in I(x^*)$. Thus $\kappa y \in C$.

*C **is a convex of** $\mathbb{R}^n$ since for $y, y' \in C$ and $s \in [0, 1]$, we have

$$g_i'(x^*)(sy + (1-s)y') = sg_i'(x^*)y + (1-s)g_i'(x^*)y' \leqslant s.0 + (1-s).0 = 0$$

for $\quad i \in I(x^*)$. Thus $sy + (1-s)y' \in C$.

Solved Problems

1. – Find and draw the cone of feasible directions at the point $(0, 3, 0)$ belonging to the set $x^2 + y^2 + z^2 \geqslant 9$.

Solution: Set $g(x, y, z) = 9 - (x^2 + y^2 + z^2)$. We have

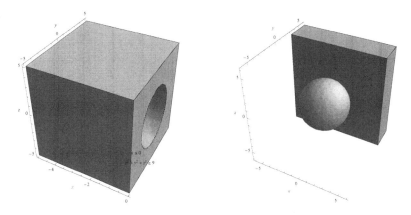

FIGURE 4.4: The set $[g \geqslant 3] \cap [x \leqslant 0]$ and $C(0, 3, 0) = [y \geqslant 3]$

$g'(x, y, z) = -2xi - 2yj - 2zk$, $g'(0, 3, 0) = -6j \neq \mathbf{0}$, $rank(g'(0, 3, 0)) = 1$.

So $(0, 3, 0)$ is a regular point and the cone of feasible directions to $[g \leqslant 0]$, with vertex at this point (see Figure 4.4), is given by

$$C(0, 3, 0) = \{(x, y, z) \in \mathbb{R}^3 : \quad g'(0, 3, 0). \begin{bmatrix} x - 0 \\ y - 3 \\ z - 0 \end{bmatrix} \leqslant 0\}.$$

We have

$$\begin{bmatrix} 0 & -6 & 0 \end{bmatrix}. \begin{bmatrix} x - 0 \\ y - 3 \\ z - 0 \end{bmatrix} \leqslant 0 \quad \Longleftrightarrow \quad 0(x - 0) - 6(y - 3) + 0(z - 1) \leqslant 0$$

$$\Longleftrightarrow \quad y \geqslant 3: \qquad C(0,3,0) = [y \geqslant 3].$$

2. – Find the cone of feasible directions at the point $(0,1,0)$ to the set

$$g(x,y,z) = (g_1(x,y,z), g_2(x,y,z)) \leqslant (1,1)$$

$$g_1(x,y,z) = x+y+z, \qquad g_2(x,y,z) = x^2 + y^2 + z^2.$$

Solution: The set $S = [g \leqslant (1,1)]$, as illustrated in Figure 4.5, is the part of the unit ball located below the plane $x+y+z \leqslant 1$.

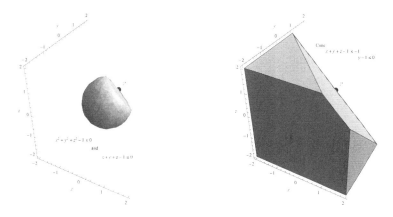

FIGURE 4.5: $[g \leqslant (1,1)]$ and $C(0,1,0)$

The point $(0,1,0) \in S$ satisfies the two constraints $g_1(x,y,z) = g_2(x,y,z) = 1$ and is a regular point since we have:

$$g'(x,y,z) = \begin{bmatrix} \frac{\partial g_1}{\partial x} & \frac{\partial g_1}{\partial y} & \frac{\partial g_1}{\partial z} \\ \frac{\partial g_2}{\partial x} & \frac{\partial g_2}{\partial y} & \frac{\partial g_2}{\partial z} \end{bmatrix} = \begin{bmatrix} 1 & 1 & 1 \\ 2x & 2y & 2z \end{bmatrix}$$

$$g'(0,1,0) = \begin{bmatrix} 1 & 1 & 1 \\ 0 & 2 & 0 \end{bmatrix} \qquad \text{has rank 2.}$$

The cone of feasible directions to the set S at the point $(0, 1, 0)$, with vertex this point, is the set of points (x, y, z) such that

$$g'(0, 1, 0). \begin{bmatrix} x - 0 \\ y - 1 \\ z - 0 \end{bmatrix} = \begin{bmatrix} 1 & 1 & 1 \\ 0 & 2 & 0 \end{bmatrix} . \begin{bmatrix} x \\ y - 1 \\ z \end{bmatrix} \leqslant \begin{bmatrix} 0 \\ 0 \end{bmatrix}$$

$$\Longleftrightarrow \qquad x + y - 1 + z \leqslant 0 \quad \text{and} \quad 2(y - 1) \leqslant 0.$$

Thus

$$C(0, 1, 0) = \{(x, y, z) \in \mathbb{R}^3 : \quad x + y + z \leqslant 1 \quad \text{and} \quad y \leqslant 1\}.$$

3. – Show that the sets

$$z \leqslant \sqrt{x^2 + y^2} \qquad \text{and} \qquad z \leqslant \frac{1}{10}(x^2 + y^2) + \frac{5}{2}$$

have a common cone of feasible directions at the point $(3, 4, 5)$.

Solution: Set

$$g_1(x, y, z) = z - \sqrt{x^2 + y^2} \qquad g_2(x, y, z) = z - \frac{1}{10}(x^2 + y^2) - \frac{5}{2}.$$

We have

$$g_1(3, 4, 5) = g_2(3, 4, 5) = 0$$

$$g_1' = \frac{1}{\sqrt{x^2 + y^2}}\Big(-xi - yj + k\Big), \qquad g_1'(3, 4, 5) = -\frac{3}{5}i - \frac{4}{5}j + k \neq \mathbf{0}$$

$$g_2'(x, y, z) = -\frac{x}{5}i - \frac{y}{5}j + k, \qquad g_2'(3, 4, 5) = -\frac{3}{5}i - \frac{4}{5}j + k \neq \mathbf{0}$$

$$rank(g_1'(3, 4, 5)) = 1 \qquad\qquad rank(g_2'(3, 4, 5)) = 1.$$

So $(3, 4, 5)$ is a regular point for the two constraints $g_1(x, y, z) = 0$ and $g_2(x, y, z) = 0$. Therefore, the cones of feasible directions at the point $(3, 4, 5)$ for the sets $[g_1(x, y, z) \leqslant 0]$ and $[g_2(x, y, z) \leqslant 0]$, with vertex $(3, 4, 5)$, are given respectively by:

$$C_1(3, 4, 5) = \{(x, y, z) \in \mathbb{R}^3 : g_1'(3, 4, 5).^t \begin{bmatrix} x - 3 & y - 4 & z - 5 \end{bmatrix} \leqslant 0\}$$

$$C_2(3, 4, 5) = \{(x, y, z) \in \mathbb{R}^3 : g_2'(3, 4, 5).^t \begin{bmatrix} x - 3 & y - 4 & z - 5 \end{bmatrix} \leqslant 0\}.$$

Clearly, since $g'_1(3, 4, 5) = g'_2(3, 4, 5)$, the two sets are equal and we have for $i = 1, 2$

$$g'_i(3, 4, 5). \begin{bmatrix} x - 3 \\ y - 4 \\ z - 5 \end{bmatrix} \leqslant 0 \iff \begin{bmatrix} -\frac{3}{5} & -\frac{4}{5} & 1 \end{bmatrix}. \begin{bmatrix} x - 3 \\ y - 4 \\ z - 5 \end{bmatrix} \leqslant 0$$

$$\iff -\frac{3}{5}(x - 3) - \frac{4}{5}(y - 4) + 1(z - 5) \leqslant 0.$$

Hence, the two given sets have a common cone of feasible directions at this point (see the illustrations in Figure 4.6) characterized by the inequality

$$-\frac{3}{5}(x - 3) - \frac{4}{5}(y - 4) + (z - 5) \leqslant 0.$$

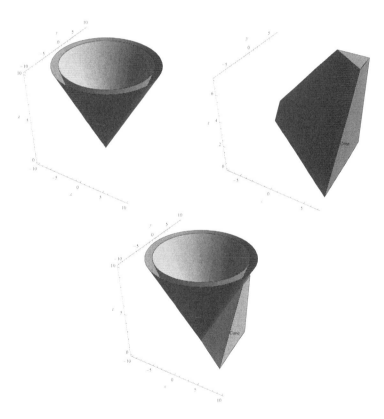

FIGURE 4.6: Sets $[g_1 \leqslant 0]$, $[g_2 \leqslant 0]$ and $C(3, 4, 5)$

4.2   Necessary Condition for Local Extreme Points/ Inequality Constraints

In what follows, we will be interested in the study of the maximization problem

$$\max\ f(x_1,\ldots,x_n) \qquad subject\ to \qquad \begin{cases} g_1(x_1,\ldots,x_n) \leqslant b_1 \\ \quad\vdots \qquad\qquad \vdots \\ g_m(x_1,\ldots,x_n) \leqslant b_m \end{cases}$$

The results established are strongly related to the fact that we are maximizing a function f under inequality constraint $g(x) \leqslant b$. To solve a minimization problem $\min\ f(x)$, we can maximize $-f(x)$, and if a constraint is given in the form $g_j(x) \geqslant b_j$, we can transform it into $-g_j(x) \leqslant -b_j$. An equality constraint $g_j(x) = b_j$ can be equivalently written as $g_j(x) \leqslant b_j$ and $-g_j(x) \leqslant -b_j$.

We have the following preliminary lemma

Lemma 4.2.1   *Let f and $g = (g_1,\ldots,g_m)$ be C^1 functions in a neighborhood of $x^* \in [g(x) \leqslant b]$. If x^* is a regular point and a local maximum point of f subject to these constraints, then we have*

$$\left(\forall y \in \mathbb{R}^n: \quad g_i'(x^*).y \leqslant 0, \quad i \in I(x^*)\right) \qquad\Longrightarrow\qquad f'(x^*).y \leqslant 0.$$

Proof. Let $y \in \mathbb{R}^n$ such that $g'(x^*).y \leqslant 0$. Because, x^* is a regular point of the set $[g(x) \leqslant b]$, then $y \in C(x^*)$, the cone of feasible directions at x^* to the set $[g(x) \leqslant b]$. So $\exists a > 0$, $\exists x \in C^1[0.a]$ such that

$$g(x(t)) \leqslant c \quad \forall t \in [0,a], \quad x(0) = x^*, \quad x'(0) = y.$$

Now, since x^* is a local maximum point of f on the set $g(x) \leqslant b$, then there exists $\delta \in (0,a)$ such that $\forall t \in (0,\delta)$,

$$f(x(t)) \leqslant f(x^*) = f(x(0)) \qquad\Longleftrightarrow\qquad \frac{f(x(t)) - f(x(0))}{t\quad 0} \leqslant 0$$

from which we deduce

$$f'(x^*).y = f'(x(t)).x'(t)\Big]_{t=0} = \frac{d}{dt}f(x(t))\Big]_{t=0} = \lim_{t\to 0^+} \frac{f(x(t)) - f(x(0))}{t - 0} \leqslant 0.$$

Remark 4.2.1 *The lemma generalizes the necessary condition for a local maximum point x^* in a convex S:*

$$f'(x^*).(x - x^*) \leqslant 0 \qquad \forall x \in S.$$

Without assuming the set $S = [g(x) \leqslant b]$ is convex, the local maximum point must satisfy an inequality on the convex cone $C(x^*)$:

$$f'(x^*).(x - x^*) \leqslant 0 \qquad \forall x \in C(x^*).$$

As a consequence of the lemma, we have the following characterization of a constrained local maximum point.

Theorem 4.2.1 *Let f and $g = (g_1, \ldots, g_m)$ be C^1 functions in a neighborhood of $x^* \in [g(x) \leqslant b]$. If x^* is a regular point and a local maximum point of f subject to these constraints, then*

$$\exists \lambda_j^* \geqslant 0, \qquad j \in I(x^*) = \{k \in \mathbb{N}, \quad 1 \leqslant k \leqslant m : \quad g_k(x) = b_k\}$$

such that

$$\frac{\partial f}{\partial x_i}(x^*) - \sum_{j \in I(x^*)} \lambda_j^* \frac{\partial g_j}{\partial x_i}(x^*) = 0, \qquad i = 1, \cdots, n.$$

The proof uses an argument of linear algebra called "Farkas-Minkowski's Lemma" [5] that says:

Farkas-Minkowski's Lemma. *Let A be an $p \times n$ real matrix and $c \in \mathbb{R}^n$. Then, the inclusion*

$$\{x \in \mathbb{R}^n : \quad Ax \geqslant 0\} \quad \subset \quad \{x \in \mathbb{R}^n : \quad c.x \geqslant 0\}$$

is satisfied if and only if

$$\exists \lambda = \langle \lambda_1, \ldots, \lambda_p \rangle \in \mathbb{R}^p, \qquad \lambda \geqslant 0, \qquad such \ that \qquad c = {}^t A \lambda.$$

Proof. Set $\qquad I(x^*) = \{i_1, i_2, \cdots, i_p\},$

$$A = -[g'_j(x^*)]_{j \in I(x^*)} = -\begin{bmatrix} \frac{\partial g_{i_1}}{\partial x_1} & \frac{\partial g_{i_1}}{\partial x_2} & \cdots & \frac{\partial g_{i_1}}{\partial x_n} \\ \frac{\partial g_{i_2}}{\partial x_1} & \frac{\partial g_{i_2}}{\partial x_2} & \cdots & \frac{\partial g_{i_2}}{\partial x_n} \\ \vdots & \vdots & \vdots & \vdots \\ \frac{\partial g_{i_p}}{\partial x_1} & \frac{\partial g_{i_p}}{\partial x_2} & \cdots & \frac{\partial g_{i_p}}{\partial x_n} \end{bmatrix}$$

$$c = -{}^t f'(x^*) = -{}^t \left[\frac{\partial f}{\partial x_1}, \cdots, \frac{\partial f}{\partial x_n} \right] = -\begin{bmatrix} \frac{\partial f}{\partial x_1} \\ \vdots \\ \frac{\partial f}{\partial x_n} \end{bmatrix}.$$

From Farkas-Minkowski's Lemma, the inclusion

$$\{y = (y_1, \cdots, y_n) \in \mathbb{R}^n : \quad Ay \geqslant 0\} = \bigcap_{i \in I(x^*)} \{y \in \mathbb{R}^n : \quad g'_i(x^*).y \leqslant 0\}$$

$$\subset \quad \{y \in \mathbb{R}^n : \quad f'(x^*).y \leqslant 0\} = \{y \in \mathbb{R}^n : \quad c.y = -f'(x^*).y \geqslant 0\}$$

is satisfied, then $\exists \lambda^* = (\lambda_1^*, \ldots, \lambda_p^*) \in \mathbb{R}^p, \quad \lambda^* \geqslant 0$ such that

$$-{}^t f'(x^*) = c = {}^t A \lambda^* = -\sum_{k=1}^{p} \lambda_k^* \, {}^t g'_{i_k}(x^*) \quad \Longleftrightarrow \quad f'(x^*) = \sum_{k=1}^{p} \lambda_k^* \, g'_{i_k}(x^*).$$

So we are led to solve the system

$$\frac{\partial f}{\partial x_i}(x) - \sum_{j \in I(x^*)} \lambda_j \frac{\partial g_j}{\partial x_i}(x) = 0 \quad i = 1, \ldots, n, \quad \lambda_j \geqslant 0 \quad j \in I(x^*)$$

$$g_j(x) - b_j = 0 \qquad \forall j \in I(x^*)$$

$$g_j(x) - b_j < 0 \qquad \forall j \notin I(x^*).$$

To find a practical way to solve the system, we introduce the **complementary slackness conditions**

$$\lambda_j \geqslant 0, \qquad \text{with} \qquad \lambda_j = 0 \quad \text{if} \quad g_j(x) < b_j, \qquad j = 1, \ldots, m$$

When $g_j(x^*) = b_j$, we say that the constraint $g_j(x) \leqslant b_j$ is **active** or **binding** at x^*.

When $g_j(x^*) < b_j$, we say that the constraint $g_j(x) \leqslant b_j$ is **inactive** or **slack** at x^*.

We introduce the **Lagrangian** function

$$\mathcal{L}(x, \lambda) = f(x) - \lambda_1(g_1(x) - b_1) - \ldots - \lambda_m(g_m(x) - b_m)$$

where $\lambda_1, \cdots, \lambda_m$ are the **generalized Lagrange multipliers**.

Then, we reformulate the previous theorem as follows:

Theorem 4.2.2 *Let f and $g = (g_1, \ldots, g_m)$ be C^1 functions in a neighborhood of $x^* \in [g(x) \leqslant b]$. If x^* is a regular point and a local maximum point of f subject to these constraints, then $\exists! \lambda^* = (\lambda_1^*, \ldots, \lambda_m^*)$ such that the following **Karush-Kuhn-Tucker (KKT)** conditions hold at (x^*, λ^*):*

$$\begin{cases} \dfrac{\partial \mathcal{L}}{\partial x_i}(x^*, \lambda^*) = \dfrac{\partial f}{\partial x_i}(x^*) - \displaystyle\sum_{j=1}^{m} \lambda_j^* \dfrac{\partial g_j}{\partial x_i}(x^*) = 0 \qquad i = 1, \ldots, n \\[4mm] \lambda_j^* \geqslant 0, \qquad with \qquad \lambda_j^* = 0 \quad if \quad g_j(x^*) < b_j, \qquad j = 1, \ldots, m. \end{cases}$$

Remark 4.2.2 *The numbers λ_j^*, $j \in I(x^*)$ are unique. Indeed, suppose there exist $\lambda = \langle \lambda_1, \cdots, \lambda_p \rangle$ and $\lambda' = \langle \lambda_1', \cdots, \lambda_p' \rangle$ solutions of*

$$c = {}^t A\lambda \qquad and \qquad c = {}^t A\lambda'$$

then ${}^t A(\lambda - \lambda') = 0$, which we can write

$$\sum_{j \in I(x^*)} (\lambda_j - \lambda_j') g_j'(x^*) = 0.$$

Since the vectors $g_j'(x^*)$ are linearly independent, deduce that

$$(\lambda_j - \lambda_j') = 0 \qquad for \ each \quad j \in I(x^*).$$

Remark 4.2.3 *If $I(x^*) = \emptyset$ then the Karush-Kuhn-Tucker conditions reduce to $\nabla f(x^*) = 0$ which is expected since then the point x^* belongs to the interior of the set of the constraints. On the other hand, this shows that the Kuhn-Tucker conditions are not sufficient for optimality. In fact, when x^* is an interior point, it could be a local maximum, a local minimum or a saddle point.*

First, let us practice writing the KKT conditions through simple examples.

Example 1. Solve the problem

$$\max \ (x-2)^3 \qquad \text{subject to} \qquad g(x) = -x \leqslant 0.$$

Solution: Since f, g, are C^1 in $\mathbb{R}$, consider the Lagrangian

$$\mathcal{L}(x, \alpha) = (x-2)^3 - \alpha(-x)$$

and write the KKT conditions:

$$\begin{cases} (1) & \dfrac{\partial \mathcal{L}}{\partial x} = 3(x-2)^2 + \alpha = 0 \\[2mm] (2) & \alpha \geqslant 0 \quad \text{with} \quad \alpha = 0 \quad \text{if} \quad -x < 0 \end{cases}$$

From (1), the only possible solution is $(x^*, \alpha^*) = (2, 0)$ and 2 is an interior point to the constraint set $[0, +\infty)$. Thus we have a critical point for the Lagrangian without $x^* = 2$ being the maximum point solution of the problem since $y = (x-2)^3$ is increasing on $\mathbb{R}$ $(y' = 3(x-2)^2)$. Therefore, it doesn't attain its maximal value on $[0, +\infty)$.

Example 2. Solve the problem

$$\max \ (x-2)^3 \qquad \text{subject to} \qquad g_1(x) = -x \leqslant 0, \qquad g_2(x) = x \leqslant 3.$$

Solution: i) The set of constraints is the set reduced to the closed interval $[0, 3]$. The problem consists of maximizing the real function $y = (x-2)^3 = f(x)$ which is increasing on $\mathbb{R}$ $(f'(x) = 3(x-2)^2)$. Therefore, it attains its maximal value on $[0, 3]$, at $x = 3$.

ii) *Writing the KKT conditions.* Note that f, g_1, g_2, are C^1 in $\mathbb{R}$. Consider the Lagrangian

$$\mathcal{L}(x, \alpha, \beta) = (x-2)^3 - \alpha(-x) - \beta(x-3).$$

The KKT conditions are

$$\begin{cases} (1) & \dfrac{\partial \mathcal{L}}{\partial x} = 3(x-2)^2 + \alpha - \beta = 0 \\[2mm] (2) & \alpha \geqslant 0 \qquad \text{with} \qquad \alpha = 0 \quad \text{if} \quad -x < 0 \\[2mm] (3) & \beta \geqslant 0 \qquad \text{with} \qquad \beta = 0 \quad \text{if} \quad x < 3 \end{cases}$$

iii) *Solving the system.* We proceed by discussing whether a constraint is active or not.

∗ If $x < 3$, then $\beta = 0$, and with (1), we have

$$3(x-2)^2 + \alpha = 0 \qquad \Longrightarrow \qquad \alpha = -3(x-2)^2 \leqslant 0.$$

With (2), we deduce that $\alpha = 0$, and then by (1), we obtain $x = 2$, which is an interior point to the set of the constraint $[0,3]$. Thus a candidate point is

$$x = 2 \quad \text{with} \quad (\alpha, \beta) = (0,0).$$

∗∗ If $x = 3$, then $x > 0$, and by (2), we deduce that $\alpha = 0$. Then, by (1), we have

$$3(x-2)^2 + \alpha = 3(3-2)^2 + 0 - \beta = 0 \qquad \Longrightarrow \qquad \beta = 3 > 0.$$

Thus we obtain
$$x = 3 \quad \text{with} \quad (\alpha, \beta) = (0,3).$$

Note that only the constraint $g_2(x) = x$ is active at 3. We have

$$g_2'(x) = [\,1\,] \neq 0 \qquad rank(g_2'(3)) = 1.$$

Thus point 3 is regular and, therefore, it is a candidate point.

vi) *Conclusion.* $x = 2$ is not the optimal point since $f(2) = 0 < 1 = f(3)$. Because, f is increasing, then 3 is the maximum point. We can also conclude by using the extreme value theorem since f is continuous on the closed bounded constraint set $(g_1, g_2) \leqslant (0,3)$.

Example 3. *Distance problem.* For $(a, b) \in \mathbb{R}^2$ with $a^2 + b^2 > 1$, solve the problem

$$\min f(x, y) = \|(x, y) - (a, b)\|^2 \qquad \text{subject to} \qquad g(x, y) = x^2 + y^2 \leqslant 1.$$

Solution: i) The problem describes the shortest distance of the point (a, b) to the unit disk (here (a, b) is located outside the unit disk). This distance is attained by the extreme value theorem since f is continuous on the constraint set $[g \leqslant 1]$, which is a closed and bounded subset of $\mathbb{R}^2$. The case $(a, b) = (2, 3)$ is illustrated in Figure 4.7 and a graphical solution is described in Figure 4.8 using level curves.

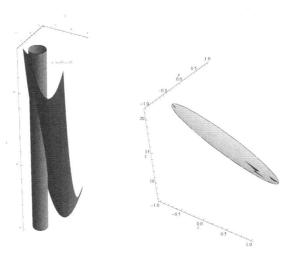

FIGURE 4.7: Graph of $z = (x - 2)^2 + (y - 3)^2$ on $x^2 + y^2 \leqslant 1$

ii) *KKT conditions.* f and g being C^1, introduce, for the corresponding maximization problem, the Lagrangian

$$\mathcal{L}(x, y, \lambda) = -(x - a)^2 - (y - b)^2 - \lambda(x^2 + y^2 - 1).$$

The necessary conditions to satisfy are:

$$\begin{cases} (i) \quad \mathcal{L}_x = -2(x - a) - 2\lambda x = 0 \quad \Longleftrightarrow \quad x(1 + \lambda) = a \\[2mm] (ii) \quad \mathcal{L}_y = -2(y - b) - 2\lambda y = 0 \quad \Longleftrightarrow \quad y(1 + \lambda) = b \\[2mm] (iii) \quad \lambda \geqslant 0 \qquad \text{with} \qquad \lambda = 0 \quad \text{if} \quad x^2 + y^2 < 1 \end{cases}$$

$*$ If $x^2 + y^2 < 1$ then $\lambda = 0$, and then (i) and (ii) yield $(x, y) = (a, b)$ which leads to a contradiction since $a^2 + b^2 > 1$.

** If $x^2 + y^2 = 1$ then from (i) and (ii), we deduce that $(x, y) = (\dfrac{a}{1+\lambda}, \dfrac{b}{1+\lambda})$. By substitution in $x^2 + y^2 = 1$, we get

$$\Big(\frac{a}{1+\lambda}\Big)^2 + \Big(\frac{b}{1+\lambda}\Big)^2 = 1, \quad \lambda \geqslant 0 \qquad \Longleftrightarrow \qquad \lambda = \sqrt{a^2 + b^2} - 1.$$

Thus, the only solution of the system is the point

$$(x^*, y^*) = (\frac{a}{\sqrt{a^2 + b^2}}, \frac{b}{\sqrt{a^2 + b^2}})$$

where the constraint is active. Finally, the point is regular since we have

$$g'(x, y) = \begin{bmatrix} 2x & 2y \end{bmatrix}, \qquad \text{and} \qquad rank g'(x^*, y^*) = 1.$$

Therefore, the point is candidate for optimality.

Conclusion. Now, since it is guaranteed that the maximum value is attained, it must be at the candidate point found. Hence,

$$\max_{x^2 + y^2 \leqslant 1} f(x, y) = f(x^*, y^*) = \sqrt{a^2 + b^2} - 1.$$

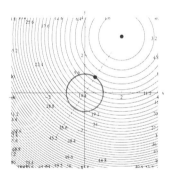

FIGURE 4.8: Minimal value of $z = (x - 2)^2 + (y - 3)^2$ on $x^2 + y^2 = 1$

Remark 4.2.4 *The conclusion of the Karush-Kuhn-Tucker theorem is also true when the extreme point x^* satisfies any one of the following regularity conditions (see [14], [5]):*

i) Linear constraints: $g_j(x)$ is linear, $j = 1, \cdots, m$.

ii) Slater's condition: $g_j(x)$ is convex and there exists $\bar{x}$ such that $g_j(\bar{x}) < b_j$, $j = 1, \cdots, m$ (with f concave).

iii) Concave programming (with f concave): $g_j(x)$ is convex and there exists $\bar{x}$ such that for any $j = 1, \ldots, m$,

$$g_j(\bar{x}) \leqslant b_j \qquad \text{and} \qquad g_j(\bar{x}) < b_j \quad \text{if} \quad g_j \quad \text{is not linear.}$$

iv) The rank condition: The constraints $g_{i_1}, \ldots, g_{i_p}$, $(p \leqslant m)$, are **binding**. The rank of the matrix

$$\begin{bmatrix} g'_{i_1}(x^*) \\ \vdots \\ g'_{i_p}(x^*) \end{bmatrix}$$

is equal to p.

This last case is the one we consider here in our study. These four conditions are not equivalent to one another. For example, the uniqueness of the Lagrange multipliers is established under the rank condition iv).

Example 4. (Non-uniqueness of Lagrange multipliers).
Solve the problem

$$\max f(x, y) = x^{1/2} y^{1/4} \qquad \text{subject to}$$

$$2x + y \leqslant 3, \qquad x + 2y \leqslant 3, \qquad x + y \leqslant 2 \quad \text{with} \quad x \geqslant 0, \quad y \geqslant 0.$$

Solution: To simplify calculations, we will transform the problem to an equivalent one as we did for distance problems, where the square distance is considered instead of the distance itself. Here, to avoid the powers, we will use the logarithmic function.

i) The constraint set, sketched in Figure 4.9, and defined by:

$$S = \{(x, y) \in \mathbb{R}^+ \times \mathbb{R}^+ : \quad 2x + y \leqslant 3, \quad x + 2y \leqslant 3, \quad x + y \leqslant 2\}$$

is a closed bounded subset of $\mathbb{R}^2$. f is continuous on S, then, by the extreme value theorem,

$$\exists (x_*, y_*) \in S \qquad \text{such that} \qquad f(x_*, y_*) = \max_{(x,y) \in S} f(x, y).$$

Note that, we have

$$f(0, y) = f(x, 0) = 0 \qquad \forall x \geqslant 0, \quad y \geqslant 0$$

$$f(x, y) > 0 \qquad \forall x > 0, \quad y > 0.$$

So $f(x_*, y_*) = \max\limits_{(x,y) \in S} f(x,y) > 0$. Therefore, at the maximum point, the constraints $x \geqslant 0$ and $y \geqslant 0$ cannot be binding.

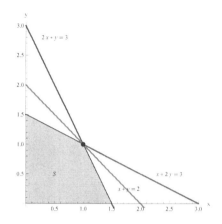 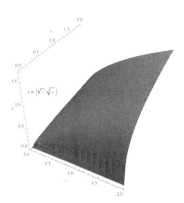

FIGURE 4.9: Graph of $z = x^{1/2} y^{1/4}$ on S

ii) Set $\quad \Omega = (0, +\infty) \times (0, +\infty)$.

As a consequence of i), we have

$$\max\limits_{(x,y) \in S} f(x,y) = \max\limits_{(x,y) \in S^*} f(x,y)$$

where

$$S^* = \{(x,y) \in \Omega: \quad 2x + y \leqslant 3, \quad x + 2y \leqslant 3, \quad x + y \leqslant 2\}.$$

Set

$$F(x,y) = \ln f(x,y) = \frac{1}{2}\ln(x) + \frac{1}{4}\ln(y) \qquad (x,y) \in \Omega$$

F is well defined and we have

$$\max\limits_{(x,y) \in S} f(x,y) = \max\limits_{(x,y) \in S^*} f(x,y) = \max\limits_{(x,y) \in S^*} e^{\ln F(x,y)}$$

$$\max\limits_{(x,y) \in S^*} F(x,y) = \ln\left(\max\limits_{(x,y) \in S^*} f(x,y)\right) = \ln f(x_*, y_*)$$

since the functions $t \longmapsto \ln t$ and $t \longmapsto e^t$ are increasing.

Note that S^* is a bounded subset of $\mathbb{R}^2$ but not closed. Thus, we cannot apply the extreme value theorem to conclude about the existence of a solution to the problem

$$\max_{(x,y)\in S^*} F(x,y).$$

iii) Since F and the constraints are C^1 in Ω, to solve the problem, we write the KKT conditions for the associated Lagrangian

$$\mathcal{L}(x,y,\lambda_1,\lambda_2,\lambda_3) = \frac{1}{2}\ln x + \frac{1}{4}\ln y - \lambda_1(2x+y-3) - \lambda_2(x+2y-3) - \lambda_3(x+y-2),$$

The necessary conditions to satisfy are:

$$\begin{cases} (i) \quad \mathcal{L}_x = \dfrac{1}{2x} - 2\lambda_1 - \lambda_2 - \lambda_3 = 0 \\[3mm] (ii) \quad \mathcal{L}_y = \dfrac{1}{4y} - \lambda_1 - 2\lambda_2 - \lambda_3 = 0 \\[3mm] (iii) \quad \lambda_1 \geqslant 0 \qquad \text{with} \quad \lambda_1 = 0 \quad \text{if} \quad 2x+y<3 \\[3mm] (iv) \quad \lambda_2 \geqslant 0 \qquad \text{with} \quad \lambda_2 = 0 \quad \text{if} \quad x+2y<3 \\[3mm] (v) \quad \lambda_3 \geqslant 0 \qquad \text{with} \quad \lambda_3 = 0 \quad \text{if} \quad x+y<2 \end{cases}$$

* If $x+y<2$ then $\lambda_3 = 0$, and discuss the cases

    o   if $x+2y<3$ then $\lambda_2 = 0$, and

$$\lambda_1 = \frac{1}{4x} = \frac{1}{4y} > 0 \qquad \Longrightarrow \qquad y = x.$$

Because $\lambda_1 > 0$ then $2x+y = 3$. Thus, we have $(x,y) = (1,1)$. But $x+y = 1+1 = 2$: contradiction.

    oo if $x+2y = 3$, then

        −   if $2x+y<3$, then $\lambda_1 = 0$, and

$$\lambda_2 = \frac{1}{2x} = \frac{1}{8y} > 0 \qquad \Longrightarrow \qquad x = 4y.$$

Because $x+2y = 3$, then, we have $(x,y) = (2,1/2)$. But $x+y = 2+1/2 > 2$: contradiction.

        −   if $2x+y = 3$, then $(x,y) = (1,1)$. But $x+y = 1+1 = 2$: contradiction.

** If $x + y = 2$, then by drawing the constraint set, we see that the only point satisfying $x + y = 2$ is $(x, y) = (1, 1)$ for which we have also $2x + y = 3$ and $x + 2y = 3$ with

$$2\lambda_1 + \lambda_2 + \lambda_3 = 1 \qquad \lambda_1 + 2\lambda_2 + \lambda_3 = 1 \quad \Longleftrightarrow \quad \lambda_1 = \lambda_2 \qquad \lambda_3 = 1 - 3\lambda_1.$$

iv) *Conclusion.* The only point candidate is $(1, 1)$, and it is the maximum point since we know that such a point exists.

However, we see that we do not have uniqueness of the Lagrange multipliers, but still we can apply the KKT conditions since the constraints are linear.

Note also, that the rank condition is not satisfied since we have

$$g(x, y) = (2x + y, \; x + 2y, \; x + y) \qquad\qquad g(1, 1) = (3, 3, 3)$$

$$g'(x, y) = \begin{bmatrix} 2 & 1 \\ 1 & 2 \\ 1 & 1 \end{bmatrix} \qquad\qquad rank(g'(1, 1)) = 2 \neq 3$$

and the three constraints are active at $(1, 1)$; see Figure 4.10.

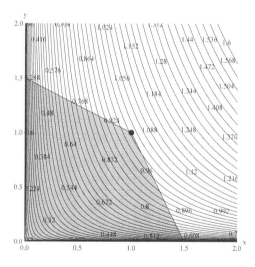

FIGURE 4.10: Maximal value of $z = x^{1/2}\, y^{1/4}$ on S

Mixed Constraints

Some maximization problems take the form

$$\max \ f(x) \quad \text{subject to} \quad \begin{cases} g_j(x) = b_j, & j = 1, \cdots, r \quad (r < n) \\ h_k(x) \leqslant c_k, & k = 1, \cdots, s \end{cases}$$

We have:

Theorem 4.2.3 *Let f, $g = (g_1, \ldots, g_r)$, and $h = (h_1, \ldots, h_s)$ be C^1 functions in a neighborhood of $x^* \in [g(x) = b] \cap [h(x) \leqslant c]$. If x^* is a regular point and a local maximum point of f subject to these constraints, then $\exists! (\lambda^*, \mu^*)$, $\lambda^* = (\lambda_1^*, \ldots, \lambda_r^*)$, $\mu^* = (\mu_1^*, \ldots, \mu_s^*)$ such that the following* **Karush-Kuhn-Tucker (KKT) conditions** *hold at (x^*, λ^*, μ^*):*

$$\begin{cases} \dfrac{\partial \mathcal{L}}{\partial x_i}(x^*, \lambda^*, \mu^*) = \dfrac{\partial f}{\partial x_i}(x^*) - \sum_{j=1}^{r} \lambda_j^* \dfrac{\partial g_j}{\partial x_i}(x^*) - \sum_{k=1}^{s} \mu_k^* \dfrac{\partial h_k}{\partial x_i}(x^*) = 0 \\ \qquad i = 1, \ldots, n \\[2mm] \dfrac{\partial \mathcal{L}}{\partial \lambda_j}(x^*, \lambda^*, \mu^*) = -(g_j(x^*) - b_j) = 0 \qquad j = 1, \ldots, r. \\[2mm] \mu_k^* \geqslant 0, \quad \text{with} \quad \mu_k^* = 0 \quad \text{if} \quad h_k(x^*) < c_k, \quad k = 1, \ldots, s, \end{cases}$$

where

$$\mathcal{L}(x, \lambda, \mu) = f(x) - \sum_{j=1}^{r} \lambda_j (g_j(x) - b_j) - \sum_{k=1}^{s} \mu_k (h_k(x) - c_k).$$

Proof. The maximization problem is equivalent to

$$\max \ f(x) \quad \text{subject to} \quad \begin{cases} g_j(x) \leqslant b_j, & j = 1, \ldots, r \\ -g_j(x) \leqslant -b_j, & j = 1, \ldots, r \\ h_k(x) \leqslant c_k, & k = 1, \ldots, s \end{cases}$$

By applying the KKT conditions with the Lagrangian

$$\mathcal{L}^*(x,\tau,\kappa,\mu) = f(x) - \sum_{j=1}^{r} \tau_j(g_j(x) - b_j) - \sum_{j=1}^{r} \kappa_j(-g_j(x) + b_j) - \sum_{k=1}^{s} \mu_k(h_k(x) - c_k)$$

there exist unique multipliers $\tau_j^*, \kappa_j^*, \mu_k^*$ such that the necessary conditions are satisfied:

$$\begin{cases} \dfrac{\partial \mathcal{L}^*}{\partial x_i}(x^*, \tau^*, \kappa^*, \mu^*) = \dfrac{\partial f}{\partial x_i}(x^*) - \sum_{j=1}^{r} \tau_j^* \dfrac{\partial g_j}{\partial x_i}(x^*) \\[4mm] \qquad + \sum_{j=1}^{r} \kappa_j^* \dfrac{\partial g_j}{\partial x_i}(x^*) - \sum_{k=1}^{s} \mu_k^* \dfrac{\partial h_k}{\partial x_i}(x^*) = 0 \qquad i = 1, \ldots, n \\[4mm] \tau_j^* \geqslant 0, \qquad \text{with} \qquad \tau_j^* = 0 \quad \text{if} \quad g_j(x^*) < b_j, \qquad j = 1, \ldots, r \\[2mm] \kappa_j^* \geqslant 0, \qquad \text{with} \qquad \kappa_j^* = 0 \quad \text{if} \quad -g_j(x^*) < -b_j, \qquad j = 1, \ldots, r \\[2mm] \mu_k^* \geqslant 0, \qquad \text{with} \qquad \mu_k^* = 0 \quad \text{if} \quad h_k(x^*) < c_k, \qquad k = 1, \ldots, s. \end{cases}$$

Setting

$$\lambda^* = \tau^* - \kappa^* \quad \text{and}$$

$$\mathcal{L}(x, \lambda, \mu) = \mathcal{L}^*(x, \tau, \kappa, \mu)$$

$$= f(x) - \sum_{j=1}^{r} (\tau_j - \kappa_j)(g_j(x) - b_j) - \sum_{k=1}^{s} \mu_k(h_k(x) - c_k)$$

we deduce that (x^*, λ^*, μ^*) is also a solution of the KKT conditions corresponding to Lagrangian $\mathcal{L}$. Moreover, for $j = 1, \ldots, r$, λ_j changes sign

$$\lambda_j = \tau_j - \kappa_j = -\kappa_j \leqslant 0 \qquad \text{if} \qquad g_j(x) < b_j$$
$$\lambda_j = \tau_j - \kappa_j = \tau_j \geqslant 0 \qquad \text{if} \qquad g_j(x) > b_j.$$

Uniqueness of λ^* and μ^*. Suppose λ^* and μ^* are not uniquely defined, then we would have for some $\lambda' \neq \lambda$ and $\mu' \neq \mu$

$$\frac{\partial f}{\partial x_i}(x^*) - \sum_{j=1}^{r} \lambda_j \frac{\partial g_j}{\partial x_i}(x^*) - \sum_{k=1}^{s} \mu_k \frac{\partial h_k}{\partial x_i}(x^*) = 0$$

$$\frac{\partial f}{\partial x_i}(x^*) - \sum_{j=1}^{r} \lambda'_j \frac{\partial g_j}{\partial x_i}(x^*) - \sum_{k=1}^{s} \mu'_k \frac{\partial h_k}{\partial x_i}(x^*) = 0.$$

Subtracting the two equalities and using the fact that x^* is a regular point, we obtain a contradiction:

$$\sum_{j=1}^{r}(\lambda'_j - \lambda_j)\nabla g_j(x^*) - \sum_{k=1}^{s}(\mu'_k - \mu_k)\nabla h_k(x^*) = 0 \quad \Longrightarrow \quad (\lambda', \mu') = (\lambda, \mu).$$

Nonnegativity constraints

Some maximization problems take the form

$$\max \ f(x) \qquad \text{subject to} \qquad \begin{cases} g_j(x) \leqslant b_j, & j = 1, \ldots, m \\ x_1 \geqslant 0, \ldots, x_n \geqslant 0. \end{cases}$$

We introduce the following n new constraints:

$$g_{m+1}(x) = -x_1 \leqslant 0, \ldots\ldots\ldots\ldots, g_{m+n}(x) = -x_n \leqslant 0.$$

The maximization problem is equivalent to

$$\max \ f(x) \qquad \text{subject to} \qquad \begin{cases} g_j(x) \leqslant b_j, & j = 1, \ldots, m \\ g_j(x) \leqslant 0, & j = m+1, \ldots, m+n. \end{cases}$$

By applying the KKT conditions, for a regular point x, with the Lagrangian

$$\mathcal{L}^*(x, \lambda, \mu) = f(x) - \sum_{j=1}^{m} \lambda_j(g_j(x) - b_j) - \sum_{k=1}^{n} \mu_k(-x_k)$$

there exist unique multipliers λ_j, μ_k such that

$$
\begin{cases}
\dfrac{\partial \mathcal{L}^*}{\partial x_i}(x, \lambda, \mu) = \dfrac{\partial f}{\partial x_i}(x) - \displaystyle\sum_{j=1}^{m} \lambda_j \dfrac{\partial g_j}{\partial x_i}(x) + \mu_i = 0 \qquad i = 1, \ldots, n \\[4mm]
\lambda_j \geqslant 0, \qquad \text{with} \qquad \lambda_j = 0 \quad \text{if} \quad g_j(x) < b_j, \qquad j = 1, \ldots, m \\[3mm]
\mu_k \geqslant 0, \qquad \text{with} \qquad \mu_k = 0 \quad \text{if} \quad x_k > 0, \qquad k = 1, \ldots, n.
\end{cases}
$$

We deduce then:

Theorem 4.2.4 *Let f and $g = (g_1, \ldots, g_r)$ be C^1 functions in a neighborhood of $x^* \in [g(x) \leqslant b] \cap [x \geqslant 0]$. If x is a regular point and a local maximum point of f subject to these constraints, then $\exists! \lambda^* = (\lambda_1^*, \ldots, \lambda_m^*)$ such that the following **Karush-Kuhn-Tucker (KKT) conditions** hold at (x, λ):*

$$
\begin{cases}
\dfrac{\partial \mathcal{L}}{\partial x_i}(x, \lambda) = \dfrac{\partial f}{\partial x_i}(x) - \displaystyle\sum_{j=1}^{m} \lambda_j \dfrac{\partial g_j}{\partial x_i}(x) \leqslant 0 \qquad (= 0 \quad \text{if} \quad x_i > 0), \\[4mm]
\quad i = 1, \ldots, n \\[3mm]
\lambda_j \geqslant 0, \qquad \text{with} \qquad \lambda_j = 0 \qquad \text{if} \qquad g_j(x) < b_j, \qquad j = 1, \ldots, m
\end{cases}
$$

where the Lagrangian is

$$
\mathcal{L}(x, \lambda) = f(x) - \sum_{j=1}^{m} \lambda_j \, (g_j(x) - b_j).
$$

Solved Problems

1. – *Importance of KKT hypotheses.* Show that the KKT conditions fail to hold at the optimal solution of the problem

$$\max \; f(x,y) = x^2 + y \qquad \text{subject to} \qquad \begin{cases} g_1(x,y) = (x-2)^2 = 0 \\[2mm] g_2(x,y) = (y+1)^3 \leqslant 0. \end{cases}$$

Solution: i) The set of constraints, graphed in Figure 4.11, is

$$S = \{(x,y) : \quad g_1(x,y) = 0 \quad \text{and} \quad g_2(x,y) \leqslant 0\}$$
$$= \{(x,y) : \quad x = 2 \quad \text{and} \quad y \leqslant -1\}.$$

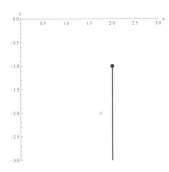

FIGURE 4.11: Constraint set S

ii) The Karush-Kuhn-Tucker conditions for the Lagrangian

$$\mathcal{L}(x,y,\alpha,\beta) = x^2 + y - \alpha((x-2)^2) - \beta((y+1)^3)$$

are

$$
\begin{cases}
(1) & \mathcal{L}_x = 2x - 2\alpha(x-1) = 0 \\[2mm]
(2) & \mathcal{L}_y = 1 - 3\beta(y+1)^2 = 0 \\[2mm]
(3) & \beta \geqslant 0 \quad \text{with} \quad \beta = 0 \quad \text{if} \quad (y+1)^3 < 0
\end{cases}
$$

* If $(y+1)^3 < 0$, then $\beta = 0$. We get a contradiction with (2) which leads to $1 = 0$.

* If $(y+1)^3 = 0$, then $y = -1$, and by (3) again, we obtain $1 = 0$ which is not possible.

Thus, KKT conditions have no solution.

ii) The problem has a solution at $(2, -1)$ since, we have

$$f(x,y) = x^2 + y = 2^2 + y \leqslant 4 + (-1) = f(2,-1) \qquad \forall (x,y) \in S.$$

Thus

$$\max_S f(x,y) = f(2,-1) = 3.$$

Note that the point is not a candidate for the KKT conditions. This is because it doesn't satisfy the constraint qualification under which the KKT conditions are established. In particular, the rank condition is not satisfied. Indeed, the two constraints are active at $(2,-1)$, but we have $rank\left(\begin{bmatrix} g_1'(2,-1) \\ g_2'(2,-1) \end{bmatrix}\right) = 0$ since

$$\begin{bmatrix} g_1'(x,y) \\ g_2'(x,y) \end{bmatrix} = \begin{bmatrix} 2(x-2) & 0 \\ 0 & 3(y+1)^2 \end{bmatrix} \qquad \begin{bmatrix} g_1'(2,-1) \\ g_2'(2,-1) \end{bmatrix} = \begin{bmatrix} 0 & 0 \\ 0 & 0 \end{bmatrix}.$$

2. –KKT conditions are not sufficient. Consider the problem

$$\min \ f(x,y) = 2 - y - (x-1)^2 \quad \text{subject to} \quad y - x = 0, \quad x+y-2 \leqslant 0$$

i) Sketch the feasible set and write down the necessary KKT conditions.

ii) Find the point(s) solution of the KKT conditions and check their regularity.

iii) What can you conclude about the solution of the minimization problem?

iv) Does this contradict the theorem on the necessary conditions for a constrained candidate point?

Solution: i) The set of the constraints is the set of points on the line $y = x$ included in the region below the line $y = 2 - x$, as shown in Figure 4.12.

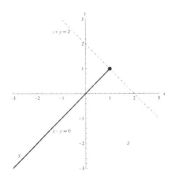

FIGURE 4.12: Constraint set S

Writing the Karush-Kuhn-Tucker conditions. First, transform the problem into a maximization one as:

$$\max \ -f(x, y) = y - 2 + (x - 1)^2 \quad \text{subject to} \quad y - x = 0, \quad x + y - 2 \leqslant 0$$

Note that f, and the constraints g_1 and g_2 are C^∞ in $\mathbb{R}^2$ where

$$g_1(x, y) = y - x \qquad\qquad g_2(x, y) = x + y - 2.$$

Thus, the Lagrangian associated is

$$\mathcal{L}(x, y, \alpha, \beta) = y - 2 + (x - 1)^2 - \alpha(y - x) - \beta(x + y - 2)$$

and the Karush-Kuhn-Tucker conditions are

$$\begin{cases} (1) & \mathcal{L}_x = 2(x - 1) + \alpha - \beta = 0 \\[2mm] (2) & \mathcal{L}_y = 1 - \alpha - \beta = 0 \\[2mm] (3) & \mathcal{L}_\alpha = -(y - x) = 0 \\[2mm] (4) & \beta \geqslant 0 \quad \text{with} \quad \beta = 0 \quad \text{if} \quad x + y - 2 < 0. \end{cases}$$

ii) ***Solving the KKT conditions.***

∗ If $x + y - 2 < 0$ then $\beta = 0$ and

$$\begin{cases} 2(x-1) + \alpha = 0 \\ 1 - \alpha = 0 \\ y - x = 0 \end{cases} \implies (x,y) = (\frac{1}{2}, \frac{1}{2}) \quad \text{and} \quad (\alpha, \beta) = (1, 0).$$

** If $x + y - 2 = 0$ then

$$\begin{cases} 2(x-1) + \alpha - \beta = 0 \\ 1 - \alpha - \beta = 0 \\ y - x = 0 \\ x + y - 2 = 0 \end{cases} \implies (x,y) = (1,1) \quad \text{and} \quad (\alpha, \beta) = (1/2, 1/2).$$

So, there are two solutions: $(1/2, 1/2)$ and $(1, 1)$ for the KKT conditions.

Regularity of the point $(1/2, 1/2)$. Only the constraint $g_1(x, y) = y - x$ is active at $(1/2, 1/2)$ and we have

$$\begin{bmatrix} g_1'(x,y) \end{bmatrix} = \begin{bmatrix} -1 & 1 \end{bmatrix} \qquad rank(\begin{bmatrix} g_1'(1/2, 1/2) \end{bmatrix}) = rank(\begin{bmatrix} -1 & 1 \end{bmatrix}) = 1.$$

The point $(1/2, 1/2)$ is a regular point.

Regularity of the point $(1, 1)$. The two constraints are active at $(1, 1)$. We have

$$\begin{bmatrix} g_1'(x,y) \\ g_2'(x,y) \end{bmatrix} = \begin{bmatrix} -1 & 1 \\ 1 & 1 \end{bmatrix} \qquad rank(\begin{bmatrix} g_1'(1,1) \\ g_2'(1,1) \end{bmatrix}) = rank(\begin{bmatrix} -1 & 1 \\ 1 & 1 \end{bmatrix}) = 2.$$

Thus the point $(1, 1)$ is a regular point.

iii) ***Conclusion.*** The two points are candidates for optimality. Comparing the values taken by f at these points gives:

$$f(1,1) = 1, \qquad f(\frac{1}{2}, \frac{1}{2}) = 2 - \frac{1}{2} - \frac{1}{4} = \frac{5}{4} > 1,$$

we deduce that, only $(1, 1)$ is the candidate for minimality. However, it is not the minimum point. Indeed, we have

$$f(x, x) = 2 - x - (x-1)^2 \longrightarrow -\infty \quad \text{as} \quad x \longrightarrow -\infty.$$

Therefore, f doesn't attain its minimal value.

iv) This doesn't contradict the theorem since KKT conditions indicate only where to find the possible points when they exist.

3. – Positivity constraints. Solve the problem by two methods:

$$\max \ f(x,y) = 3 + x - y + xy \quad \text{subject to} \quad \begin{cases} y - x^2 \geqslant 0, & y \leqslant 4 \\ x \geqslant 0, & y \geqslant 0 \end{cases}$$

i) using the extreme value theorem.

ii) using the KKT conditions.

Solution: i) EVT method. The constraints set, graphed in Figure 4.13, is

$$S = \{(x,y) \ / \ 0 \leqslant x \leqslant 2, \quad x^2 \leqslant y \leqslant 4\}.$$

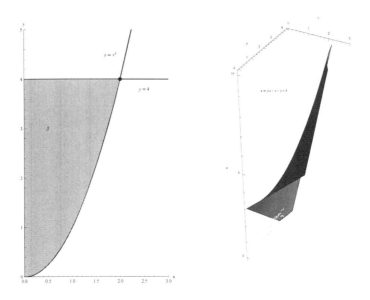

FIGURE 4.13: Graph of $z = 3 + x - y + xy$ on S

f is continuous (because it is a polynomial) on the set S, which is a bounded and closed subset of $\mathbb{R}^2$. So f attains its absolute extreme points on S (by the extreme value theorem), either at the critical points located in $\overset{\circ}{S}$ or on ∂S.

* *Critical points of* f: f has no critical point in the interior of S because

$$\nabla f(x,y) = \langle 1 + y, -1 + x \rangle = \langle 0, 0 \rangle \quad \Longleftrightarrow \quad (x, y) = (1, -1) \notin S.$$

** *Extreme values on* ∂S :

Let L_1, L_2 and L_3 the three parts of the boundary of S defined by:

$$L_1 = \{(x, x^2),\ 0 \leqslant x \leqslant 2\}, \qquad L_2 = \{(x, 4),\ 0 \leqslant x \leqslant 2\}$$

$$L_3 = \{(0, y),\ 0 \leqslant y \leqslant 4\}$$

– *On* L_1, we have $f(x, x^2) = 3 + x - x^2 + x^3 = g(x)$,

$$g'(x) = 3x^2 - 2x + 1.$$

x	0		2
$g'(x)$		+	
$g(x)$	3	↗	9

TABLE 4.1: Variations of $g(x) = 3 + x - x^2 + x^3$ on $[0, 2]$

Then, using Table 4.1, we deduce that

$$\max_{L_1} f = f(2, 4) = 9 \qquad\qquad \min_{L_1} f = f(0, 0) = 3.$$

– *On* L_2, we have: $f(x, 4) = 5x - 1 = h(x)$, $\quad h'(x) = 5$.

x	0		2
$h'(x)$		+	
$h(x)$	-1	↗	9

TABLE 4.2: Variations of $h(x) = 5x - 1$ on $[0, 2]$

From Table 4.2, the extreme values on this side are

$$\max_{L_2} f = f(2, 4) = 9 \qquad\qquad \min_{L_2} f = f(0, 4) = -1.$$

– *On* L_3, we have: $f(0, y) = 3 - y = \varphi(y)$, $\quad \varphi'(y) = -1$.
Hence, we obtain from Table 4.3,

$$\max_{L_3} f = f(0, 0) = 3 \qquad\qquad \min_{L_3} f = f(0, 4) = -1.$$

y	0		4
$\varphi'(y)$		$-$	
$\varphi(y)$	3	$\searrow$	-1

TABLE 4.3: Variations of $\varphi(y) = 3 - y$ on $[0, 4]$

$* * * Conclusion:$

The maximal value of f on S is 9 and is attained at the point $(2, 4)$.
The minimal value of f on S is -1 and is attained at the point $(0, 4)$.

ii) KKT conditions. Consider the Lagrangian

$$\mathcal{L}(x, y, \lambda, \mu) = 3 + x - y + x y - \lambda(x^2 - y) - \mu(y - 4).$$

The Karush-Kuhn-Tucker conditions are

$$
\begin{cases}
(1) & \mathcal{L}_x = 1 + y - 2\lambda x \leqslant 0 \quad (= 0 \quad \text{if} \quad x > 0) \\[2mm]
(2) & \mathcal{L}_y = -1 + x + \lambda - \mu \leqslant 0 \quad (= 0 \quad \text{if} \quad y > 0) \\[2mm]
(3) & \lambda \geqslant 0 \quad \text{with} \quad \lambda = 0 \quad \text{if} \quad y < x^2 \\[2mm]
(4) & \mu \geqslant 0 \quad \text{with} \quad \mu = 0 \quad \text{if} \quad y < 4
\end{cases}
$$

Solving the KKT conditions.

$*$ If $y < 4$ then $\mu = 0$ and

    – Suppose $y < x^2$, then $\lambda = 0$ and by (1), we get $1 + y \leqslant 0$ which contradicts $y \geqslant 0$.

    – Suppose $y = x^2$.

        ○ if $x > 0$, then $y = x^2 > 0$. From (1) and (2), we get

$$1 + x^2 - 2\lambda x = 0 \quad \text{and} \quad \lambda = 1 - x \implies 3x^2 - 2x + 1 = 0$$

with no solution.

        ○ if $x = 0$, then $y = x^2 = 0$, and (1) leads to a contradiction.

** If $y = 4$

– Suppose $y < x^2$ then $\lambda = 0$ and by (1), we get $1 + 4 \leqslant 0$ which is not possible.

– Suppose $y = x^2$. We deduce that $x = 2$ or $x = -2$. The second value is not possible since $x \geqslant 0$. For $x = 2 > 0$, we insert the values $x = 2$ and $y = 4$ in (1) and (2), and obtain

$$\begin{cases} 5 - 4\lambda = 0 \\ 1 + \lambda - \mu = 0 \end{cases} \implies (\lambda, \mu) = (\frac{5}{4}, \frac{9}{4}).$$

Note that, both constraints are active at $(2, 4)$, and if $g(x, y) = (x^2 - y, y - 4)$, then

$$g'(x, y) = \begin{bmatrix} 2x & -1 \\ 0 & 1 \end{bmatrix} \qquad rank(g'(2, 4)) = rank(\begin{bmatrix} 4 & -1 \\ 0 & 1 \end{bmatrix}) = 2.$$

Thus, $(2, 4)$ is a regular point and, therefore, a candidate point. Moreover, $(2, 4)$ solves the problem since the maximal value of f is attained on S by the EVT; see Figure 4.14.

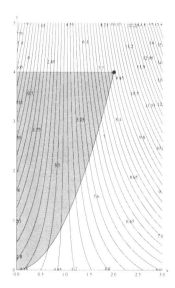

FIGURE 4.14: Maximal value of $z = 3 + x - y + xy$ on S

Introduction to the Theory of Optimization in Euclidean Space

4. – Application. Find $(x,y) \in S = \{(x,y) : \ x + y \leqslant 0, \ x^2 - 4 \leqslant 0\}$ that lies closest to the point $(2,3)$ by following the steps below:

i) Formulate the problem as an optimization problem.

ii) Illustrate the problem graphically (Hint: use level curves).

iii) Write down the KKT conditions.

iv) Find all points that satisfy the KKT conditions. Check whether or not each point is regular.

v) What can you conclude about the solution of the problem?

Solution: i) The square of the distance between (x,y) and $(2,3)$ is given by $(x-2)^2 + (y-3)^2$. To find the point $(x,y) \in S$ that lies closest to the point $(2,3)$ is equivalent to solve the minimization problem

$$\min \ (x-2)^2 + (y-3)^2 \qquad \text{subject to} \qquad \begin{cases} g_1(x,y) = x + y \leqslant 0 \\[2mm] g_2(x,y) = x^2 - 4 \leqslant 0 \end{cases}$$

or to maximize the objective function f below subject to the two constraints:

$$\max \ f(x,y) = -(x-2)^2 - (y-3)^2 \ \text{subject to} \ \begin{cases} g_1(x,y) = x + y \leqslant 0 \\[2mm] g_2(x,y) = x^2 - 4 \leqslant 0 \end{cases}$$

ii) The feasible set, graphed in Figure 4.15, is also described by

$$S = \{(x,y) : \quad y \leqslant -x, \quad -2 \leqslant x \leqslant 2\}$$

The level curves of f, with equations: $(x-2)^2 + (y-3)^2 = k$ where $k \geqslant 0$, are circles centered at $(2,3)$ with radius $\sqrt{k}$; see Figure 4.16. If we increase the values of the radius, the values of f decrease. The first circle that will intersect the set S will be the circle with radius equal to the distance of the point $(2,3)$ to the line $y = -x$. So, only the first constraint will be active in solving the optimization problem.

iii) **Writing the KKT conditions.** Consider the Lagrangian

$$\mathcal{L}(x,y,\lambda,\beta) = -(x-2)^2 - (y-3)^2 - \lambda(x+y) - \beta(x^2 - 4)$$

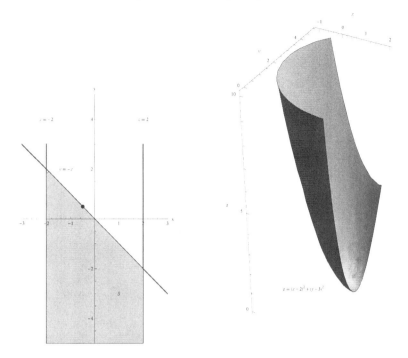

FIGURE 4.15: Graph of $z = (x-2)^2 + (y-3)^2$ on S

The KKT conditions are

$$
\begin{cases}
(1) & \dfrac{\partial \mathcal{L}}{\partial x} = -2(x-2) - \lambda - 2\beta x = 0 \\[2mm]
(2) & \dfrac{\partial \mathcal{L}}{\partial y} = -2(y-3) - \lambda = 0 \\[2mm]
(3) & \lambda \geqslant 0 \quad \text{with} \quad \lambda = 0 \quad \text{if} \quad x+y < 0 \\[2mm]
(4) & \beta \geqslant 0 \quad \text{with} \quad \beta = 0 \quad \text{if} \quad x^2 - 4 < 0
\end{cases}
$$

iv) *Solving the KKT conditions.*

∗ If $x + y < 0$ then $\lambda = 0$ and

$$
\begin{cases}
-2(x-2) - 2\beta x = 0 \\
-2(y-3) = 0
\end{cases}
\quad \implies \quad x(1+\beta) = 2 \quad \text{and} \quad y = 3
$$

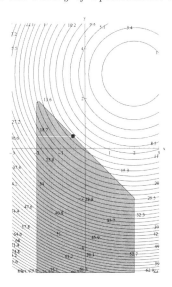

FIGURE 4.16: Minimal value of $z = (x-2)^2 + (y-3)^2$ on S

then with (4) we have

$$\beta \geqslant 0 \qquad \Longrightarrow \qquad x = \frac{2}{1+\beta} > 0.$$

This contradicts $x + y < 0$.

** If $x + y = 0$ then

– Suppose $x^2 - 4 < 0$ then $\beta = 0$ and

$$\begin{cases} -2(x-2) - \lambda = 0 \\ \\ -2(y-3) - \lambda = 0 \end{cases} \qquad \Longrightarrow \qquad y = x + 1.$$

With $x+y = 0$, we deduce that $(x, y) = (-1/2, 1/2)$. Note that $(-1/2)^2 - 4 < 0$ is satisfied and $\lambda = 5 > 0$.

– Suppose $x^2 - 4 = 0$. We deduce that $x = 2$ or $x = -2$. Then, inserting in (1) and (2), we obtain

$$(x, y) = (2, -2) \qquad \Longrightarrow \qquad \begin{cases} \lambda + 4\beta = 0 \\ \\ 10 - \lambda = 0 \end{cases} \qquad \Longrightarrow \qquad (\lambda, \beta) = (10, -5/2)$$

$$(x, y) = (-2, 2) \quad \implies \quad \begin{cases} 8 - \lambda + 4\beta = 0 \\ \\ \lambda = 0 \end{cases} \quad \implies \quad (\lambda, \beta) = (0, -2)$$

contradicting $\beta \geqslant 0$.
So, the only point solution of the system is

$$(x^*, y^*) = (-1/2, 1/2) \qquad \text{with} \qquad (\lambda, \beta) = (5, 0).$$

Regularity of the candidate point $(-1/2, 1/2)$. Note that only the constraint $g_1(x, y) = x + y$ is active at $(-1/2, 1/2)$. We have

$$\begin{bmatrix} g_1'(x, y) \end{bmatrix} = \begin{bmatrix} 1 & 1 \end{bmatrix} \qquad rank(\begin{bmatrix} g_1'(-1/2, 1/2) \end{bmatrix}) = rank(\begin{bmatrix} 1 & 1 \end{bmatrix}) = 1.$$

Thus the point $(-1/2, 1/2)$ is a regular point.

iv) ***Conclusion.*** The constraint set is an unbounded closed convex and we have

$$|f(x, y)| = \|(x, y) - (2, 3)\|^2 \geqslant (\|(x, y)\| - \|(2, 3)\|)^2$$
$$\implies \quad \lim_{\|(x,y)\| \to +\infty} f(x, y) = +\infty.$$

By Theorem 2.4.2, there exists a minimum point for f on S. Thus, the candidate found solves the problem.

5. – ***Mixed constraints.*** Solve the problem

$$\max x^2 + y^2 + z^2 \qquad \text{subject to} \qquad \begin{cases} 2x^2 + y^2 + z^2 = 1 \\ \\ x + y + z \leqslant 0. \end{cases}$$

Solution: Set $U(x, y, z) = x^2 + y^2 + z^2$ and

$$g(x, y, z) = x + y + z \qquad\qquad h(x, y, z) = 2x^2 + y^2 + z^2 - 1.$$

First, the maximization problem has a solution by the extreme-value theorem. Indeed, U is continuous on the set

$$S = \{(x, y, z): \quad x + y + z \leqslant 0, \quad 2x^2 + y^2 + z^2 = 1\}$$

which is a closed and bounded subset of $\mathbb{R}^3$ as the intersection of the ellipsoid $\dfrac{x^2}{(1/\sqrt{2})^2} + y^2 + z^2 = 1$ with the region below the plane $x + y + z = 0$. The plane passes through the center of the ellipsoid; see Figure 4.17.

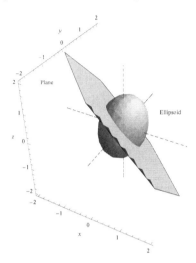

FIGURE 4.17: S is the part of the ellipsoid below the plane

Next, the functions U, g and h are C^1 around each point $(x, y, z) \in \mathbb{R}^3$. We then may deduce the solution by using the Karusk-Kuhn-Tucker conditions. The Lagrangian is given by

$$\mathcal{L}(x, y, \lambda, \mu) = x^2 + y^2 + z^2 - \lambda(x + y + z) - \mu(2x^2 + y^2 + z^2 - 1),$$

and the necessary conditions to satisfy are:

$$\begin{cases} (i) \quad \mathcal{L}_x = 2x - \lambda - 4\mu x = 0 \quad \Longleftrightarrow \quad 2x(1 - 2\mu) = \lambda \\[2mm] (ii) \quad \mathcal{L}_y = 2y - \lambda - 2\mu y = 0 \quad \Longleftrightarrow \quad 2y(1 - \mu) = \lambda \\[2mm] (iii) \quad \mathcal{L}_z = 2z - \lambda - 2\mu z = 0 \quad \Longleftrightarrow \quad 2z(1 - \mu) = \lambda \\[2mm] (iv) \quad \mathcal{L}_\mu = -(2x^2 + y^2 + z^2 - 1) = 0 \\[2mm] (v) \quad \lambda \geqslant 0 \quad \text{with} \quad \lambda = 0 \quad \text{if} \quad x + y + z < 0. \end{cases}$$

* If $x + y + z < 0$, then $\lambda = 0$, and from (i), (ii), (iii) and (iv), we deduce that

$$\begin{cases} x = 0 & \text{or} & \mu = 1/2 \\ y = 0 & \text{or} & \mu = 1 \\ z = 0 & \text{or} & \mu = 1 \\ 2x^2 + y^2 + z^2 = 1 \end{cases}$$

We obtain the points

$$(0, 0, -1), \qquad (0, -1, 0) \qquad \text{with} \qquad \mu = 1$$

$$(-\frac{1}{\sqrt{2}}, 0, 0) \qquad \text{with} \qquad \mu = \frac{1}{2}$$

The active constraint at these points satisfies: $h'(x, y, z) = (4x, 2y, 2z)$,

$$rank\Big(h'(0, -1, 0)\Big) = rank\Big(h'(0, 0, -1)\Big) = rank\Big(h'(-\frac{1}{\sqrt{2}}, 0, 0)\Big) = 1.$$

Thus, the points are regular and candidate for optimality.

* If $x + y + z = 0$, then

- Suppose $x = 0$. We deduce from (iv) that

$$y^2 + z^2 = 1 \qquad \text{and} \qquad y + z = 1$$

and deduce the two candidate points

$$(x, y, z) = (0, \frac{1}{\sqrt{2}}, -\frac{1}{\sqrt{2}}) \quad \text{or} \quad (0, -\frac{1}{\sqrt{2}}, \frac{1}{\sqrt{2}}) \quad \text{with} \quad (\lambda, \mu) = (0, 1/2).$$

The two constraints are active at theses points and satisfy

$$\begin{bmatrix} g(x, y, z) & h(x, y, z) \end{bmatrix} = \begin{bmatrix} x + y + z & 2x^2 + y^2 + z^2 - 1 \end{bmatrix}$$

$$\begin{bmatrix} g'(x, y, z) \\ h'(x, y, z) \end{bmatrix} = \begin{bmatrix} 1 & 1 & 1 \\ 4x & 2y & 2z \end{bmatrix}$$

$$rank \begin{bmatrix} g'(0, \frac{1}{\sqrt{2}}, -\frac{1}{\sqrt{2}}) \\ h'(0, \frac{1}{\sqrt{2}}, -\frac{1}{\sqrt{2}}) \end{bmatrix} = rank \begin{bmatrix} 1 & 1 & 1 \\ 0 & 2\sqrt{2} & -2\sqrt{2} \end{bmatrix} = 2$$

$$rank \begin{bmatrix} g'(0, -\frac{1}{\sqrt{2}}, \frac{1}{\sqrt{2}}) \\ h'(0, -\frac{1}{\sqrt{2}}, \frac{1}{\sqrt{2}}) \end{bmatrix} = rank \begin{bmatrix} 1 & 1 & 1 \\ 0 & -2\sqrt{2} & 2\sqrt{2} \end{bmatrix} = 2$$

The points satisfy the constraint qualification. They are regular points and candidates for optimality.

• Suppose $x \neq 0$. Then

– if $\mu = 1/2$, then $\lambda = 0$. By (ii) and (iii), we have $y = z = 0$. Thus, from $x + y + z = 0$, we deduce $x = 0$: contradiction with $x \neq 0$.

– if $\mu \neq 1/2$, then from (i), we have $\lambda \neq 0$. Moreover, by (ii) and (iii), we have $\mu \neq 1$. So, by dividing each side of (ii) by each side of (iii), we obtain $y = z$. Then we deduce that

$$x + 2y = 0 \qquad 2x^2 + 2y^2 = 1 \qquad \Longrightarrow \qquad y = \pm\frac{1}{\sqrt{10}}$$

$$2y(1 - \mu) = \lambda = 2(-2y)(1 - 2\mu) \qquad \Longrightarrow \qquad \mu = \frac{3}{5}.$$

With $\lambda \geqslant 0$, the only possible point is:

$$(x, y, z) = \left(-\frac{2}{\sqrt{10}}, \frac{1}{\sqrt{10}}, \frac{1}{\sqrt{10}}\right) \qquad \text{with} \qquad (\lambda, \mu) = \left(\frac{4}{5\sqrt{10}}, \frac{3}{5}\right).$$

It is clear also that the constraint qualification condition is satisfied, so the point is regular.

$$rank \begin{bmatrix} g'\left(-\frac{2}{\sqrt{10}}, \frac{1}{\sqrt{10}}, \frac{1}{\sqrt{10}}\right) \\ h'\left(-\frac{2}{\sqrt{10}}, \frac{1}{\sqrt{10}}, \frac{1}{\sqrt{10}}\right) \end{bmatrix} = rank \begin{bmatrix} 1 & 1 & 1 \\ -\frac{8}{\sqrt{10}} & \frac{2}{\sqrt{10}} & \frac{2}{\sqrt{10}} \end{bmatrix} = 2.$$

Conclusion: Finally, comparing the values of f at the candidate points

$$f\left(-\frac{2}{\sqrt{10}}, \frac{1}{\sqrt{10}}, \frac{1}{\sqrt{10}}\right) = \frac{3}{5} \qquad f\left(-\frac{1}{\sqrt{2}}, 0, 0\right) = \frac{1}{2}$$

$$f(0, 0, -1) = f(0, -1, 0) = f\left(0, \frac{1}{\sqrt{2}}, -\frac{1}{\sqrt{2}}\right) = f\left(0, -\frac{1}{\sqrt{2}}, \frac{1}{\sqrt{2}}\right) = 1$$

we deduce that f attains its maximum value subject to the constraints at

$$(0, 0, -1), \qquad (0, -1, 0), \qquad \left(0, \frac{1}{\sqrt{2}}, -\frac{1}{\sqrt{2}}\right) \qquad \text{and} \qquad \left(0, -\frac{1}{\sqrt{2}}, \frac{1}{\sqrt{2}}\right).$$

4.3   Classification of Local Extreme Points-Inequality Constraints

To classify a candidate point x^* for optimality of the problem

$$local \max\,(\min)\,f(x) \qquad \text{subject to} \qquad g(x) \leqslant b$$

with $g = (g_1, \ldots, g_m)$ and $b = (b_1, \ldots, b_m)$, we proceed as in the case of equality constraints by comparing the values taken by the Lagrangian

$$\mathcal{L}(x, \lambda) = f(x) - \lambda_1(g_1(x) - b_1) - \cdots - \lambda_m(g_m(x) - b_m),$$

at points close to x^*. Then, since, $x^* \in [g(x) \leqslant b]$ means that

$$x^* \in \bigcap_{j \notin I(x^*)} [g_j(x) < b_j] = \mathcal{O} \quad \text{and} \quad x^* \in \bigcap_{j \in I(x^*)} [g_j(x) = b_j],$$

we remark that by working in a neighborhood of x^* included in the open set $\mathcal{O}$, we bring ourselves to solving a local optimization problem of type equality constraints

$$local \max\,(\min)\,f(x) \qquad \text{subject to} \qquad g_j(x) = b_j, \qquad j \in I(x^*).$$

Consequently, we can apply the second derivative test established for equality constraints by considering in the test only the active constraints at that point. In what follows, suppose we have:

Hypothesis (H) f and $g = (g_1, \ldots, g_m)$ be C^2 functions in a neighborhood of x^* in $\mathbb{R}^n$ such that:

$$g_j(x^*) = b_j \qquad \text{if} \qquad j \in I(x^*) = \{i_1, \ldots, i_p\} \qquad p < n$$

$$\lambda_j = 0 \qquad \text{if} \qquad g_j(x^*) < b_j \qquad j \notin I(x^*)$$

$$rank(G'(x^*)) = p, \qquad G(x) = (g_{i_1}(x), \ldots, g_{i_p}(x)),$$

$$\nabla_x \mathcal{L}(x^*, \lambda^*) = 0 \qquad \text{for a unique vector} \qquad \lambda^* = (\lambda_1^*, \ldots, \lambda_m^*).$$

For $r = p+1, \ldots, n$, let $\mathbb{B}_r(x^*)$ be the bordered Hessian determinant

$$\mathbb{B}_r(x^*) = \begin{vmatrix} 0 & \cdots & 0 & \frac{\partial g_{i_1}}{\partial x_1}(x^*) & \cdots & \frac{\partial g_{i_1}}{\partial x_r}(x^*) \\ \vdots & \ddots & \vdots & \vdots & \ddots & \vdots \\ 0 & \cdots & 0 & \frac{\partial g_{i_p}}{\partial x_1}(x^*) & \cdots & \frac{\partial g_{i_p}}{\partial x_r}(x^*) \\ \frac{\partial g_{i_1}}{\partial x_1}(x^*) & \cdots & \frac{\partial g_{i_p}}{\partial x_1}(x^*) & \mathcal{L}_{x_1 x_1}(x^*, \lambda^*) & \cdots & \mathcal{L}_{x_1 x_r}(x^*, \lambda^*) \\ \vdots & \ddots & \vdots & \vdots & \ddots & \vdots \\ \frac{\partial g_{i_1}}{\partial x_r}(x^*) & \cdots & \frac{\partial g_{i_p}}{\partial x_r}(x^*) & \mathcal{L}_{x_r x_1}(x^*, \lambda^*) & \cdots & \mathcal{L}_{x_r x_r}(x^*, \lambda^*) \end{vmatrix}$$

The variables are **renumbered** in order to make the first p columns in the matrix $G'(x^*)$ linearly independent.

Theorem 4.3.1 *Sufficient conditions for a local constrained extreme point.*

If assumptions (H) hold, then

$$(i) \quad \lambda \leqslant 0, \quad (-1)^p \mathbb{B}_r(x^*) > 0 \quad \forall r = p+1, \ldots, n$$
$$\implies \quad x^* \text{ is a strict local minimum point}$$

$$(ii) \quad \lambda \geqslant 0, \quad (-1)^r \mathbb{B}_r(x^*) > 0 \quad \forall r = p+1, \ldots, n$$
$$\implies \quad x^* \text{ is a strict local maximum point.}$$

Proof. The proof follows the one seen for the case of equality constraints. We outline here the key modification that allows us to conclude with the previous proof. We assume that $I(x^*) \neq \{1, \ldots, m\}$ to avoid the case of equality constraints. Note that the positivity of λ is not assumed in the hypothesis H in order to include both the maximization and minimization problems as explained below. The Lagrangian introduced is used to link values of f and g for comparison. Then depending on its positivity or negativity on the plan tangent of the active constraints at that point, we identify whether we have a minimum or a maximum point.

Step 0: Suppose that we assign for the problems

$$\max f : \quad L(x, \alpha) = f(x) - \alpha.(g(x) - b) \quad \alpha \geqslant 0$$
$$\min f : \quad L(x, \beta) = -f(x) - \beta.(g(x) - b) \quad \beta \geqslant 0$$

then

$$-L(x,\beta) = f(x) - (-\beta).(g(x) - b) \qquad -\beta \leqslant 0.$$

So, to consider the two problems simultaneously, we can introduce the Lagrangian

$$\mathcal{L}(x,\lambda) = f(x) - \lambda.(g(x) - b)$$

with $\lambda \geqslant 0$ (resp. $\leqslant$) for the maximization (resp. minimization) problem.

Step 1: We have

$$[g(x) \leqslant b] = \Big(\bigcap_{j \notin I(x^*)} [g_j(x) < b_j] \Big) \bigcap \Big(\bigcap_{j \in I(x^*)} [g_j(x) = b_j] \Big) \subset \mathcal{O}.$$

Thus x^* belongs to the open set $\mathcal{O}$. So, one can find $\rho_0 > 0$ such that $B_{\rho_0}(x^*) \subset \mathcal{O}$. Then, for $h \in \mathbb{R}^n$ such that $x^* + h \in B_{\rho_0}(x^*)$, we have from Taylor's formula, for some $\tau \in (0,1)$,

$$\mathcal{L}(x^*+h,\lambda^*) = \mathcal{L}(x^*,\lambda^*) + \sum_{i=1}^{n} \mathcal{L}_{x_i}(x^*,\lambda^*)h_i + \frac{1}{2}\sum_{i=1}^{n}\sum_{j=1}^{n} \mathcal{L}_{x_i x_j}(x^*+\tau h,\lambda^*)h_i h_j.$$

By assumptions, we have

$$\mathcal{L}_{x_i}(x^*,\lambda^*) = 0 \qquad\qquad i = 1,\ldots,n$$

$$\mathcal{L}(x,\lambda) = f(x) - \sum_{j \in I(x^*)} \lambda_j(g_j(x) - b_j)$$

$$g_{i_1}(x^*) - b_{i_1} = g_{i_2}(x^*) - b_{i_2} = \ldots = g_{i_p}(x^*) - b_{i_p} = 0$$

then, we have

$$\mathcal{L}(x^*,\lambda^*) = f(x^*) - \lambda_{i_1}^*(g_{i_1}(x^*) - b_{i_1}) - \ldots - \lambda_{i_p}^*((g_{i_p}(x^*) - b_{i_p}) = f(x^*)$$

$$\mathcal{L}(x^*+h,\lambda^*) = f(x^*+h) - \lambda_{i_1}^*(g_{i_1}(x^*+h) - b_{i_1}) - \ldots - \lambda_{i_p}^*(g_{i_p}(x^*+h) - b_{i_p})$$

from which we deduce

$$f(x^*+h) - f(x^*) = \sum_{k \in I(x^*)} \lambda_k^*[g_k(x^*+h) - b_k] + \frac{1}{2}\sum_{i=1}^{n}\sum_{j=1}^{n} \mathcal{L}_{x_i x_j}(x^*+\tau h,\lambda^*)h_i h_j.$$

Using Taylor's formula for each g_k, $k \in I(x^*)$, we obtain

$$g_k(x^*+h) - b_k = g_k(x^*+h) - g_k(x^*) = \sum_{j=1}^{n} \frac{\partial g_k}{\partial x_j}(x^* + \tau_k h)h_j \qquad \tau_k \in (0,1).$$

Step 2: Consider the $(p + n) \times (p + n)$ bordered Hessian matrix $\mathbf{B}(\mathbf{x}^0, \mathbf{x}^1, \ldots, \mathbf{x}^p)$ with

$$\mathbf{G}(\mathbf{x}^1, \cdots, \mathbf{x}^p) = \left(\frac{\partial g_{i_k}}{\partial x_j}(\mathbf{x}^k) \right)_{p \times n} = \begin{bmatrix} \frac{\partial g_{i_1}}{\partial x_{i_1}}(\mathbf{x}^1) & \cdots & \frac{\partial g_{i_1}}{\partial x_n}(\mathbf{x}^1) \\ \vdots & \ddots & \vdots \\ \frac{\partial g_{i_p}}{\partial x_1}(\mathbf{x}^p) & \cdots & \frac{\partial g_{i_p}}{\partial x_n}(\mathbf{x}^p). \end{bmatrix}$$

The remaining steps of the equality constraints' proof work is shown using the above notations.

Remark 4.3.1 *If we introduce the notations:*

$$Q(h) = Q(h_1, \ldots, h_n) = \sum_{i=1}^{n} \sum_{j=1}^{n} \mathcal{L}_{x_i x_j}(x^*, \lambda^*) h_i h_j$$

the $(p+n) \times (p+n)$ bordered matrix $\begin{bmatrix} 0_{p \times p} & G'(x^*) \\ {}^t G'(x^*) & [\mathcal{L}_{x_i x_j}(x^*, \lambda^*)]_{n \times n} \end{bmatrix}$

$$M = \{ h \in \mathbb{R}^n : G'(x^*).h = 0 \}$$

the theorem says that

$$Q(h) > 0 \qquad \forall h \in M, \qquad h \neq 0$$
$$\implies \qquad x^* \text{ is a strict local constrained minimum}$$

$$Q(h) < 0 \qquad \forall h \in M, \qquad h \neq 0$$
$$\implies \qquad x^* \text{ is a strict local constrained maximum.}$$

It suffices then to study the positivity (negativity) of the quadratic form on the tangent plan M to the constraints $g_k(x) = b_k$, $k \in I(x^*)$ at x^*.

Example 1. Solve the problem

$$local \max (\min) f(x, y) = xy \qquad \text{subject to} \qquad g(x, y) = x + y \leqslant 2.$$

Solution: Consider the Lagrangian

$$\mathcal{L}(x, y, \lambda) = f(x, y) - \lambda(g(x, y) - 2) = xy - \lambda(x + y - 2)$$

and the system

$$
\begin{cases}
(i) & \mathcal{L}_x = y - \lambda = 0 \\
(ii) & \mathcal{L}_y = x - \lambda = 0 \\
(iii) & \lambda = 0 \quad \text{if} \quad x + y < 2.
\end{cases}
$$

From (i) and (ii), we deduce that $\lambda = x = y$.

* If $x + y < 2$, then $\lambda = 0$. Thus $(0,0)$ is a candidate point, that is an interior point of $[g \leqslant 2]$. To explore its nature, we use the second derivatives test for unconstrained problems. We have

$$
H_f(x,y) = \begin{bmatrix} 0 & 1 \\ 1 & 0 \end{bmatrix}, \qquad D_1(0,0) = 0 \quad \text{and} \quad D_2(0,0) = -1 < 0.
$$

Then, $(0,0)$ is a saddle point.

** If $x + y = 2$, then $(x,y) = (1,1)$ is a candidate point with $\lambda = 1$.

First, $(1,1)$ is a regular point since $g'(x,y) = \langle 1,1 \rangle$ and $rank[g'(1,1)] = 1$. Next, since $n = 2$ and $p = 1$, we have to consider the sign of the bordered Hessian determinant:

$$
(-1)^2 \mathbb{B}_2(1,1) = \begin{vmatrix} 0 & g_x(1,1) & g_y(1,1) \\ g_x(1,1) & \mathcal{L}_{xx}(1,1,1) & \mathcal{L}_{xy}(1,1,1)^* \\ g_y(1,1) & \mathcal{L}_{xy}(1,1,1) & \mathcal{L}_{yy}(1,1,1) \end{vmatrix} = \begin{vmatrix} 0 & 1 & 1 \\ 1 & 0 & 1 \\ 1 & 1 & 0 \end{vmatrix} = 2 > 0.
$$

We conclude that the point $(1,1)$ is a local maximum to the problem.

Finally, we also have

Theorem 4.3.2 *Necessary conditions for a local constrained extreme points*

If assumptions (H) hold, then

    (i) x^* *is a local minimum point* $\Longrightarrow$ $H_{\mathcal{L}} = (\mathcal{L}_{x_i x_j}(x^*, \lambda^*))_{n \times n}$ *is*
         positive semi definite on M: ${}^t y H_{\mathcal{L}} y \geqslant 0$ $\forall y \in M$

    (ii) x^* *is a local maximum point* $\Longrightarrow$ $H_{\mathcal{L}} = (\mathcal{L}_{x_i x_j}(x^*, \lambda^*))_{n \times n}$ *is*
         is negative semi definite on M: ${}^t y H_{\mathcal{L}} y \leqslant 0$ $\forall y \in M$

where $M = \{h \in \mathbb{R}^n : G'(x^*).h = 0\}$ *is the tangent plan to the constraints* $g_k(x) = b_k$, $k \in I(x^*)$ *at* x^*.

Proof. Let $x(t) \in C^2[0, a]$, $a > 0$, be a curve on the constraint set $g(x) \leqslant b$ passing through x^* at $t = 0$. Suppose that x^* is a local maximum point for f subject to the constraint $g(x) \leqslant b$. Then,

$$f(x^*) \geqslant f(x(t)) \qquad \forall t \in [0, a).$$

or

$$\widetilde{f}(0) = f(x^*) \geqslant f(x(t)) = \widetilde{f}(t) \qquad \forall t \in [0, a).$$

So $\widetilde{f}$ is a one variable function that has a local maximum at $t = 0$. Consequently, it satisfies $\widetilde{f}'(0) \leqslant 0$ and $\widetilde{f}''(0) \leqslant 0$ or equivalently

$$\nabla f(x^*).x'(0) \leqslant 0 \qquad \text{and} \qquad \frac{d^2}{dt^2} f(x(t)) \Big|_{t=0} \leqslant 0.$$

We have

$$\frac{d^2}{dt^2} f(x(t)) \;=\; {}^t x'(0) H_f(x^*) x'(0) \;+\; \nabla f(x^*).x''(0).$$

Moreover, differentiating the relations $g_k(x(t)) = b_k$, $k \in I(x^*)$ twice and denoting $\Lambda^* = \langle \lambda_{i_1}^*, \ldots, \lambda_{i_p}^* \rangle$, we obtain

$${}^t x'(0) H_G(x^*) x'(0) \;+\; \nabla G(x^*) x''(0) = 0$$

$$\implies \quad {}^t x'(0) {}^t \Lambda^* H_G(x^*) x'(0) \;+\; {}^t \Lambda^* \nabla G(x^*) x''(0) = 0.$$

Hence

$$0 \geqslant \frac{d^2}{dt^2} f(x(t)) \Big|_{t=0} \;=\; [{}^t x'(0) H_f(x^*) x'(0) \;+\; \nabla f(x^*) x''(0)] \;-$$

$$[{}^t x'(0) {}^t \Lambda H_G(x^*) x'(0) \;+\; {}^t \Lambda \nabla G(x^*) x''(0)]$$

$$= \; {}^t x'(0) [H_f(x^*) - {}^t \Lambda H_G(x^*)] x'(0) \;+\; [\nabla f(x^*) \;+\; {}^t \Lambda \nabla G(x^*)] x''(0)$$

$$= \; {}^t x'(0) [H_{\mathcal{L}}(x^*)] x'(0) \quad \text{since} \quad \nabla f(x^*) \;+\; {}^t \Lambda \nabla G(x^*) = 0$$

and the result follows since $x'(0)$ is an arbitrary element of M.

Example 2. Suppose that $(4, 0)$ is a candidate satisfying the KKT conditions where only the constraint g is active and such that $g'(4, 0) = \begin{bmatrix} -1 & 0 \end{bmatrix}$ and the Hessian of the associated Lagrangian is $\mathcal{H}_{\mathcal{L}(., -8)}(4, 0) = \begin{bmatrix} -2 & 0 \\ 0 & 14 \end{bmatrix}$. Can $(4, 0)$ be a local maximum or minimum to the constrained optimization problem?

Solution: The point $(4, 0)$ is regular since $rank(g_1'(4, 0)) = 1$.

We have $p = 1 < 2 = n$. Then we can consider the following determinant $(r = p + 1 = 2)$. (Note that the first column vector of $g_1'(4, 0)$ is linearly independent, so we do not have to renumber the variables).

$$\mathbb{B}_2(4,0) = \begin{vmatrix} 0 & -1 & 0 \\ -1 & -2 & 0 \\ 0 & 0 & 14 \end{vmatrix} = -14.$$

We have $(-1)^1 \mathbb{B}_2(4, 0) = 14 > 0$ and $\lambda = -8 < 0$. So the second derivatives test is satisfied and $(4, 0)$ is a strict local minimum. This shows also that the Hessian is positive definite under the constraint. Indeed, we can check this directly:

$$g_1'(4,0)\begin{bmatrix} h \\ k \end{bmatrix} = \begin{bmatrix} -1 & 0 \end{bmatrix} \cdot \begin{bmatrix} h \\ k \end{bmatrix} = -h + (0)k = 0$$

Thus, $M = \{ \begin{bmatrix} 0 \\ k \end{bmatrix} \quad k \in \mathbb{R} \}$ and

$$\begin{bmatrix} 0 & k \end{bmatrix} \begin{bmatrix} -2 & 0 \\ 0 & 14 \end{bmatrix} \begin{bmatrix} 0 \\ k \end{bmatrix} = 14k^2 \geqslant 0 \qquad \forall \begin{bmatrix} 0 \\ k \end{bmatrix} \in M.$$

1. – Solve the problem

$$local \min \ f(x,y) = x^2 + y^2 \quad \text{s.t} \quad \begin{cases} x + 2y \geqslant 3 \\ 2x - y \geqslant 1 \end{cases}$$

Solution: The problem is equivalent to the maximization problem.

$$local \max \ -f(x,y) = -(x^2 + y^2) \quad \text{s.t} \quad \begin{cases} -(x + 2y) \leqslant -3 \\ -(2x - y) \leqslant -1 \end{cases}$$

Consider the Lagrangian

$$\mathcal{L}(x, y, \lambda_1, \lambda_2) = -f(x,y) - \lambda_1(-(x + 2y) + 3) - \lambda_2(-(2x - y) + 1)$$

$$= -(x^2 + y^2) + \lambda_1(x + 2y - 3) + \lambda_2(2x - y - 1).$$

The constraints are linear, so we can look for the candidate points by writing the Karush-Kuhn-Tucker conditions:

$$\begin{cases} (i) & \mathcal{L}_x = -2x + \lambda_1 + 2\lambda_2 = 0 \\[2mm] (ii) & \mathcal{L}_y = -2y + 2\lambda_1 - \lambda_2 = 0 \\[2mm] (iii) & \lambda_1 \geqslant 0 \quad \text{with} \quad \lambda_1 = 0 \quad \text{if} \quad x + 2y > 3 \\[2mm] (iv) & \lambda_2 \geqslant 0 \quad \text{with} \quad \lambda_2 = 0 \quad \text{if} \quad 2x - y > 1. \end{cases}$$

We distinguish several cases:

• If $2x - y > 1$, then $\lambda_2 = 0$. From (i) and (ii), we deduce that $\lambda_1 = 2x = y$. But $2x - 2x = 0 \not> 1$. So, no solution.

- If $2x - y = 1$, then

    –   If $x + 2y > 3$, then $\lambda_1 = 0$. From (i) and (ii), we deduce that $\lambda_2 = x = 2y$.

    With $2x - y = 1$, we deduce $(x, y) = (2/3, 1/3)$. But $2/3 + 2(1/3) \not> 3$. So, no solution.

    –   If $x + 2y = 3$, then with $2x - y = 1$, we have $(x, y) = (1, 1)$ and (λ_1, λ_2) are such that

$$\begin{cases} \lambda_1 + 2\lambda_2 = 2 \\[2mm] 2\lambda_1 - \lambda_2 = 2 \end{cases} \qquad \Longleftrightarrow \qquad (\lambda_1, \lambda_2) = (\frac{6}{5}, \frac{2}{5}).$$

Hence, the only solution point is

$$(x, y) = (1, 1) \qquad \text{with} \qquad (\lambda_1, \lambda_2) = (\frac{6}{5}, \frac{2}{5}).$$

Regularity of the point. The two constraints are active at the point. We have $g = (g_1, g_2) = (-(x + 2y), -(2x - y))$,

$$g'(x, y) = \begin{bmatrix} \frac{\partial g_1}{\partial x} & \frac{\partial g_1}{\partial y} \\[2mm] \frac{\partial g_2}{\partial x} & \frac{\partial g_2}{\partial y} \end{bmatrix} = \begin{bmatrix} -1 & -2 \\ -2 & 1 \end{bmatrix} \qquad \Longrightarrow \qquad rank(g'(1, 1)) = 2.$$

Classification of the point. Since $n = 2$, $p = 2$, $p \not< n$, then we can't apply the second derivatives test. Let us use comparison to conclude. We have

$$\mathcal{L}(x, y, \frac{6}{5}, \frac{2}{5}) = -(x^2 + y^2) + \frac{6}{5}(x + 2y - 3) + \frac{2}{5}(2x - y - 1)$$

$$= -x^2 - y^2 + 2x + 2y - 4 = -(x - 1)^2 - (y - 1)^2 - 2 \leqslant -2$$

and, on the set of the constraints, we have

$$-f(x, y) \leqslant \mathcal{L}(x, y, \frac{6}{5}, \frac{2}{5}) = -f(x, y) + \frac{6}{5}(x + 2y - 3) + \frac{2}{5}(2x - y - 1)$$

Thus,

$$-f(x, y) \leqslant -2 = -f(1, 1) \qquad \forall (x, y) \quad x + 2y \geqslant 3, \qquad 2x - y \geqslant 1.$$

Hence, $(1, 1)$ is the minimum point solution; see Figure 4.18 for a geometric interpretation of the solution.

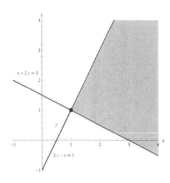 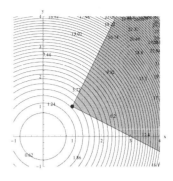

FIGURE 4.18: Local minimum of $z = x^2 + y^2$ on $x + 2y \geqslant 3$ and $2x - y \geqslant 1$

2. – Classify the solutions of the problem

$$local \max (\min) f(x, y) = x^2 y + 3y - 4 \quad \text{s.t} \quad g(x, y) = 4 - xy \leqslant 0$$

Solution: i) Consider the Lagrangian

$$\mathcal{L}(x, y, \lambda) = f(x, y) - \lambda(g(x, y) - 0) = x^2 y + 3y - 4 - \lambda(4 - xy)$$

and write the conditions

$$\begin{cases} (1) \quad \mathcal{L}_x = 2xy - \lambda(-y) = 0 \quad \Longleftrightarrow \quad y(2x + \lambda) = 0 \\ \\ (2) \quad \mathcal{L}_y = x^2 + 3 - \lambda(-x) = 0 \\ \\ (3) \quad \lambda = 0 \quad \text{if} \quad xy > 4. \end{cases}$$

$*$ If $xy > 4$, then $\lambda = 0$, and with (2), we have $x^2 + 3 = 0$ which has no solution.

$**$ If $xy = 4$, then $x \neq 0$ and $y \neq 0$. By (1), we deduce that $\lambda = -2x$, which inserted in (2), we obtain $3 - x^2 = 0$. Thus, we have two solutions:

$$(x, y) = (\sqrt{3}, \frac{4}{\sqrt{3}}) \quad \text{with} \quad \lambda = -2\sqrt{3}$$

$$(x, y) = (-\sqrt{3}, -\frac{4}{\sqrt{3}}) \quad \text{with} \quad \lambda = 2\sqrt{3}.$$

ii) *Constraint qualification.* Note that g is C^1 in $\mathbb{R}^2$ and any point of the set of the constraints $g = 0$ is an interior point and regular; see Figure 4.19. Indeed, we have

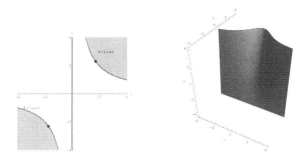

FIGURE 4.19: Graph of $z = x^2y + 3y - 4$ on $xy \geqslant 4$

$$g'(x,y) = \begin{bmatrix} -y & -x \end{bmatrix} \qquad rank(g'(x,y)) = 1 \quad \text{for } (x,y) \in [g = 0]$$

since $g'(x,y) = 0 \iff (x,y) = (0,0)$ and $(0,0) \notin [g = 0]$. In particular $(\sqrt{3}, \frac{4}{\sqrt{3}})$ and $(-\sqrt{3}, -\frac{4}{\sqrt{3}})$ are regular points of $[g = 0]$. Therefore, they are candidate points; see in Figure 4.20, the variations of the values of the function close to these points.

iii) *Classification.* With $m = 1$ (the number of the active constraints), $n = 2$ (the dimension of the space), then r taking values from $m+1$ to $n = 2$, must be equal to $r = 2$. So, consider the following determinant

$$\mathbb{B}_2(x,y) = \begin{vmatrix} 0 & g_x & g_y \\ g_x & \mathcal{L}_{xx} & \mathcal{L}_{xy} \\ g_y & \mathcal{L}_{yx} & \mathcal{L}_{yy} \end{vmatrix} = \begin{vmatrix} 0 & -y & -x \\ -y & 2y & 2x+\lambda \\ -x & 2x+\lambda & 0 \end{vmatrix}$$

$*$ At $(\sqrt{3}, 4/\sqrt{3})$, we have $\lambda = 2\sqrt{3}$, then

$$\mathbb{B}_2(\sqrt{3}, 4/\sqrt{3}) = \begin{vmatrix} 0 & -4/\sqrt{3} & -\sqrt{3} \\ -4/\sqrt{3} & 8/\sqrt{3} & 0 \\ -\sqrt{3} & 0 & 0 \end{vmatrix} = -\sqrt{3} \begin{vmatrix} -4/\sqrt{3} & -8/\sqrt{3} \\ 8/\sqrt{3} & 0 \end{vmatrix} = -8\sqrt{3}$$

Because $\lambda = -2\sqrt{3} \leqslant 0$ and $(-1)^1\mathbb{B}_2(\sqrt{3}, 4/\sqrt{3}) > 0$, we deduce that $(\sqrt{3}, 4/\sqrt{3})$ is a local minimum.

$*$ At $(-\sqrt{3}, -4/\sqrt{3})$, we have $\lambda = 2\sqrt{3}$, then

$$\mathbb{B}_2(-\sqrt{3}, -4/\sqrt{3}) = \begin{vmatrix} 0 & 4/\sqrt{3} & \sqrt{3} \\ 4/\sqrt{3} & -8/\sqrt{3} & 0 \\ \sqrt{3} & 0 & 0 \end{vmatrix} = \sqrt{3} \begin{vmatrix} 4/\sqrt{3} & \sqrt{3} \\ 8/\sqrt{3} & 0 \end{vmatrix} = 8\sqrt{3}$$

Because $\lambda = 2\sqrt{3} \geqslant 0$ and $(-1)^2\mathbb{B}_2(-\sqrt{3}, -4/\sqrt{3}) > 0$, we deduce that $(-\sqrt{3}, -4/\sqrt{3})$ is a local maximum.

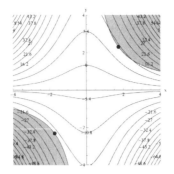

FIGURE 4.20: Local extrema of $z = x^2y + 3y - 4$ on $xy \geqslant 4$

3. – Solve the problem

$$local \min \; f(x,y,z) = x^2 + y^2 + z^2 \quad \text{s.t} \quad \begin{cases} g_1(x,y,z) = x + 2y + z \geqslant 30 \\ \\ g_2(x,y,z) = 2x - y - 3z \leqslant 10 \end{cases}$$

Solution: Note that f, g_1 and g_2 are C^1 in $\mathbb{R}^3$. The problem is equivalent to the maximization problem

$$local \max \; -f = -(x^2 + y^2 + z^2) \quad \text{s.t} \quad \begin{cases} -g_1 = -(x + 2y + z) \leqslant -30 \\ \\ g_2 = 2x - y - 3z \leqslant 10 \end{cases}$$

Consider the Lagrangian

$$\mathcal{L}(x,y,z,\lambda_1,\lambda_2) = -(x^2 + y^2 + z^2) + \lambda_1(x + 2y + z - 30) - \lambda_2(2x - y - 3z - 10).$$

Because the constraints are linear, the local candidate points satisfy the KKT conditions:

$$\begin{cases} (i) & \mathcal{L}_x = -2x + \lambda_1 - 2\lambda_2 = 0 \\ \\ (ii) & \mathcal{L}_y = -2y + 2\lambda_1 + \lambda_2 = 0 \\ \\ (iii) & \mathcal{L}_z = -2z + \lambda_1 + 3\lambda_2 = 0 \\ \\ (iv) & \lambda_1 \geqslant 0 \quad \text{with} \quad \lambda_1 = 0 \quad \text{if} \quad x + 2y + z > 30 \\ \\ (v) & \lambda_2 \geqslant 0 \quad \text{with} \quad \lambda_2 = 0 \quad \text{if} \quad 2x - y - 3z < 10. \end{cases}$$

From the first three equations, we deduce that

$$x = \frac{1}{2}\lambda_1 - \lambda_2 \qquad\qquad y = \lambda_1 + \frac{1}{2}\lambda_2 \qquad\qquad z = \frac{1}{2}\lambda_1 + \frac{3}{2}\lambda_2.$$

We distinguish several cases:

* If $x + 2y + z = 30$ and $2x - y - 3z = 10$, then inserting the expressions of x, y and z above into the two equations gives

$$\begin{cases} 3\lambda_1 + \frac{3}{2}\lambda_2 = 30 \\[2mm] -\frac{3}{2}\lambda_1 - 7\lambda_2 = 10 \end{cases} \quad\Longleftrightarrow\quad \lambda_1 = \frac{54}{5}, \qquad \lambda_2 = -\frac{8}{5}$$

which contradicts $\lambda_2 \geqslant 0$.

** If $x + 2y + z = 30$ and $2x - y - 3z < 10$, then $\lambda_2 = 0$ and

$$(x, y, z) = \lambda_1 \left(\frac{1}{2}, 1, \frac{1}{2}\right)$$

which inserted into the equation $x + 2y + z = 30$ gives $\lambda_1 = 10$ and $(x, y, z) = (5, 10, 5)$. We have $2x - y - 3z = 2(5) - 10 - 3(5) = -15 < 10$. So the point

$$(x, y, z) = (5, 10, 5) \qquad\qquad \lambda_1 = 10, \qquad \lambda_2 = 0$$

is a candidate for optimality.

Now, let us study the nature of the point $(5, 10, 5)$. For this, we use the second derivatives test since f, g_1 and g_2 are C^2 around this point. Since $n = 3$ and $p = 1$ (only the constraint g_1 is active), then r takes the values $p + 1 = 2$ to $n = 3$. First, we consider the matrix

$$g'(x, y, z) = \begin{bmatrix} -\frac{\partial g_1}{\partial x} & -\frac{\partial g_1}{\partial y} & -\frac{\partial g_1}{\partial z} \end{bmatrix} = \begin{bmatrix} -1 & -2 & -1 \end{bmatrix}$$

Then $rank(g'(x, y, z)) = 1$. Moreover, the first column vector of $g'(5, 10, 5)$ is linearly independent, so we don't have to renumber the variables.

Next, we have to consider the sign of the following bordered Hessian determinants:

$$\mathbb{B}_2(5, 10, 5) = \begin{vmatrix} 0 & -\frac{\partial g_1}{\partial x} & -\frac{\partial g_1}{\partial y} \\[2mm] -\frac{\partial g_1}{\partial x} & \mathcal{L}_{xx} & \mathcal{L}_{xy} \\[2mm] -\frac{\partial g_1}{\partial y} & \mathcal{L}_{yx} & \mathcal{L}_{yy} \end{vmatrix} = \begin{vmatrix} 0 & -1 & -2 \\ -1 & -2 & 0 \\ -2 & 0 & -2 \end{vmatrix} = 10.$$

$$\mathbb{B}_3(5,10,5) = \begin{vmatrix} 0 & -\frac{\partial g_1}{\partial x} & -\frac{\partial g_1}{\partial y} & -\frac{\partial g_1}{\partial z} \\ -\frac{\partial g_1}{\partial x} & \mathcal{L}_{xx} & \mathcal{L}_{xy} & \mathcal{L}_{xz} \\ -\frac{\partial g_1}{\partial y} & \mathcal{L}_{yx} & \mathcal{L}_{yy} & \mathcal{L}_{yz} \\ -\frac{\partial g_1}{\partial z} & \mathcal{L}_{zx} & \mathcal{L}_{zy} & \mathcal{L}_{zz} \end{vmatrix} = \begin{vmatrix} 0 & -1 & -2 & -1 \\ -1 & -2 & 0 & 0 \\ -2 & 0 & -2 & 0 \\ -1 & 0 & 0 & -2 \end{vmatrix} = -24.$$

Here, the partial derivatives of g_1 are evaluated at the point $(5,10,5)$ and the second partial derivatives of $\mathcal{L}$ are evaluated at the point $(5,10,5,10,0)$. We have

$$(-1)^2 \mathbb{B}_2(5,10,5) = 10 > 0 \qquad \text{and} \qquad (-1)^3 \mathbb{B}_3(5,10,5) = 24 > 0.$$

We conclude that the point $(5,10,5)$ is a local maximum to the maximization problem, or equivalently, a local minimum to the minimization problem.

∗ ∗ ∗   If $x + 2y + z > 30$ and $2x - y - 3z = 10$ then $\lambda_1 = 0$ and

$$(x,y,z) = \lambda_2(1, -\frac{1}{2}, -\frac{3}{2})$$

which inserted into the equation $2x - y - 3z = 10$ gives $\lambda_2 = -10/7 < 0$: contradiction.

∗ ∗ ∗∗   If $x + 2y + z > 30$ and $2x - y - 3z < 10$ then $\lambda_1 = 0$ and $\lambda_2 = 0$. So $(x,y,z) = (0,0,0)$ which contradicts the first above inequality.

Conclusion: The minimization problem has one local minimum at the point $(5,10,5)$.

4. – Classify the candidates of the problem

$$local \max(\min) f(x,y,z) = x + y + z \quad \text{s.t} \quad \begin{cases} g_1 = x^2 + y^2 + z^2 = 1 \\ g_2 = x - y - z \geqslant 1. \end{cases}$$

Solution: i) Note that f, g_1 and g_2 are C^∞ in $\mathbb{R}^3$ and consider the Lagrangian

$$\mathcal{L}(x,y,z,\lambda_1,\lambda_2) = f(x,y,z) - \lambda_1(g_1(x,y,z) - 1) - \lambda_2(1 - g_2(x,y,z))$$
$$= x + y + z - \lambda_1(x^2 + y^2 + z^2 - 1) + \lambda_2(x - y - z - 1)$$

and let us look for the solutions of the system

$$\begin{cases} (1) & \mathcal{L}_x = 1 - 2x\lambda_1 + \lambda_2 = 0 \\ (2) & \mathcal{L}_y = 1 - 2y\lambda_1 - \lambda_2 = 0 \\ (3) & \mathcal{L}_z = 1 - 2z\lambda_1 - \lambda_2 = 0 \\ (4) & \mathcal{L}_{\lambda_1} = -(x^2 + y^2 + z^2 - 1) = 0 \\ (5) & \lambda_2 = 0 \quad \text{if} \quad x - y - z > 1. \end{cases}$$

From the first three equations, we deduce that

$$-\lambda_2 = 1 - 2x\lambda_1 \qquad \lambda_1(x + y) = 1 \qquad \lambda_1(x + z) = 1.$$

Note that $\lambda_1 = 0$ is not possible because we would have from (1) : $\lambda_2 = -1$, and from (2) : $\lambda_2 = 1$. So $\lambda_1 \neq 0$ and we have

$$x + y = x + z = \frac{1}{\lambda_1} \qquad \Longrightarrow \qquad y = z.$$

* If $x - y - z > 1$, then $\lambda_2 = 0$. We deduce that $\frac{1}{\lambda_1} = 2x$, thus $x + y = x + z = \frac{1}{\lambda_1} = 2x$. So $x = y = z$, which inserted into (4) gives $3x^2 = 1$. Hence, we have two points

$$(x, y, z) = (\frac{1}{\sqrt{3}}, \frac{1}{\sqrt{3}}, \frac{1}{\sqrt{3}}) \quad \text{or} \quad (-\frac{1}{\sqrt{3}}, -\frac{1}{\sqrt{3}}, -\frac{1}{\sqrt{3}}).$$

But, they do not satisfy $x - y - z > 1$.

* If $x - y - z = 1$, then with $y = z$ and (4), we have

$$\begin{cases} x = 1 + 2y \\ 2y(3y + 2) = 0 \end{cases} \quad \Longleftrightarrow \quad (x, y) = (1, 0) \quad \text{or} \quad (x, y) = (-\frac{1}{3}, -\frac{2}{3}).$$

We deduce then

$$(x, y, z) = (1, 0, 0) \qquad \text{with} \qquad \lambda_1 = 1, \qquad \lambda_2 = 1$$

$$(x, y, z) = (-\frac{1}{3}, -\frac{2}{3}, -\frac{2}{3}) \qquad \text{with} \qquad \lambda_1 = -1, \qquad \lambda_2 = -\frac{1}{3}.$$

ii) *Regularity of the points.* We have

$$g'(x,y,z) = \begin{bmatrix} \frac{\partial g_1}{\partial x} & \frac{\partial g_1}{\partial y} & \frac{\partial g_1}{\partial z} \\ -\frac{\partial g_2}{\partial x} & -\frac{\partial g_2}{\partial y} & -\frac{\partial g_2}{\partial z} \end{bmatrix} = \begin{bmatrix} 2x & 2y & 2z \\ -1 & 1 & 1 \end{bmatrix}$$

$$g'(1,0,0) = \begin{bmatrix} 2 & 0 & 0 \\ -1 & 1 & 1 \end{bmatrix} \qquad g'(-\frac{1}{3},-\frac{2}{3},-\frac{2}{3}) = \begin{bmatrix} -\frac{2}{3} & -\frac{4}{3} & -\frac{4}{3} \\ -1 & 1 & 1 \end{bmatrix}.$$

Then

$$rank(g'(1,0,0)) = rank(g'(-\frac{1}{3},-\frac{2}{3},-\frac{2}{3})) = 2.$$

The two points are regular. Moreover, we remark that the first two column vectors are linearly independent and we will not renumber the variables.

iii) *Classification of the points.* Now, let us study the nature of the points $(1,0,0)$ and $(-\frac{1}{3},-\frac{2}{3},-\frac{2}{3})$. For this we use the second derivatives test since f, g_1 and g_2 are C^2 around these points. We have to consider the sign of the following bordered Hessian determinant:

$$\mathbb{B}_3(x,y,z) = \begin{vmatrix} 0 & 0 & \frac{\partial g_1}{\partial x} & \frac{\partial g_1}{\partial y} & \frac{\partial g_1}{\partial z} \\ 0 & 0 & -\frac{\partial g_2}{\partial x} & -\frac{\partial g_2}{\partial y} & -\frac{\partial g_2}{\partial z} \\ \frac{\partial g_1}{\partial x} & -\frac{\partial g_2}{\partial x} & \mathcal{L}_{xx} & \mathcal{L}_{xy} & \mathcal{L}_{xz} \\ \frac{\partial g_1}{\partial y} & -\frac{\partial g_2}{\partial y} & \mathcal{L}_{yx} & \mathcal{L}_{yy} & \mathcal{L}_{yz} \\ \frac{\partial g_1}{\partial z} & -\frac{\partial g_2}{\partial z} & \mathcal{L}_{zx} & \mathcal{L}_{zy} & \mathcal{L}_{zz} \end{vmatrix}$$

$$= \begin{vmatrix} 0 & 0 & 2x & 2y & 2z \\ 0 & 0 & -1 & 1 & 1 \\ 2x & -1 & -2\lambda_1 & 0 & 0 \\ 2y & 1 & 0 & -2\lambda_1 & 0 \\ 2z & 1 & 0 & 0 & -2\lambda_1 \end{vmatrix}.$$

The first partial derivatives of g_1 and g_2 are evaluated at (x,y,z). The second partial derivatives of $\mathcal{L}$ are evaluated at $(x,y,z,\lambda_1,\lambda_2)$.

$*$ At $(1,0,0)$ with $\lambda_1 = 1$ and $\lambda_2 = 1$, we have

$$\mathbb{B}_3(1,0,0) = \begin{vmatrix} 0 & 0 & 2 & 0 & 0 \\ 0 & 0 & -1 & 1 & 1 \\ 2 & -1 & -2 & 0 & 0 \\ 0 & 1 & 0 & -2 & 0 \\ 0 & 1 & 0 & 0 & -2 \end{vmatrix} = -16 \qquad (-1)^3 \mathbb{B}_3 = 16 > 0.$$

We conclude that the point $(1,0,0)$ is a local maximum to the constrained optimization problem $(\lambda_2 \geqslant 0, \; (-1)^3 \mathbb{B}_3 > 0)$.

$**$ At $(-\frac{1}{3}, -\frac{2}{3}, -\frac{2}{3})$ with $\lambda_1 = -1$ and $\lambda_2 = -\frac{1}{3}$, we have

$$\mathbb{B}_3(-\frac{1}{3}, -\frac{2}{3}, -\frac{2}{3}) = \begin{vmatrix} 0 & 0 & -\frac{2}{3} & -\frac{4}{3} & -\frac{4}{3} \\ 0 & 0 & -1 & 1 & 1 \\ -\frac{2}{3} & -1 & 2 & 0 & 0 \\ -\frac{4}{3} & 1 & 0 & 2 & 0 \\ -\frac{4}{3} & 1 & 0 & 0 & 2 \end{vmatrix} = 16 \qquad (-1)^2 \mathbb{B}_3 = 16 > 0.$$

We conclude that the point $(-\frac{1}{3}, -\frac{2}{3}, -\frac{2}{3})$ is a local minimum to the constrained optimization problem $(\lambda_2 \leqslant 0, \; (-1)^2 \mathbb{B}_3 > 0)$.

iii) The set of the constraints is a closed bounded set of $\mathbb{R}^2$ as it is the intersection of the unit sphere $[g_1 = 1]$ and the region above the plane $[g_2 = 1]$. By the extreme value theorem, f attains its extreme values on this set of the constraints. Therefore, the local points found in ii) are also the global extreme points. Hence, we have

$$\max_{g_1=1, \; g_2 \geqslant 1} f = f(1,0,0) = 1 \qquad \min_{g_1=1, \; g_2 \geqslant 1} f = f(-\frac{1}{3}, -\frac{2}{3}, -\frac{2}{3}) = -\frac{5}{3}.$$

5. – Classify the candidates of the problem

$$local\max(\min) f(x,y,z) = x + y + z \quad \text{s.t} \quad \begin{cases} g_1 = x^2 + y^2 + z^2 = 1 \\[2mm] g_2 = x - y - z \leqslant 1. \end{cases}$$

Solution: i) Note that f, g_1 and g_2 are C^∞ in $\mathbb{R}^3$ and consider the Lagrangian

$$\mathcal{L}(x,y,z,\lambda_1,\lambda_2) = f(x,y,z) - \lambda_1(g_1(x,y,z) - 1) - \lambda_2(g_2(x,y,z) - 1)$$
$$= x + y + z - \lambda_1(x^2 + y^2 + z^2 - 1) - \lambda_2(x - y - z - 1).$$

We look for the solutions of the system

$$\begin{cases} (1) \quad \mathcal{L}_x = 1 - 2x\lambda_1 - \lambda_2 = 0 \\[2mm] (2) \quad \mathcal{L}_y = 1 - 2y\lambda_1 + \lambda_2 = 0 \\[2mm] (3) \quad \mathcal{L}_z = 1 - 2z\lambda_1 + \lambda_2 = 0 \end{cases} \qquad \begin{cases} (4) \quad \mathcal{L}_{\lambda_1} = -(x^2 + y^2 + z^2 - 1) = 0 \\[2mm] (5) \quad \lambda_2 = 0 \quad \text{if} \quad x - y - z < 1. \end{cases}$$

From the first three equations, we deduce that

$$\lambda_2 = 1 - 2x\lambda_1 \qquad \lambda_1(x+y) = 1 \qquad \lambda_1(x+z) = 1.$$

Note that $\lambda_1 = 0$ is not possible because we would have from (1) : $\lambda_2 = -1$, and from (2) : $\lambda_2 = 1$. So $\lambda_1 \neq 0$ and we have

$$x + y = x + z = \frac{1}{\lambda_1} \qquad \Longrightarrow \qquad y = z.$$

$*$ If $x - y - z < 1$, then $\lambda_2 = 0$. We deduce that $\dfrac{1}{\lambda_1} = 2x$, thus $x + y = x + z = \dfrac{1}{\lambda_1} = 2x$. So $x = y = z$, which inserted into (4) gives $3x^2 = 1$. Hence, we have two solutions

$$(x, y, z) = (\frac{1}{\sqrt{3}}, \frac{1}{\sqrt{3}}, \frac{1}{\sqrt{3}}) \qquad \text{with} \qquad \lambda_1 = \frac{\sqrt{3}}{2}, \qquad \lambda_2 = 0$$

$$(x, y, z) = (-\frac{1}{\sqrt{3}}, -\frac{1}{\sqrt{3}}, -\frac{1}{\sqrt{3}}) \qquad \text{with} \qquad \lambda_1 = -\frac{\sqrt{3}}{2}, \qquad \lambda_2 = 0.$$

$*$ If $x - y - z = 1$, then with $y = z$ and (4), we have

$$\begin{cases} x = 1 + 2y \\ 2y(3y + 2) = 0 \end{cases} \qquad \Longleftrightarrow \qquad (x, y) = (1, 0) \quad \text{or} \quad (x, y) = (-\frac{1}{3}, -\frac{2}{3}).$$

We deduce then

$$(x, y, z) = (1, 0, 0) \qquad \text{with} \qquad \lambda_1 = 1, \qquad \lambda_2 = -1$$

$$(x, y, z) = (-\frac{1}{3}, -\frac{2}{3}, -\frac{2}{3}) \qquad \text{with} \qquad \lambda_1 = -1, \qquad \lambda_2 = \frac{1}{3}.$$

ii) *Regularity of the points.* We have

$$g'(x, y, z) = \begin{bmatrix} \frac{\partial g_1}{\partial x} & \frac{\partial g_1}{\partial y} & \frac{\partial g_1}{\partial z} \\ \frac{\partial g_2}{\partial x} & \frac{\partial g_2}{\partial y} & \frac{\partial g_2}{\partial z} \end{bmatrix} = \begin{bmatrix} 2x & 2y & 2z \\ 1 & -1 & -1 \end{bmatrix}$$

$$g'_2(\frac{1}{\sqrt{3}}, \frac{1}{\sqrt{3}}, \frac{1}{\sqrt{3}}) = \begin{bmatrix} \frac{2}{\sqrt{3}} & \frac{2}{\sqrt{3}} & \frac{2}{\sqrt{3}} \end{bmatrix} = -g'_2(-\frac{1}{\sqrt{3}}, -\frac{1}{\sqrt{3}}, -\frac{1}{\sqrt{3}})$$

$$rank(g_2'(\frac{1}{\sqrt3},\frac{1}{\sqrt3},\frac{1}{\sqrt3})) = rank(g_2'(-\frac{1}{\sqrt3},-\frac{1}{\sqrt3},-\frac{1}{\sqrt3})) = 1.$$

$$g'(1,0,0) = \begin{bmatrix} 2 & 0 & 0 \\ 1 & -1 & -1 \end{bmatrix} \qquad g'(-\frac{1}{3},-\frac{2}{3},-\frac{2}{3}) = \begin{bmatrix} -\frac{2}{3} & -\frac{4}{3} & -\frac{4}{3} \\ 1 & -1 & -1 \end{bmatrix}.$$

$$rank(g'(1,0,0)) = rank(g'(-\frac{1}{3},-\frac{2}{3},-\frac{2}{3})) = 2.$$

The four points are regular. Moreover, we will not have to renumber the variables since the first two column vectors of each derivative above are linearly independent.

iii) *Classification of the points* $(\pm\frac{1}{\sqrt3},\pm\frac{1}{\sqrt3},\pm\frac{1}{\sqrt3})$.

Here $n = 3$, $p = 1$, thus we have to consider the sign of the following bordered Hessian determinants:

$$\mathbb{B}_2 = \begin{vmatrix} 0 & 2x & 2y \\ 2x & -2\lambda_1 & 0 \\ 2y & 0 & -2\lambda_1 \end{vmatrix} \qquad \mathbb{B}_3 = \begin{vmatrix} 0 & 2x & 2y & 2z \\ 2x & -2\lambda_1 & 0 & 0 \\ 2y & 0 & -2\lambda_1 & 0 \\ 2z & 0 & 0 & -2\lambda_1 \end{vmatrix}.$$

We have

$$\mathbb{B}_2(\frac{1}{\sqrt3},\frac{1}{\sqrt3},\frac{1}{\sqrt3}) = \frac{8}{\sqrt3} \qquad \mathbb{B}_3(\frac{1}{\sqrt3},\frac{1}{\sqrt3},\frac{1}{\sqrt3}) = -12$$

$$(-1)^r\mathbb{B}_r(\frac{1}{\sqrt3},\frac{1}{\sqrt3},\frac{1}{\sqrt3}) > 0 \qquad r = 2,3.$$

Thus, the point is a local maximum since $\lambda_1 = 1 > 0$ and $(-1)^2\mathbb{B}_2 > 0$, $(-1)^3\mathbb{B}_3 > 0$.

$$\mathbb{B}_2(-\frac{1}{\sqrt3},-\frac{1}{\sqrt3},-\frac{1}{\sqrt3}) = -\frac{8}{\sqrt3} \qquad \mathbb{B}_3(-\frac{1}{\sqrt3},-\frac{1}{\sqrt3},-\frac{1}{\sqrt3}) = -12$$

$$(-1)^1\mathbb{B}_r(-\frac{1}{\sqrt3},-\frac{1}{\sqrt3},-\frac{1}{\sqrt3}) > 0 \qquad r = 2,3$$

Thus, the point is a local minimum since $\lambda_1 = -1 < 0$ and $(-1)^1\mathbb{B}_2 > 0$, $(-1)^1\mathbb{B}_3 > 0$.

iv) *Classification of the points* $(1,0,0)$, $(-\frac{1}{3},-\frac{2}{3},-\frac{2}{3})$.

Here $n = 3$, $p = 2$, thus we have to consider the sign of the following bordered Hessian determinant:

$$\mathbb{B}_3(x,y,z) = \begin{vmatrix} 0 & 0 & 2x & 2y & 2z \\ 0 & 0 & 1 & -1 & -1 \\ 2x & 1 & -2\lambda_1 & 0 & 0 \\ 2y & -1 & 0 & -2\lambda_1 & 0 \\ 2z & -1 & 0 & 0 & -2\lambda_1 \end{vmatrix}.$$

$*$ At $(1,0,0)$ with $\lambda_1 = 1$ and $\lambda_2 = -1$, we have $\mathbb{B}_3(1,0,0) = -16$. We conclude that the point cannot be a local maximum because $\lambda_2 = -1 \not\geq 0$. It cannot also be a local minimum because the Hessian is not semi definite positive at the point on the tangent plane

$$M = \left\{ \begin{bmatrix} h \\ k \\ l \end{bmatrix} : \begin{bmatrix} 2 & 0 & 0 \\ 1 & -1 & -1 \end{bmatrix} \begin{bmatrix} h \\ k \\ l \end{bmatrix} = \begin{bmatrix} 0 \\ 0 \end{bmatrix} \right\} = \left\{ k \begin{bmatrix} 0 \\ 1 \\ -1 \end{bmatrix} : k \in \mathbb{R} \right\}$$

$$\begin{bmatrix} 0 & k & -k \end{bmatrix} \begin{bmatrix} -2 & 0 & 0 \\ 0 & -2 & 0 \\ 0 & 0 & -2 \end{bmatrix} \begin{bmatrix} 0 \\ k \\ -k \end{bmatrix} = -4k^2 \leqslant 0 \quad \text{on } M.$$

$**$ At $(-\frac{1}{3}, -\frac{2}{3}, -\frac{2}{3})$ with $\lambda_1 = -1$ and $\lambda_2 = \frac{1}{3}$, we have $\mathbb{B}_3(-\frac{1}{3}, -\frac{2}{3}, -\frac{2}{3}) = 16$. We conclude that the point cannot be a local minimum because $\lambda_2 = 1/3 \not\leq 0$. It cannot also be a local maximum because the Hessian is not semi definite negative at the point on the tangent plane

$$M = \left\{ \begin{bmatrix} h \\ k \\ l \end{bmatrix} : \begin{bmatrix} -\frac{2}{3} & -\frac{4}{3} & -\frac{4}{3} \\ 1 & -1 & -1 \end{bmatrix} \begin{bmatrix} h \\ k \\ l \end{bmatrix} = \begin{bmatrix} 0 \\ 0 \end{bmatrix} \right\} = \left\{ k \begin{bmatrix} 0 \\ 1 \\ -1 \end{bmatrix} : k \in \mathbb{R} \right\}$$

$$\begin{bmatrix} 0 & k & -k \end{bmatrix} \begin{bmatrix} 2 & 0 & 0 \\ 0 & 2 & 0 \\ 0 & 0 & 2 \end{bmatrix} \begin{bmatrix} 0 \\ k \\ -k \end{bmatrix} = 4k^2 \geqslant 0 \quad \text{on } M.$$

v) The set of the constraints is a closed bounded set of $\mathbb{R}^2$ as it is the intersection of the unit sphere $[g_1 = 1]$ and the region below the plane $[g_2 = 1]$. By the extreme value theorem, f attains its extreme values on this set of the constraints. Hence, we have

$$\max_{g_1=1,\, g_2\leqslant 1} f(x,y,z) = \sqrt{3} \qquad \text{and} \qquad \min_{g_1=1,\, g_2\leqslant 1} f(x,y,z) = -\sqrt{3}.$$

4.4 Global Extreme Points-Inequality Constraints

When the Lagrangian is concave/convex on a convex constraint set, a solution of the Karush-Kuhn-Tucker conditions is a global maximum/minimum point.

Theorem 4.4.1 *Let $\Omega \subset \mathbb{R}^n$, Ω be an open set and $f, g_1, \ldots, g_m : \Omega \longrightarrow \mathbb{R}$ be C^1 functions. Let $S \subset \Omega$ be convex, $x^* \in \overset{\circ}{S}$ and*

$$\mathcal{L}(x, \lambda) = f(x) - \lambda_1(g_1(x) - b_1) - \ldots - \lambda_m(g_m(x) - b_m)$$

$$\exists \lambda^* = \langle \lambda_1^*, \ldots, \lambda_m^* \rangle : \quad \nabla_x \mathcal{L}(x^*, \lambda^*) = 0$$

$$\lambda_j^* = 0 \quad if \quad g_j(x^*) < b_j \quad j = 1, \ldots, m.$$

Then, we have

$$\lambda^* \geqslant 0 \quad and \quad \mathcal{L}(., \lambda^*) \text{ is concave in } x \in S \implies f(x^*) = \max_{S \cap \{x \in \Omega : \, g(x) \leqslant b\}} f(x)$$

$$\lambda^* \leqslant 0 \quad and \quad \mathcal{L}(., \lambda^*) \text{ is convex in } x \in S \implies f(x^*) = \min_{S \cap \{x \in \Omega : \, g(x) \leqslant b\}} f(x)$$

Proof. i) *First implication.* The point x^* is a critical point for the Lagrangian $\mathcal{L}(., \lambda^*)$ ($\nabla_x \mathcal{L}(x^*, \lambda^*) = 0$) and $\mathcal{L}(., \lambda^*)$ is concave on the convex set S, then x^* is a global maximum for $\mathcal{L}(., \lambda^*)$ on S (by Theorem 2.3.4). Thus, we have

$$\mathcal{L}(x^*, \lambda^*) = f(x^*) - \lambda_1^*(g_1(x^*) - b_1) - \ldots - \lambda_m^*(g_m(x^*) - b_m)$$
$$\geqslant f(x) - \lambda_1^*(g_1(x) - b_1) - \ldots - \lambda_m^*(g_m(x) - b_m) = \mathcal{L}(x, \lambda^*) \qquad \forall x \in S.$$

At x^*, we have $\quad \lambda_j^* \geqslant 0, \quad$ with $\quad \lambda_j^* = 0 \quad$ if $\quad g_j(x^*) < b_j \quad j = 1, \ldots, m$ so

$$-\lambda_j^*(g_j(x^*) - b_j) = 0 \qquad j = 1, \ldots, m,$$

and, the previous inequality reduces to

$$\mathcal{L}(x^*, \lambda^*) = f(x^*) \geqslant f(x) - \lambda_1^*(g_1(x) - b_1) - \ldots - \lambda_m^*(g_m(x) - b_m) = \mathcal{L}(x, \lambda^*).$$

For each $j = 1, \ldots, m$, we also have , $\lambda_j^* \geqslant 0$ and $g_j(x) - b_j \leqslant 0$, then $-\lambda_j^*(g_j(x) - b_j) \geqslant 0$. Therefore,

$$\mathcal{L}(x^*, \lambda^*) = f(x^*) \geqslant \mathcal{L}(x, \lambda^*) \geqslant f(x) \qquad \forall x \in S \cap \{x \in \Omega : g(x) \leqslant b\}.$$

Hence x^* solves the constrained problem.

ii) *Second implication.* This part can be deduced similarly. Moreover, it suggests, for example, when looking for candidates for a maximization problem that we keep the points with negative Lagrange multipliers and see if they are global minima points without maximizing $(-f)$ and introducing another Lagrangian.

Example 1. Solve the problem

$$\min(\max) f(x, y, z) = x^2 + y^2 + z^2 \quad \text{s.t} \quad g(x, y, z) = x - 2z \leqslant -5.$$

Solution: Form the Lagrangian using the C^∞ functions f and g on $\mathbb{R}^3$:

$$\mathcal{L}(x, y, z, \lambda) = x^2 + y^2 + z^2 - \lambda(x - 2z + 5)$$

Let us solve the system

$$\begin{cases} (i) & \mathcal{L}_x = 2x - \lambda = 0 \\[2mm] (ii) & \mathcal{L}_y = 2y = 0 \\[2mm] (iii) & \mathcal{L}_z = 2z + 2\lambda = 0 \\[2mm] (iv) & \lambda = 0 \quad \text{if} \quad x - 2z + 5 < 0. \end{cases}$$

∗ If $x - 2z + 5 < 0$, then $\lambda = 0$. From the equations $(i), (ii)$ and (iii), we deduce that $(x, y, z) = (0, 0, 0)$. But, then the inequality $x - 2z + 5 < 0$ is not satisfied.

∗∗ If $x - 2z + 5 = 0$, then using (i), (ii) and (iii), we obtain

$$\lambda = 2x = -z, \quad y = 0, \quad x - 2z + 5 = 0 \iff (x, y, z) = (-1, 0, 2) \text{ with } \lambda = -2$$

which is the only candidate point for maximality.

Now, we study the convexity/concavity of $\mathcal{L}$ in (x, y, z) when $\lambda = -2$. We have

$$H_{\mathcal{L}(.,-2)}(x, y, z) = \begin{bmatrix} 2 & 0 & 0 \\ 0 & 2 & 0 \\ 0 & 0 & 2 \end{bmatrix}$$

The leading principal minors are such that: $\quad \forall (x, y, z) \in \mathbb{R}^3$,

$$D_1(x, y, z) = 2 > 0, \qquad D_2(x, y, z) = 4 > 0, \qquad D_3(x, y, z) = 8 > 0.$$

Hence, $\mathcal{L}(., -2)$ is strictly convex in (x, y, z), and we conclude that the point $(-1, 0, 2)$ is the solution to the constrained manimization problem.

The maximization problem doesn't have a solution, since there is only one solution to the system and it is a global minimum point.

Interpretation. The problem looks for the shortest and farthest distance of the origin to the space region located below the plan $x - 2z + 5 = 0$. The shortest distance is attained on the plane.

Remark 4.4.1 *The rank condition, at the point x^*, is not assumed in the theorem. The proof uses the characterization of a C^1 convex function on a convex set only.*

Example 2. In Example 4, Section 4.2, the point $(1, 1)$ doesn't satisfy the rank condition. It solves the KKT conditions related to the problem with linear constraints:

$$\max F(x, y) = \ln x + \ln y \qquad \text{subject to}$$

$$2x + y \leqslant 3, \qquad x + 2y \leqslant 3 \quad \text{and} \quad x + y \leqslant 2 \quad \text{with} \quad x > 0, \quad y > 0.$$

Use concavity to show that $(1, 1)$ solves the problem.

Solution: i) With the Lagrangian

$$\mathcal{L}(x, y, \lambda_1, \lambda_2, \lambda_3) = \frac{1}{2} \ln x + \frac{1}{4} \ln y - \lambda_1(2x+y-3) - \lambda_2(x+2y-3) - \lambda_3(x+y-2),$$

the Hessian with respect to (x, y) is

$$H_{\mathcal{L}(.,\lambda_1,\lambda_2,\lambda_3)}(x, y) = \begin{bmatrix} -\dfrac{1}{x^2} & 0 \\ 0 & -\dfrac{1}{y^2} \end{bmatrix}$$

is strictly definite negative since the leading principal minors are such that

$$D_1(x,y) = -\frac{1}{x^2} < 0, \qquad D_2(x,y) = \frac{1}{x^2 y^2} > 0$$

for $(x,y) \in \Omega = (0,+\infty) \times (0,+\infty)$. So the Lagrangian is strictly concave in $(x,y) \in \Omega$, and $(1,1)$ is the maximum point.

Remark 4.4.2 *The concavity/convexity hypothesis is a sufficient condition. We may have a global extreme point with a Lagrangian that is neither concave nor convex (see Exercise 3).*

Example 3. In Exercise 2, Section 4.3, the points

$$(\sqrt{3}, \frac{4}{\sqrt{3}}) \quad \text{with} \quad \lambda = -2\sqrt{3} \quad \text{and} \quad (-\sqrt{3}, -\frac{4}{\sqrt{3}}) \quad \text{with} \quad \lambda = 2\sqrt{3}$$

solve respectively the local min and local max problems

$$local \max (\min) f(x,y) = x^2 y + 3y - 4 \qquad \text{s.t} \qquad g(x,y) = 4 - xy \leqslant 0.$$

Are there global extreme points?

Solution: i) Let us explore the concavity and convexity of $\mathcal{L}$ with respect to (x,y)

$$\mathcal{L}(x,y,\lambda) = x^2 y + 3y - 4 + \lambda(xy - 4)$$

The Hessian matrix of $\mathcal{L}$ in (x,y) is

$$\mathcal{H}_{\mathcal{L}} = \begin{bmatrix} \mathcal{L}_{xx} & \mathcal{L}_{xy} \\ \mathcal{L}_{yx} & \mathcal{L}_{yy} \end{bmatrix} = \begin{bmatrix} 2y & 2x+\lambda \\ 2x+\lambda & 0 \end{bmatrix}$$

When $\lambda = 2\sqrt{3}$, the principal minors are

$$\Delta_1^1 = \mathcal{L}_{yy} = 0, \qquad \Delta_1^2 = \mathcal{L}_{xx} = 2y \qquad \text{and} \qquad \Delta_2 = -(2x+\lambda)^2.$$

So $\mathcal{L}$ is neither concave nor convex in $(x,y) \in \mathbb{R}^2$.

Similarly, when $\lambda = -2\sqrt{3}$ the principal minors are

$$\Delta_1^1 = \mathcal{L}_{yy} = 0, \qquad \Delta_1^2 = \mathcal{L}_{xx} = 2y \qquad \text{and} \qquad \Delta_2 = -(2x - 2\sqrt{3})^2,$$

and $\mathcal{L}$ is neither concave nor convex in (x,y).

Therefore, we cannot use this sufficient condition to conclude anything about the global optimality of the candidate points.

ii) Note that, on the boundary of the constraint set $[g \geqslant 4]$, we have $y = 4/x$ and f takes the values

$$f(x, \frac{4}{x}) = 4x + \frac{12}{x} - 4$$

and

$$\lim_{x \to +\infty} f(x, \frac{4}{x}) = +\infty \qquad \text{and} \qquad \lim_{x \to -\infty} f(x, \frac{4}{x}) = -\infty.$$

Hence f doesn't attain an absolute maximum nor an absolute minimum value on the constraint set.

Remark. Note that

$$f(\sqrt{3}, 4/\sqrt{3}) = \frac{24}{\sqrt{3}} - 4 \qquad f(-\sqrt{3}, -4/\sqrt{3}) = -\frac{24}{\sqrt{3}} - 4.$$

With $f(\sqrt{3}, 4/\sqrt{3}) > f(-\sqrt{3}, -4/\sqrt{3})$, $(\sqrt{3}, 4/\sqrt{3})$ being a local minimum and $(-\sqrt{3}, -4/\sqrt{3})$ being a local maximum, we can see that $(\sqrt{3}, 4/\sqrt{3})$ cannot be a global minimum and $(-\sqrt{3}, -4/\sqrt{3})$ cannot be a global maximum. A constrained global extreme point would be a local one since any point of the set of the constraints $g = 4$ is an interior point and regular.

Example 4. Quadratic programming. The general quadratic program (QP) can be formulated as

$$\min \frac{1}{2} \, {}^t x Q x + {}^t x.d \qquad \text{s.t} \qquad Ax \leqslant b$$

where Q is a symmetric $n \times n$ matrix, $d \in \mathbb{R}^n$, $b \in \mathbb{R}^m$ and A an $m \times n$ matrix.

Introduce the Lagrangian

$$\mathcal{L}(x, \lambda) = -(\frac{1}{2} \, {}^t x Q x + {}^t x.d) - \lambda(Ax - b)$$

and write the KKT conditions

$$\begin{cases} \nabla_x \mathcal{L} = -Qx - d - {}^t A\lambda = 0 \\ \\ \lambda_i \geqslant 0 \qquad \text{with} \qquad \lambda_i = 0 \quad \text{if} \quad (Ax)_i < b_i. \end{cases}$$

If (x^*, λ^*) is a solution of the KKT conditions, and x^* is a candidate point where p constraints are active $(Ax)_{i_k} = b_{i_k}$, $k = 1, \ldots, p$, then the second derivatives test at the point shows whether the point is a solution or not since the $H_{\mathcal{L}}(x, \lambda^*) = Q$ is constant and the constraints are linear. Thus the positivity of the Hessian subject to these constraints is equivalent to test the bordered determinants formed from the matrix

$$\begin{bmatrix} 0 & A_p \\ {}^t A_p & Q \end{bmatrix} \qquad A_p = \begin{bmatrix} {}^t a_{i_1} \\ \vdots \\ {}^t a_{i_p} \end{bmatrix} \qquad {}^t a_{i_k} \text{ is the } i_k{}^{eme} \text{ row vector of } A$$

Remark 4.4.3 ** To sum up, solving an unconstrained or constrained optimization problem leads to solving a nonlinear system $F(x, \lambda) = 0$ that appears in different forms*

no constraints

$$f'(x) = 0$$

$$\downarrow$$

$$\mathbf{F(x, \lambda) = 0}$$

$$\nearrow \qquad\qquad\qquad \nwarrow$$

equality constraints   **inequality constraints**

$$\nabla_{x,\lambda} \mathcal{L}(x, \lambda) = 0 \qquad\qquad \left(\nabla_x \mathcal{L}(x, \lambda) = 0, \quad \lambda.(g(x) - b) = 0 \right)$$

On the other hand, solving a nonlinear equation is not easy even when F is a polynomial of degree 3 of one variable.

** The importance of the theorems studied comes from

- locating the possible candidates

- showing how to compare the values of f along the feasible directions.

These two points are the start for the development of numerical methods for approaching the solution with accuracy (see [17], [19], [8], [4]).

*** The proofs we studied for optimization problems in the Euclidean space constitute a natural step to more complex ones developed in calculus of variation where the maximum and minimum are searched in a class of functions and where the objective function is a function defined on that class (see [16], [6], [9]).

Solved Problems

1. – *Distance to an hyperplane.* Let $a \in \mathbb{R}^n$, $a \neq 0$, $b \in \mathbb{R}$ $b > 0$.

i) Solve

$$\min \ \|x\|^2 \qquad \text{subject to} \qquad {}^t a.x \geqslant b.$$

ii) Deduce the solution to the following problems.

$$\alpha) \quad \min_{-x+y \geqslant 2} \ \sqrt{5 + x^2 + y^2} \qquad \beta) \quad \max_{2x-y+2z \leqslant -1} \ -6x^2 - 6y^2 - 6z^2 + 4$$

Solution: i) Let ${}^t a = \begin{bmatrix} a_1 & \dots & a_n \end{bmatrix}$, ${}^t x = \begin{bmatrix} x_1 & \dots & x_n \end{bmatrix}$. The minimization problem looks for points in the region above the hyperplane

$${}^t a.x = b \qquad \Longleftrightarrow \qquad a_1 x_1 + a_2 x_2 + \dots + a_n x_n = b$$

that are closest to the origin. It is a nonlinear minimization problem with inequality constraints. We introduce the Lagrangian

$$\mathcal{L}(x, \lambda) = -(x_1^2 + x_2^2 + \dots + x_n^2) + \lambda(a_1 x_1 + a_2 x_2 + \dots + a_n x_n - b)$$

and write the KKT conditions

$$
\begin{cases}
\mathcal{L}_{x_1} = -2x_1 + \lambda a_1 = 0 \\[2mm]
\quad \vdots \\[2mm]
\mathcal{L}_{x_n} = -2x_n + \lambda a_n = 0 \\[2mm]
\lambda \geqslant 0 \qquad \text{with} \qquad \lambda = 0 \quad \text{if} \quad a_1 x_1 + a_2 x_2 + \dots + a_n x_n > b.
\end{cases}
$$

Finding a candidate.

* If $a_1x_1 + a_2x_2 + \ldots + a_nx_n > b$, then $\lambda = 0$. We get $x_1 = \ldots = x_n = 0$. But then, we have a contradiction with $a_1(0) + \ldots + a_n(0) = 0 \geqslant b$.

* If $a_1x_1 + a_2x_2 + \ldots + a_nx_n = b$, then

$$x_i = \frac{\lambda}{2}a_i, \qquad i = 1, \ldots, n$$

which inserted in the equation of the hyperplane, we obtain

$$a_1\left(\frac{\lambda}{2}a_1\right) + a_2\left(\frac{\lambda}{2}a_2\right) + \ldots + a_n\left(\frac{\lambda}{2}a_n\right) = b \qquad \Longleftrightarrow \qquad \left(\frac{\lambda}{2}\right) = \frac{b}{\|a\|^2}.$$

Hence, a solution to the system is

$$x_i = \frac{b}{\|a\|^2}a_i, \qquad i = 1, \ldots, n \qquad \Longleftrightarrow \qquad x = \frac{b}{\|a\|^2}a.$$

Finding the solution.

To study the concavity of $\mathcal{L}$ in x when $\lambda = 2\dfrac{b}{\|a\|^2}$, consider the Hessian matrix

$$H_{\mathcal{L}(.,\lambda)}(x) = \begin{bmatrix} \mathcal{L}_{x_1x_1} & \cdots & \mathcal{L}_{x_1x_n} \\ \vdots & \vdots & \vdots \\ \mathcal{L}_{x_nx_1} & \cdots & \mathcal{L}_{x_nx_n} \end{bmatrix} = \begin{bmatrix} -2 & \cdots & 0 \\ \vdots & \ddots & \vdots \\ 0 & \cdots & -2 \end{bmatrix}$$

The leading minor principals are equal to $D_k(x) = (-2)^k$, $k = 1, \ldots, n$.

The matrix is semi-definite negative. Thus, the point maximizes $-\|x\|^2$ subject to the constraint ${}^t ax \geqslant b$. Hence, the point solves the minimization problem and the minimal distance of the origin to this point is equal to

$$\left\| \frac{b}{\|a\|^2}a \right\| = \frac{b}{\|a\|}.$$

ii) α) Note that

$$\min_{-x+y\geqslant 2} \sqrt{5 + x^2 + y^2} = \sqrt{\left(\min_{-x+y\geqslant 2} 5 + x^2 + y^2\right)} = \sqrt{\left(5 + \min_{-x+y\geqslant 2} x^2 + y^2\right)}.$$

Moreover, we have

$$\min_{-x+y\geqslant 2} x^2 + y^2 = \min_{\begin{bmatrix} -1 & 1 \end{bmatrix}\cdot\begin{bmatrix} x \\ y \end{bmatrix}\geqslant 2} \|(x,y)\|^2.$$

Thus

$$\min_{-x+y \geqslant 2} x^2 + y^2 = \frac{2}{\left\| \begin{bmatrix} -1 \\ 1 \end{bmatrix} \right\|} = \frac{2}{\sqrt{2}} = \sqrt{2}$$

and is attained at $(x^*, y^*) = (-1, 1)$. Hence

$$\min_{-x+y \geqslant 2} \sqrt{5 + x^2 + y^2} = \sqrt{5 + \sqrt{2}}.$$

β) We have

$$\max_{2x-y+2z \leqslant -1} -6x^2 - 6y^2 - 6z^2 + 4 = 4 - 6 \left(\min_{-2x+y-2z \geqslant 1} x^2 + y^2 + z^2 \right)$$

Thus

$$\min_{-2x+y-2z \geqslant 1} x^2 + y^2 + z^2 = \min_{\begin{bmatrix} -2 & 1 & -2 \end{bmatrix} \cdot \begin{bmatrix} x \\ y \\ z \end{bmatrix} \geqslant 1} \|(x, y, z)\|^2,$$

$$\max_{2x-y+2z \leqslant -1} -6x^2 - 6y^2 - 6z^2 + 4 = 4 - \frac{6}{\left\| \begin{bmatrix} -2 \\ 1 \\ -2 \end{bmatrix} \right\|} = 4 - \frac{6}{\sqrt{9}} = 2,$$

and is attained at $(x^*, y^*, z^*) = \dfrac{1}{9}(-2, 1, -2)$.

2. – *Distance to an hyperplane with positive constraints.*

i) Let $a \in \mathbb{R}^n$, $a \neq 0$, $b \in \mathbb{R}$ $b > 0$. Solve

$$\min \|x\|^2 \qquad \text{subject to} \qquad \begin{cases} {}^t a.x \geqslant b \\ \\ x \geqslant 0 \end{cases}$$

ii) Minimize $x^2 + y^2$ over the following sets

α) $-y \leqslant -2$, $x \geqslant 0$ $\qquad\qquad$ β) $x - y \geqslant 2$, $x \geqslant 0$, $y \geqslant 0$

γ) $-x + y \geqslant 2$, $x \geqslant 0$, $y \geqslant 0$ $\qquad$ δ) $x + y \geqslant 2$, $x \geqslant 0$, $y \geqslant 0$

Sketch graphs to check the solution.

Solution: i) Let ${}^t a = \begin{bmatrix} a_1 & \ldots & a_n \end{bmatrix}$, ${}^t x = \begin{bmatrix} x_1 & \ldots & x_n \end{bmatrix}$. The minimization problem looks for points with positive coordinates in the region above the hyperplane

$$ {}^t a.x = b \qquad \Longleftrightarrow \qquad a_1 x_1 + a_2 x_2 + \ldots + a_n x_n = b, $$

and that are closest to the origin. It is a nonlinear minimization problem with inequality constraints. We introduce the Lagrangian

$$ \mathcal{L}(x, \lambda) = -(x_1^2 + x_2^2 + \ldots + x_n^2) + \lambda(a_1 x_1 + a_2 x_2 + \ldots + a_n x_n - b) $$

and write the KKT conditions

$$ \begin{cases} \mathcal{L}_{x_1} = -2x_1 + \lambda a_1 \leqslant 0 & (= 0 \quad \text{if } x_1 > 0) \\ \quad \vdots \\ \mathcal{L}_{x_n} = -2x_n + \lambda a_n \leqslant 0 & (= 0 \quad \text{if } x_n > 0) \\ \lambda \geqslant 0 \quad \text{with} \quad \lambda = 0 \quad \text{if} \quad a_1 x_1 + a_2 x_2 + \ldots + a_n x_n > b. \end{cases} $$

Finding a candidate.

* If $x_i = 0$ for each $i \in \{1, \ldots, n\}$, then, $a_1(0) + \ldots + a_n(0) = 0 \geqslant b > 0$, and we get a contradiction with.

* If $x_{i_0} > 0$ for some $i_0 \in \{1, \ldots, n\}$, then

$$ -2x_{i_0} + \lambda a_{i_0} = 0 \qquad \Longleftrightarrow \qquad x_{i_0} = \frac{\lambda}{2} a_{i_0} $$

then $\lambda > 0$ and $a_{i_0} > 0$. As a consequence, we have $a_1 x_1 + a_2 x_2 + \ldots + a_n x_n = b$. Suppose $x_i > 0$ for $i \in \{i_0, i_1, \ldots, i_p\}$, and $x_i = 0$ for $i \neq i_0, i_1, \ldots, i_p$. Then,

$$ \lambda a_j \leqslant 0 \quad \text{for } j \neq i_0, i_1, \ldots, i_p \qquad \Longleftrightarrow \qquad a_j \leqslant 0 \quad \text{for } j \neq i_0, i_1, \ldots, i_p $$

since $\lambda > 0$. Hence, we can write

$$ x_j = \frac{\lambda}{2} \max(a_j, 0) = \frac{\lambda}{2}(a_j)^+ \qquad \text{for } j \neq i_0, i_1, \ldots, i_p $$

and get a unified formula for the candidate point

$$ x^* = \frac{\lambda}{2} a^+ \qquad {}^t a^+ = \begin{bmatrix} a_1^+ & \ldots & a_n^+ \end{bmatrix} $$

Inserting the expression of x^* in the equation of the hyperplane, we obtain

$$a_1\left(\frac{\lambda}{2}a_1^+\right) + a_2\left(\frac{\lambda}{2}a_2^+\right) + \ldots + a_n\left(\frac{\lambda}{2}a_n +\right) = b \qquad \Longleftrightarrow \qquad \left(\frac{\lambda}{2}\right) = \frac{b}{\|a^+\|^2}.$$

Hence, a solution to the system is

$$x_i = \frac{b}{\|a^+\|^2}a_i^+, \qquad i = 1,\ldots,n \qquad \Longleftrightarrow \qquad x = \frac{b}{\|a^+\|^2}a^+.$$

Finding the solution.

To study the concavity of $\mathcal{L}$ in x when $\lambda = 2\dfrac{b}{\|a^+\|^2}$, consider the Hessian matrix

$$H_{\mathcal{L}(.,\lambda)}(x) \quad = \quad \begin{bmatrix} \mathcal{L}_{x_1 x_1} & \cdots & \mathcal{L}_{x_1 x_n} \\ \vdots & \vdots & \vdots \\ \mathcal{L}_{x_n x_1} & \cdots & \mathcal{L}_{x_n x_n} \end{bmatrix} \quad = \quad \begin{bmatrix} -2 & \cdots & 0 \\ \vdots & \ddots & \vdots \\ 0 & \cdots & -2 \end{bmatrix}$$

The leading minor principals are equal to $D_k(x) = (-2)^k$, $k = 1,\ldots,n$.

The matrix is semi-definite negative. Thus, the point maximizes $-\|x\|^2$ subject to the constraint $^t ax \geqslant b$ and to the positivity constraint $x \geqslant 0$. Hence, the point solves the minimization problem and the minimal distance of the origin to this point is equal to

$$\left\| \frac{b}{\|a^+\|^2}a^+ \right\| = \frac{b}{\|a^+\|}.$$

ii) Here, in filling Table 4.4, we have $b = 2$.

set	$^t a$	$^t a^+$	$\|a^+\|$	(x^*, y^*)
α	$(0,1)$	$(0,1)$	1	$(0,2)$
β	$(1,-1)$	$(1,0)$	1	$(2,0)$
γ	$(-1,1)$	$(0,1)$	1	$(0,2)$
δ	$(1,1)$	$(1,1)$	$\sqrt{2}$	$(\sqrt{2}, \sqrt{2})$

TABLE 4.4: Minima points for $x^2 + y^2$ on the four sets

One can easily check the minimal distance of the origin to the given sets from the graphics in Figure 4.21.

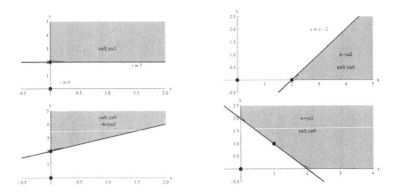

FIGURE 4.21: Closest point of the constraint set to the origin

3. – *$\mathcal{L}$ not convex nor concave* Consider the following minimization problem:

$$\min\ x^2 + y^2 \quad \text{s.t} \quad \begin{cases} y \geqslant 4 - x^2 \\[4pt] y \geqslant 3x \\[4pt] y \geqslant -3x \end{cases}$$

i) Sketch the feasible set.

ii) Write the problem as a maximization problem in the standard form, and write down the necessary KKT conditions for a point (x^*, y^*) to be a solution of the problem.

iii) Find the points that satisfy the KKT conditions. Check whether or not each point is regular.

iv) Determine whether or not the point(s) in part ii) satisfy the second-order sufficient condition.

v) Explore the concavity of the Lagrangian in $(x, y) \in \mathbb{R}^2$.

vi) What can you conclude about the solution of the problem?

vii) Give a geometric interpretation of the problem that confirms the solution you have found (Hint: use level curves).

Solution: i) The feasible set is the plane region located above the curve and the two lines, as described in Figure 4.22.

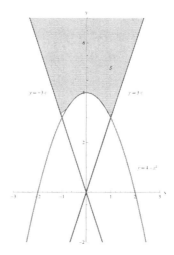

FIGURE 4.22: The constraint set S

ii) ***Writing the KKT conditions.*** The problem is equivalent to the following maximization problem

$$\max\ (-x^2 - y^2) \quad \text{subject to} \quad \begin{cases} g_1(x, y) = 4 - x^2 - y \leqslant 0 \\[2mm] g_2(x, y) = 3x - y \leqslant 0 \\[2mm] g_3(x, y) = -3x - y \leqslant 0. \end{cases}$$

Consider the Lagrangian

$$\mathcal{L}(x, y, \lambda, \beta, \gamma) = -x^2 - y^2 - \lambda(4 - x^2 - y) - \beta(3x - y) - \gamma(-3x - y).$$

The conditions are

$$\begin{cases} (1) \quad \mathcal{L}_x = -2x + 2\lambda x - 3\beta + 3\gamma = 0 \\[2mm] (2) \quad \mathcal{L}_y = -2y + \lambda + \beta + \gamma = 0 \\[2mm] (3) \quad \lambda \geqslant 0 \quad \text{with} \quad \lambda = 0 \quad \text{if} \quad 4 - x^2 - y < 0 \\[2mm] (4) \quad \beta \geqslant 0 \quad \text{with} \quad \beta = 0 \quad \text{if} \quad 3x - y < 0 \\[2mm] (5) \quad \gamma \geqslant 0 \quad \text{with} \quad \gamma = 0 \quad \text{if} \quad -3x - y < 0 \end{cases}$$

iii) *Solving the equations satisfying the KKT conditions.*

• If $4 - x^2 - y < 0$ then $\lambda = 0$ and

$$\begin{cases} -2x - 3\beta + 3\gamma = 0 \\ -2y + \beta + \gamma = 0 \end{cases}$$

then we discuss

* Suppose $3x - y < 0$, then $\beta = 0$. Thus,

$$\begin{cases} -2x + 3\gamma = 0 \\ -2y + \gamma = 0 \end{cases}$$

we get $2x = 3\gamma = 6y \Longrightarrow x = 3y$. But, then $3x - y = 3(3y) - y = 8y < 0$ and hence $\gamma < 0$ which contradicts $\gamma \geqslant 0$.

* Suppose $3x - y = 0$. We have then

$$\begin{cases} -2x - 3\beta + 3\gamma = 0 \\ -2(3x) + \beta + \gamma = 0 \end{cases} \implies 6\gamma = 20x \quad \text{and} \quad 3\beta = 8x \geqslant 0.$$

We deduce that $x \geqslant 0$ and $y \geqslant 0$. So $x = y = 0$ since $-3x - y \leqslant 0$. But, this contradicts $4 - 0^2 - 0 = 4 < 0$.

•• If $4 - x^2 - y = 0$ then

* Suppose $3x - y < 0$ then $\beta = 0$ and

$$\begin{cases} -2x + 2\lambda x + 3\gamma = 0 \\ -2y + \lambda + \gamma = 0 \end{cases}$$

− Suppose $\underline{-3x - y < 0}$ then $\gamma = 0$ and

$$\begin{cases} -2x + 2\lambda x = 0 \\ -2y + \lambda = 0 \end{cases} \iff \begin{cases} 2x(-1 + \lambda) = 0 \iff x = 0 \text{ or } \lambda = 1 \\ 2y = \lambda \geqslant 0 \end{cases}$$

○ $\lambda = 1$ leads to $y = 1/2$ and $x = \pm\sqrt{7}/2$. But, for $(x, y) = (\sqrt{7}/2, 1/2)$, the inequality $3x - y < 0$ is not satisfied, and for $(x, y) = (-\sqrt{7}/2, 1/2)$, the inequality $-3x - y < 0$ is not satisfied. So we cannot have $\lambda = 1$.

○ $x = 0$ leads to $y = 4$ and $\lambda = 8 > 0$. The two inequalities $3x - y < 0$ and $-3x - y < 0$ are satisfied at this point. Hence, the following point is a solution:

$$(x^*, y^*) = (0, 4) \quad \text{with} \quad (\lambda^*, \beta^*, \gamma^*) = (8, 0, 0) \qquad \longleftarrow$$

– Suppose $\underline{-3x - y = 0}$ then $y = -3x$ and

$$\begin{cases} -2x + 2\lambda x + 3\gamma = 0 \\ -2(-3x) + \lambda + \gamma = 0 \end{cases} \quad \Longrightarrow \quad \begin{cases} -2x + 2\lambda x + 3\gamma = 0 \\ \lambda + \gamma = -6x \end{cases}$$

From $4 - x^2 - y = 0$, we have

$$y = -3x \quad \text{and} \quad 4 - x^2 - y = 0 \quad \Longleftrightarrow \quad (x, y) = (-1, 3) \quad \text{or} \quad (4, -12).$$

The point $(4, -12)$ doesn't satisfy the inequality $3x - y < 0$, so it cannot be a solution.

The point $(-1, 3)$ satisfies the inequality $3x - y < 0$, and we have

$$\begin{cases} -2\lambda + 3\gamma = -2 \\ \lambda + \gamma = 6 \end{cases} \quad \Longrightarrow \quad (\lambda, \gamma) = (4, 2).$$

Thus, we have another candidate point:

$$(x^*, y^*) = (-1, 3) \quad \text{with} \quad (\lambda^*, \beta^*, \gamma^*) = (4, 0, 2) \qquad \longleftarrow$$

** Suppose $3x - y = 0$ then $y = 3x$. We have

$$y = 3x \quad \text{and} \quad 4 - x^2 - y = 0 \quad \Longleftrightarrow \quad (x, y) = (-4, -12) \quad \text{or} \quad (1, 3).$$

The points $(-4, -12)$ doesn't satisfy the inequality $-3x - y \leqslant 0$, so it cannot be a candidate.

The point $(1, 3)$ satisfies the inequality $-3x - y < 0$, thus $\gamma = 0$, and we have

$$\begin{cases} 2\lambda - 3\beta = 2 \\ \lambda + \beta = 6 \end{cases} \quad \Longrightarrow \quad (\lambda, \beta) = (4, 2).$$

Thus, we have another candidate point:

$$(x^*, y^*) = (1, 3) \quad \text{with} \quad (\lambda^*, \beta^*, \gamma^*) = (4, 2, 0) \qquad\qquad \longleftarrow$$

Regularity of the candidate point $(0, 4)$. Only the constraint $g_1(x, y) = 4 - x^2 - y$ is active at $(0, 4)$ and we have

$$g_1'(x, y) = \begin{bmatrix} -2x & -1 \end{bmatrix} \qquad g_1'(0, 4) = \begin{bmatrix} 0 & -1 \end{bmatrix} \qquad rank(g_1'(0, 4)) = 1.$$

Thus the point $(0, 4)$ is a regular point.

Regularity of the candidate point $(-1, 3)$. Only the constraints $g_1(x, y) = 4 - x^2 - y$ and $g_3(x, y) = -3x - y$ are active at $(-1, 3)$ and we have

$$\begin{bmatrix} g_1'(x, y) \\ g_3'(x, y) \end{bmatrix} = \begin{bmatrix} -2x & -1 \\ -3 & -1 \end{bmatrix} \qquad \begin{bmatrix} g_1'(-1, 3) \\ g_3'(-1, 3) \end{bmatrix} = \begin{bmatrix} 2 & -1 \\ -3 & -1 \end{bmatrix}.$$

Thus the point $(1, -3)$ is a regular point since $rank(\begin{bmatrix} g_1'(-1, 3) \\ g_3'(-1, 3) \end{bmatrix}) = 2.$

Regularity of the candidate point $(1, 3)$. Only the constraints $g_1(x, y) = 4 - x^2 - y$ and $g_2(x, y) = 3x - y$ are active at $(1, 3)$ and we have

$$\begin{bmatrix} g_1'(x, y) \\ g_2'(x, y) \end{bmatrix} = \begin{bmatrix} -2x & -1 \\ 3 & -1 \end{bmatrix} \qquad \begin{bmatrix} g_1'(1, 3) \\ g_2'(1, 3) \end{bmatrix} = \begin{bmatrix} -2 & -1 \\ 3 & -1 \end{bmatrix}.$$

Thus the point $(1, 3)$ is a regular point since $rank(\begin{bmatrix} g_1'(1, 3) \\ g_2'(1, 3) \end{bmatrix}) = 2.$

iv) With $p = 2$ (the number of active constraints) at the points $(3, -1)$ and $(3, 1)$, $n = 2$ (the dimension of the space), then $p = n$. The second derivatives test cannot be applied since it is established for $p < n$.

For the point $(0, 4)$, we have $p = 1 < 2 = n$. We consider the following determinant $(r = p + 1 = 2)$ (Note that the first column vector of $[g_1'(x, y)]$ is linearly dependent, so we have to renumber the variables)

$$\mathbb{B}_2(x, y) = \begin{vmatrix} 0 & \frac{\partial g_1}{\partial y} & \frac{\partial g_1}{\partial x} \\ \frac{\partial g_1}{\partial y} & \mathcal{L}_{yy} & \mathcal{L}_{yx} \\ \frac{\partial g_1}{\partial x} & \mathcal{L}_{xy} & \mathcal{L}_{xx} \end{vmatrix} = \begin{vmatrix} 0 & -1 & -2x \\ -1 & -2 & 0 \\ -2x & 0 & -2 + 2\lambda \end{vmatrix}$$

∗ At $(0,4)$, we have $\lambda = 8$,

$$\mathbb{B}_2(0,4) = \begin{vmatrix} 0 & -1 & 0 \\ -1 & -2 & 0 \\ 0 & 0 & 14 \end{vmatrix} = -14.$$

We have $(-1)^2 \mathbb{B}_2(0,4) = -14 < 0$. So the second derivatives test is not satisfied at $(0,4)$.

v) Let us explore the concavity and convexity of $\mathcal{L}$ with respect to (x,y) where the Hessian matrix of $\mathcal{L}$ in (x,y) is

$$\mathcal{H}_{\mathcal{L}} = \begin{bmatrix} \mathcal{L}_{xx} & \mathcal{L}_{xy} \\ \mathcal{L}_{yx} & \mathcal{L}_{yy} \end{bmatrix} = \begin{bmatrix} -2 & 0 \\ 0 & -2+2\lambda \end{bmatrix}$$

When $\lambda = 8$ or 4, the principal minors are

$$\Delta_1^1 = \mathcal{L}_{yy} = -2+2\lambda > 0 \qquad \Delta_1^2 = \mathcal{L}_{xx} = -2 < 0 \qquad \Delta_2 = 4(1-\lambda) < 0.$$

Therefore, $\mathcal{L}$ is neither concave, nor concave in (x,y).

vi) We have a situation where the theorems studied remain inconclusive. To conclude, we proceed by comparison. Since, the candidate points are on the boundary of the constraint set, let us study directly the values of the objective function on these points.

On the lines $y = \pm 3x$, with $|x| \geqslant 1$, the function $f(x,y) = x^2 + y^2$ takes the values

$$f(x,\pm 3x) = x^2 + (\pm 3x)^2 = 10x^2 \geqslant 10 = f(1,\pm 3) \qquad \forall\, |x| \geqslant 1.$$

On the parabola $x^2 = 4 - y$, with $|x| \leqslant 1$, we have

$$f(x, 4-x^2) = x^2 + (4-x^2)^2 = x^4 - 8x^2 + 16 + x^2 = x^4 - 7x^2 + 16 = \varphi(x)$$

$$\varphi'(x) = 4x^3 - 14x = 2x(2x^2 - 7) = 0 \iff x = 0,\ \pm\sqrt{7/2}.$$

By the extreme value theorem, φ attains its extreme values on the closed bounded interval $[-1,1]$ at the critical points inside the interval $(-1,1)$ or at the end points. Therefore, we have

$$\min_{[-1,1]} \varphi(x) = \min\{\varphi(-1), \varphi(0), \varphi(1)\} = \min\{10, 16, 10\} = 10.$$

Thus,

$$f(x, 4-x^2) = \varphi(x) \geqslant 10 = f(\pm 1, 3) \qquad \forall\, |x| \leqslant 1.$$

So we can conclude that the minimum value attained by f on the set of the constraints is 10.

vii) The feasible set is

$$S = \{(x,y): \quad 4 - x^2 - y \leqslant 0, \quad 3x - y \leqslant 0, \quad -3x - y \leqslant 0\}$$

The level curves of f, with equations : $x^2 + y^2 = k$ where $k \geqslant 0$, are circles centered at $(0,0)$ with radius $\sqrt{k}$; see Figure 4.23.
If we increase the values of the radius, the values of f increase. The value $k = 10$ is the first one at which the level curve intersects the constraints $g_1 = g_2 = 0$ and $g_1 = g_3 = 0$. Thus the value 10 is the minimal value of f reached at $(\pm 1, 3)$.

Moreover, the objective function $f(x,y) = x^2 + y^2$ is the square of the distance between (x,y) and $(0,0)$. So our problem is to find the point(s) in the feasible region that are closest to $(0,0)$.

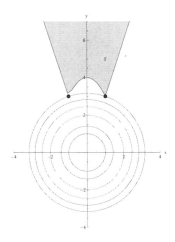

FIGURE 4.23: Level curves of f and the closest points of S to the origin

4. – The data in Table 4.5 can be found in [8]. Here we consider boundary conditions to illustrate an inequality constrained problem.

The Body Fat Index (BFI) measures the fitness of an individual. It is a function of the body density ρ (in units of kilograms per liter) according to Brozek's formula,

$$BFI = \frac{457}{\rho} - 414.2.$$

However the accurate measurement of ρ is costing. An alternative solution is to try to describe the dependence of the BFI with respect of five variables x_1, x_2, x_3, x_4, x_5 in the form

$$f : x \longmapsto BFI = y = f(x) = a_1x_1 + a_2x_2 + a_3x_3 + a_4x_4 + a_5x_5$$

The variables are easier to measure and represent

$x_1 = weight(lb.)$	$x_2 = height(in.)$	$x_3 = abdomen(cm.)$
$x_4 = wrist(cm.)$	$x_5 = neck(cm.)$	$y = BFI$

Using the following table of measurements, we assume the average of each category $\bar{x}_i$ of measurements satisfying:

$(*)$ $\qquad\qquad\qquad a_1\bar{x}_1 + a_2\bar{x}_2 + a_3\bar{x}_3 + a_4\bar{x}_4 + a_5\bar{x}_5 \leqslant \bar{y}$

hoping to find a model when $BFI \leqslant \bar{y} = 15.23$.

i)   Use a software to find a linear function f which best fits the given data, in the sense of least-squares, i.e., find a that minimizes the sum of the square errors

$$\sum_{i=1}^{10}(f(x^i) - y_i)^2 = \sum_{i=1}^{10}(a_1x_1^i + a_2x_2^i + a_3x_3^i + a_4x_4^i + a_5x_5^i - y_i)^2 \quad \text{s.t}$$

$x^i = (x_1^i, x_2^i, x_3^i, x_4^i, x_5^i)$ are the measurements for the ieme individual.

ii) Formulate the constrained problem using matrices. Use Maple to check that the Hessian of the resulting objective function is definite positive on the convex described by $(*)$.

Solution: We use Maple software for solving the problem.

i) *Finding the linear regression of best fit.*
    We solve the "least square problem" "LS" with ten linear residuals. The objective function is

$$\varphi(a_1, a_2, a_3, a_4, a_5) = \frac{1}{2}\Big((154.25a_1 + 67.75a_2 + 85.2a_3 + 17.1a_4 + 36.2a_5 - 12.7)^2$$

$+(173.25a_1 + 72.25a_2 + 83a_3 + 18.2a_4 + 38.5a_5 - 6.9)^2$
$+(154a_1 + 66.25a_2 + 87.9a_3 + 16.6a_4 + 34a_5 - 24.6)^2$
$+(184.75a_1 + 72.25a_2 + 86.4a_3 + 18.2a_4 + 37.4a_5 - 10.9)^2$
$+(184.25a_1 + 71.25a_2 + 100a_3 + 17.7a_4 + 34.4a_5 - 27.8)^2$

x_1	x_2	x_3	x_4	x_5	y
154.25	67.75	85.2	17.1	36.2	12.6
173.25	72.25	83	18.2	38.5	6.9
154	66.25	87.9	16.6	34	24.6
184.75	72.25	86.4	18.2	37.4	10.9
184.25	71.25	100	17.7	34.4	27.8
210.25	74.75	94.4	18.8	39	20.6
181	69.75	90.7	17.7	36.4	19
176	72.5	88.5	18.8	37.8	12.8
191	74	82.5	18.2	38.1	5.1
198.25	73.5	88.6	19.2	42.1	12

TABLE 4.5: Measurements involved in BFI

$+(210.25a_1 + 74.75a_2 + 94.4a_3 + 18.8a_4 + 39a_5 - 20.6)^2$
$+(181a_1 + 69.75a_2 + 90.7a_3 + 17.7a_4 + 36.4a_5 - 19)^2$
$+(176a_1 + 72.5a_2 + 88.5a_3 + 18.8a_4 + 37.8a_5 - 12.8)^2$
$+(191a_1 + 74a_2 + 82.5a_3 + 18.2a_4 + 38.1a_5 - 5.1)^2$
$+(198.25a_1 + 73.5a_2 + 88.6a_3 + 19.2a_4 + 42.1a_5 - 12)^2\big)$

$with(Optimization) : LSSolve($
$[154.25a_1 + 67.75a_2 + 85.2a_3 + 17.1a_4 + 36.2a_5 - 12.6, 173.25a_1 + 72.25a_2 + 83a_3 + 18.2a_4 + 38.5a_5 - 6.9,$
$154a_1 + 66.25a_2 + 87.9a_3 + 16.6a_4 + 34a_5 - 24.6, 184.75a_1 + 72.25a_2 + 86.4a_3 + 18.2a_4 + 37.4a_5 - 10.9,$
$184.25a_1 + 71.25a_2 + 100a_3 + 17.7a_4 + 34.4a_5 - 27.8, 210.25a_1 + 74.75a_2 + 94.4a_3 + 18.8a_4 + 39a_5 - 20.6,$
$181a_1 + 69.75a_2 + 90.7a_3 + 17.7a_4 + 36.4a_5 - 19, 176a_1 + 72.5a_2 + 88.5a_3 + 18.8a_4 + 37.8a_5 - 12.8,$
$191a_1 + 74a_2 + 82.5a_3 + 18.2a_4 + 38.1a_5 - 5.1, 198.25a_1 + 73.5a_2 + 88.6a_3 + 19.2a_4 + 42.1a_5 - 12],$
$\{180.7a_1 + 71.425a_2 + 88.72a_3 + 18.05a_4 + 37.39a_5 \le 15.23\})$
$[15.0549945448635683, [a_1 = 0.474753096134219e - 1, a_2 = -1.03634130223772,$
$a_3 = 1.22920301075594, a_4 = -1.86308283592359, a_5 = .140089140413700]]$
Thus
$$f(x_1, x_2, x_3, x_4, x_5) \approx 0.474x_1 - 1.036x_2 + 1.229x_3 - 1.863x_4 + .140x_5$$

f can be used to predict an individual's body fat index, based upon the five measurements types.

Comments. - Least square problems are solved by the LSSolve command.

- When the residuals in the objective function and the constraints are all linear, which is the case here, then an active set method is used. This is an approximate method [19],[22], [17].

-The LSSolve command uses various methods implemented in a built in library provided by a group of numerical algorithms.

ii) *Finding the linear regression of best fit using matrices.*
Let $G = (x_1^i, x_2^i, x_3^i, x_4^i, x_5^i)_{i=1,...,10} \in M_{10;5}$ be the matrix whose rows are the vectors x^i, or equivalently, the matrix whose columns are the five first columns entries of the table. Let

c be the last column entry of the table. Denote

$$a =^t (a_1, a_2, a_3, a_4, a_5), \qquad A =^t (180.7, 71.425, 88.72, 18.05, 37.39), \qquad b = 15.23$$

then

$$\varphi(a) = \varphi(a_1, a_2, a_3, a_4, a_5) = \frac{1}{2}\left(((G.a - c)_1)^2 + \ldots + ((G.a - c)_{10})^2\right) = \frac{1}{2}\|G.a - c\|^2$$

and the problem can be expressed as

$$\min \frac{1}{2}\|G.a - c\|^2 \qquad \text{subject to} \qquad Aa \leqslant b.$$

Following Maple's instructions, we enter the data using matrices

with(Optimization):
$c := Vector([12.6, 6.9, 24.6, 10.9, 27.8, 20.6, 19, 12.8, 5.1, 12],$
$datatype = float)$:
$G := Matrix([[154.25, 67.75, 85.2, 17.1, 36.2], [173.25, 72.25, 83, 18.2, 38.5],$
$[154, 66.25, 87.9, 16.6, 34], [184.75, 72.25, 86.4, 18.2, 37.4],$
$[184.25, 71.25, 100, 17.7, 34.4], [210.25, 74.75, 94.4, 18.8, 39],$
$[181, 69.75, 90.7, 17.7, 36.4], [176, 72.5, 88.5, 18.8, 37.8],$
$[191, 74, 82.5, 18.2, 38.1], [198.25, 73.5, 88.6, 19.2, 42.1]],$
$datatype = float)$:
with(Statistics) :
$A := Mean(G)$:
$b := Mean(c)$:
$A := Matrix([[180.7, 71.425, 88.72, 18.05, 37.39]], datatype = float)$:
$b := Vector([15.23], datatype = float)$:
$lc := [A, b]$:
$LSSolve([c, G], lc)$:

$$\left[15.0549945448635683, \left[\begin{array}{c} 0.0474753096134219 \\ -1.03634130223772 \\ 1.22920301075594 \\ -1.86308283592359 \\ 0.140089140413700 \end{array} \right] \right]$$

Hence, we obtain the same coefficients a_i.

The Hessian of φ is

$$\varphi(a) = \frac{1}{2}\|G.a - c\|^2 = \frac{1}{2}(G.a - c).(G.a - c)$$

$$= \frac{1}{2}(\|G.a\|^2 - 2^t c.G.a + \|c\|^2) = \frac{1}{2}(^t a^t GG.a - 2^t c.G.a + \|c\|^2)$$

$$\varphi'(a) =^t GG.a - G.c \qquad \varphi''(a) =^t GG$$

Checking that the Hessian is definite positive.
with(LinearAlgebra)
$H := Multiply(Transpose(G), G)$:
$IsDefinite(H)$
true

4.5   Dependence on Parameters

The cost to produce an output Q is equal to $rK + wL$ where r and w are respectively the prices of the input capital K and labor L. The firm would like the output to obey the Cobb-Douglas production function $Q = cK^a L^b$ ($r > 0$, $w > 0$, $c > 0$, $a + b < 1$). Thus, to minimize the cost of production, the problem is expressed as:

$$\min rK + wL \qquad \text{subject to} \qquad cK^a L^b = Q$$

with $(K, L) \in (0, +\infty) \times (0, +\infty)$. Using Lagrange's multiplier method, the unique solution is (see Example 1, Section 3.4)

$$K^* = \lambda \frac{aQ}{r} \qquad L^* = \lambda \frac{bQ}{w} \qquad \lambda^* = \left(\frac{Q}{c}\right)^{\frac{1}{a+b}} \left(\frac{aQ}{r}\right)^{\frac{a}{a+b}} \left(\frac{bQ}{w}\right)^{\frac{b}{a+b}}.$$

One can see the dependence of the extreme point on the parameters r, w, c, a, b. In general, it is not easy to express explicitly the solution with respect of many parameters. On the other hand, changing the parameters and solving a new optimization problem is costing or difficult. An alternative solution is to have an estimate on how much the optimal value changes compared to an initial situation.

To set the main result of this section, we suppose the objective function f and the constraint function g depending on a parameter $r \in \mathbb{R}^k$, i.e.

$$f(x, r) = f(x_1, \ldots, x_n, r_1, \ldots, r_k),$$
$$g(x, r) = g(x_1, \ldots, x_n, r_1, \ldots, r_k) \qquad g = (g_1, \ldots, g_n),$$
$$I(x(r)) = \{i \in \{1, \cdots, m\} : \quad g_i(x(r), r) < 0\},$$

Consider the problem (P_r)

$$f^*(r) = local \max \ f(x, r) \qquad (\text{resp. } local \min) \qquad \text{s.t} \qquad g(x, r) \leqslant 0$$

and introduce the Lagrangian

$$\mathcal{L}(x, \lambda, r) = f(x, r) - \lambda_1 g_1(x, r) - \ldots - \lambda_m g_m(x, r).$$

Hypothesis (H_r). f and g are C^2 functions in a neighborhood of x^* and for each $r \in B_\delta(\bar{r}) \subseteq \mathbb{R}^k$ such that:

$$g_i(x^*, \bar{r}) = 0 \qquad \text{if} \qquad i \in I(x^*) = \{i_1, i_2, \cdots, i_p\} \qquad p < n$$

$$\lambda_j = 0 \qquad \text{if} \qquad g_j(x^*) < b_j \qquad j \notin I(x^*)$$

$$rank(G'(x^*, \bar{r})) = p, \qquad G'(x^*, \bar{r}) = (g_{i_1}(x^*, \bar{r}), \ldots, g_{i_p}(x^*, \bar{r}))$$

$$\nabla_x \mathcal{L}(x^*, \lambda^*, \bar{r}) = 0 \qquad \text{for a unique vector} \qquad \lambda^* = (\lambda_1^*, \ldots, \lambda_m^*).$$

Theorem 4.5.1. *Assume that* (H_r) *holds and*

- $x^* = x(\bar{r})$ *solves* $(P_{\bar{r}})$

- *the second derivatives test for strict maximality with* $\lambda^* \geqslant 0$
  (resp. minimality with $\lambda^* \leqslant 0$*) is satisfied when* $r = \bar{r}$.

Then

- $\exists \eta \in (0, \delta]$ *such that* $x(.) : r \longmapsto x(r)$ *and* $\lambda(.) : r \longmapsto \lambda(r)$
  are $C^1(B_\eta(\bar{r}))$

- $f^* : r \longmapsto f(x(r), r)$ *is* C^1 *on* $B_\eta(\bar{r})$ *and*

$$\frac{\partial f^*}{\partial r_j}(r) = \frac{\partial \mathcal{L}}{\partial r_j}(x(r), \lambda(r), r) \qquad j = 1, \ldots, k.$$

Remark 4.5.1 *As a consequence of the regularity of the optimal value
function* f^*, *we have the following approximation*

$$f^*(r) \approx f^*(\bar{r}) + \sum_{j=1}^{k} \frac{\partial f^*}{\partial r_j}(\bar{r})(r_j - \bar{r}_j)$$

where

$$\frac{\partial f^*}{\partial r_j}(\bar{r}) = \left[\frac{\partial \mathcal{L}}{\partial r_j}(x, \lambda, r) \right]_{x=x^*, \ \lambda=\lambda^*, \ r=\bar{r}} \qquad j = 1, \ldots, k.$$

Thus, we can estimate the change of f^* *when the parameter* r *remains
close to* $\bar{r}$.

Proof. We write the proof for the constrained case.

Step 1.   If we assume that a C^1 regularity of $x(r)$ and $\lambda(r)$ is established, then we have

$$\mathcal{L}(x(r), \lambda(r), r) = f(x(r), r) - \lambda(r)g_1(x(r), r) - \ldots - \lambda_m(r)g_m(x(r), r)$$

$$= f(x(r), r) - \sum_{i \in I(x(r))} \lambda_i(r)g_i(x(r), r) - \sum_{i \notin I(x(r))} \lambda_i(r)g_i(x(r), r)$$

$$= f(x(r), r) = f^*(r)$$

because $g_i(x(r), r) = 0$ for $i \in I(x(r))$ and $\lambda_i(r) = 0$ for $i \notin I(x(r))$; then using the Chain rule formula, we obtain

$$\frac{\partial f^*}{\partial r_j}(r) = \frac{\partial(\mathcal{L}(x(r), \lambda(r), r))}{\partial r_j}$$

$$= \sum_{i=1}^{n} \frac{\partial \mathcal{L}}{\partial x_i}(x(r), \lambda(r), r))\frac{\partial x_i}{\partial r_j}(r) + \sum_{t=1}^{m} \frac{\partial \mathcal{L}}{\partial \lambda_t}(x(r), \lambda(r), r))\frac{\partial \lambda_t}{\partial r_j}$$

$$+ \sum_{l=1}^{k} \frac{\partial \mathcal{L}}{\partial r_l}(x(r), \lambda(r), r))\frac{\partial r_l}{\partial r_j}$$

$$= \sum_{i=1}^{n} \left[\frac{\partial f}{\partial x_i}(x(r), r) - \lambda_1(r)\frac{\partial g_1}{\partial x_i}(x(r), r) - \ldots\ldots - \lambda_m(r)\frac{\partial g_m}{\partial x_i}(x(r), r)\right]\frac{\partial x_i}{\partial r_j}(r)$$

$$+ \sum_{t=1}^{m} \left[- g_t(x(r), r)\frac{\partial \lambda_t}{\partial r_j}(r)\right] + \frac{\partial \mathcal{L}}{\partial r_j}(x(r), \lambda(r), r).$$

Since $x(r)$ optimizes $f(x, r)$ subject to the constraints $g(x, r) \leqslant 0$, then the necessary condition gives, for each $i = 1, \ldots, n$,

$$\left[\frac{\partial f}{\partial x_i}(x(r), r) - \lambda_1(r)\frac{\partial g_1}{\partial x_i}(x(r), r) - \ldots\ldots - \lambda_m(r)\frac{\partial g_m}{\partial x_i}(x(r), r)\right] = 0.$$

Now, since we have $g_t(x(r), r) = 0$ for $t \in I(x(r))$ and $\lambda_t(r) = 0$ for $t \notin I(x(r))$, then

$$\sum_{t=1}^{m} \left[- g_t(x(r), r)\frac{\partial \lambda_t}{\partial r_j}(r)\right] = 0.$$

Hence

$$\frac{\partial f^*}{\partial r_j}(r) = \frac{\partial \mathcal{L}}{\partial r_j}(x(r), \lambda(r), r).$$

Step 2.   To prove the theorem, it remains to check the definiteness of $x(r)$ and its regularity. For this, we will need the implicit function theorem recalled at the end of the proof. First, set

$$\lambda^p(r) = \big(\lambda_{i_1}(r), \ldots, \lambda_{i_p}(r)\big), \qquad\qquad \lambda^{*p}(r) = \big(\lambda^*_{i_1}(r), \ldots, \lambda^*_{i_p}(r)\big)$$

$$\mathcal{U}(x, \lambda^p(r), r) = f(x,r) - \lambda_{i_1}(r)g_{i_1}(x,r) - \ldots\ldots - \lambda_{i_p}(r)g_{i_p}(x,r)$$

$$F(x, \lambda^p, r) = \nabla_{x, \lambda^p}\mathcal{U}(x, \lambda^p, r)$$

$$= \Big(\frac{\partial f}{\partial x_1}(x,r) - \sum_{k=1}^{p}\lambda_{i_k}(r)\frac{\partial g_{i_k}}{\partial x_1}(x,r), \ldots\ldots,$$

$$\frac{\partial f}{\partial x_n}(x,r) - \sum_{k=1}^{p}\lambda_{i_k}(r)\frac{\partial g_{i_k}}{\partial x_n}(x,r), -g_{i_1}(x,r), \ldots, -g_{i_p}(x,r)\Big).$$

Consider the following equation system

$$F(x, \lambda^p, r) = 0$$

By assumption, we have

– F is C^1 function in the open set $A = \Omega \times \mathbb{R}^p \times B(\bar{r}, \delta)$ where Ω is an open neighborhood of x^*

– $F(x^*, \lambda^{*p}, \bar{r}) = 0$

– $(x^*, \lambda^{*p}, \bar{r}) \in \Omega \times \mathbb{R}^p \times B(\bar{r}, \delta)$, so $(x^*, \lambda^{*p}, \bar{r})$ is an interior point

– $det(\nabla_{x, \lambda^p} F(x^*, \lambda^{*p}, \bar{r})) = det \begin{bmatrix} (\mathcal{L}_{x_i x_j}) & -G'(x^*, \bar{r}) \\ -{}^t G'(x^*, \bar{r}) & \mathbf{0} \end{bmatrix}$

$$= (-1)^{2p} B_n(x^*, \lambda^{*p}, \bar{r}) \neq 0, \ B_n: \text{Bordered Hessian determinant.}$$

Then, by the implicit function theorem, there exists open balls

$$B_{\epsilon_1}(x^*) \subset \mathbb{R}^n, \qquad B_{\epsilon_2}(\lambda^{*p}) \subset \mathbb{R}^p, \qquad B_\eta(\bar{r}) \subset \mathbb{R}^k, \qquad \epsilon_1, \epsilon_2, \eta > 0$$

with

$$B_{\epsilon_1}(x^*) \times B_{\epsilon_2}(\lambda^{*p}) \times B_\eta(\bar{r}) \subseteq A,$$

$$det(\nabla_{x, \lambda^p} F(x, \lambda^p, r)) \neq 0 \qquad \text{in} \qquad B_{\epsilon_1}(x^*) \times B_{\epsilon_2}(\lambda^{*p}) \times B_\eta(\bar{r})$$

such that

$$\forall r \in B_\eta(\bar{r}), \quad \exists!(x, \lambda^p) \in B_{\epsilon_1}(x^*) \times B_{\epsilon_2}(\lambda^{*p}): \quad F(x, \lambda^p, r) = 0$$

$$(x, \lambda^p): \quad B_\eta(\bar{r}) \longrightarrow B_{\epsilon_1}(x^*) \times B_{\epsilon_2}(\lambda^{*p})$$

$$r \longmapsto (x(r), \lambda^p(r)) \qquad\qquad \text{are } C^1 \text{ functions.}$$

Remark 4.5.2 * *In the theorem above, the local* max(min) *problem can be replaced by the* max(min) *problem, provided we assume, for example,*

- $\forall r \in B(\bar{r}, \delta)$, $x \longmapsto \mathcal{L}(x, \lambda^*, r)$ *is strictly concave (resp. convex)*

* *For the unconstrained case, $\mathcal{L}$ is reduced to f, $F(x,r) = \nabla_x f(x,r)$ and $\det(\nabla_x F(x^*, \bar{r})) = \det H_f(x^*, \bar{r})$.*

Example 1. Suppose that when a firm produces and sells x units of a commodity, it has a revenue $R(x) = x$, while the cost is $C(x) = x^2$.

  i) Find the optimal choice of units of the commodity that maximize profit.

  ii) Find the approximate change of the optimal profit if the revenue changes to $0.99x$.

Solution: i) The profit is given by

$$P(x) = R(x) - C(x) = x - x^2 \qquad \text{with } x > 0.$$

Since the set of the constraints $S = (0, +\infty)$ is an open set and the profit function is regular, the optimal point, if it exists, is a critical point solution of the equation

$$\frac{dP}{dx} = 1 - 2x = 0 \qquad \Longleftrightarrow \qquad x = \frac{1}{2}.$$

Moreover, we have

$$\frac{d^2P}{dx^2} = -2 \qquad \text{and} \qquad \frac{d^2P}{dx^2} < 0 \quad \forall x \in S.$$

Then P is strictly concave on the convex set S. Hence, the only critical point $x = 1/2$ is a global maximum point. Thus $x^* = 1/2$ units should be produced to achieve maximum profit.

ii) Introduce the new profit function with the new revenue rx where $r > 0$:

$$P(x, r) = rx - C(x) = rx - x^2 \qquad \text{with } x > 0.$$

Proceeding as in i), one can verify that

  1. For, r close to 1, we have $\dfrac{d^2P}{dx^2}(x,r) = -2 < 0$. Thus $P(.,r)$ is concave in x.

  2. The second order condition for strict maximality is satisfied when $r = 1$.

3. $P(1/2, 1) = \max_{S} P(x, 1) = \dfrac{1}{2} - \dfrac{1}{4} = \dfrac{1}{2}$.

As a consequence,

– $\exists \eta > 0$ such that the function $P^*(r) = \max\limits_{x \in S} P(x, r)$ is defined for any $r \in (1 - \eta, 1 + \eta)$

– P^* is C^1 and

$$\frac{dP^*}{dr}(1) = \left[\frac{\partial P}{\partial r}(x, r)\right]_{x=1/2,\ r=1} = x\Big]_{x=1/2,\ r=1} = \frac{1}{4}.$$

We can write the following approximation

$$P^*(r) \approx P^*(1) + \frac{dP^*}{dr}(1)(r-1) = \frac{1}{4} + \frac{1}{2}(r-1) \qquad \text{for } r \text{ close to 1.}$$

In particular, for $r = 0.99$, the objective function P^* takes the following approximate value:

$$P^*(0.99) \approx 0.25 + 0.5(0.99 - 1) = 0.25 - 0.5(0.01) = 0.245$$

and the approximate change in the maximum value of the maximum profit function is

$$P^*(0.99) - P^*(1) \approx -0.5(0.01) = -0.005.$$

* Note that, for this example, we have easily the exact value of the objective function P^*; see Figure 4.24. Indeed, we have

$$P^*(r) = P(x^*(r), r) = P\left(\frac{r}{2}, r\right) = r\frac{r}{2} - \left(\frac{r}{2}\right)^2 = \frac{r^2}{4}$$

from which we deduce

$$P^*(0.99) = \frac{(0.99)^2}{4} = 0.245025$$

We also have the following equality

$$\frac{dP^*}{dr} = \frac{r}{2} = x^*(r) = \left[\frac{\partial P(x, r)}{\partial r}\right]_{x=x^*(r)}.$$

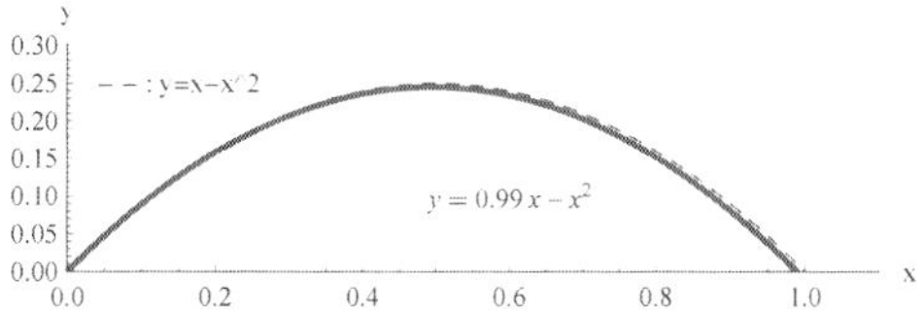

FIGURE 4.24: Highest profit for $r = 1$ and $r = 0.99$

Remark 4.5.3 *In particular, when $r = b$, $f(x,r) = f(x)$ and $g(x,r) = g(x) - b$, we have*

$$\frac{\partial f^*}{\partial b_j}(\bar{b}) = \left[\frac{\partial \mathcal{L}}{\partial b_j}(x, \lambda, b)\right]_{x=x^*,\ \lambda=\lambda^*,\ b=\bar{b}} = \lambda_j(\bar{b}) \qquad j = 1, \ldots, m.$$

This tells us that the Lagrange multiplier $\lambda_j = \lambda_j(\bar{b})$ for the jth constraint is the rate of change at which the optimal value function changes with respect of the parameter b_j at the point $\bar{b}$.

Using the linear approximation formula,

$$f^*(b) - f^*(\bar{b}) \simeq \frac{\partial f^*}{\partial b_1}(\bar{b})(b_1 - \bar{b}_1) + \cdots\cdots + \frac{\partial f^*}{\partial b_m}(\bar{b})(b_m - \bar{b}_m)$$

$$= \lambda_1(\bar{b})(b_1 - \bar{b}_1) + \cdots\cdots + \lambda_m(\bar{b})(b_m - \bar{b}_m),$$

the change in the optimal value function is estimated, when one or more components of the resource vector are slightly changed.

Example 2. For b close to 3, estimate

$$f^*(b) = local \max f(x, y, z) = xy + yz + xz \quad \text{subject to} \quad x + y + z = b$$

knowing that (see Example 2, Section 3.3)

$$f^*(3) = f(1, 1, 1) = 3, \qquad \lambda(3) = 2, \qquad (-1)^r \mathbb{B}_r(1, 1, 1) > 0 \quad \text{for} \quad r = 2, 3.$$

Solution: We can deduce that $f^* \in C^1(3 - \eta, 3 + \eta)$ for some $\eta > 0$, and write the linear approximation

$$f^*(b) \approx f^*(3) + \frac{\partial f^*}{\partial b}(3)(b - 3) \qquad \text{for } b \text{ close to 3.}$$

If we denote by

$$\mathcal{L}(x, y, z, \lambda, b) = xy + yz + xz - \lambda(x + y + z - b)$$

the Lagrangian associated with the new constrained maximization problem, then we have

$$\frac{\partial f^*}{\partial b}(3) = \frac{\partial \mathcal{L}}{\partial b}(x(b), y(b), z(b), \lambda(b), b)\Big]_{b=3} = \lambda(b)\Big]_{b=3} = 2$$

$$f^*(b) \approx 3 + 2(b - 3) \qquad \text{for } b \text{ close to 3.}$$

Solved Problems

1. – *Irregular value function.* i) Show that the value function

$$f^*(r) = \max_{x \in [-1,1]} (x - r)^2$$

is not differentiable on $\mathbb{R}$. Is there a contradiction with the theorem?

ii) Can you expect a regularity for the value function

$$g^*(r) = \min_{x \in [-1,1]} (x - r)^2.$$

Solution: This example shows that the optimal value function is not necessarily regular. Indeed, set

$$y = f(x, r) = (x - r)^2 \qquad f^*(r) = \max_{x \in [-1,1]} f(x, r).$$

We have

$$y' = \frac{dy}{dx} = f_x(x, r) = 2(x - r).$$

We distinguish different cases:

$* \ r \in (-1, 1)$: From Table 4.6, we deduce the maximum value.

x	-1		r		1
$y' = 2(x - r)$		$-$		$+$	
$y = (x - r)^2$	$(1 + r)^2$	$\searrow$	0	$\nearrow$	$(1 - r)^2$

TABLE 4.6: Variations of $y = (x - r)^2$ when $r \in (-1, 1)$

$$\max_{x \in [-1,1]} (x - r)^2 = \max\left((1 + r)^2, (1 - r)^2\right) = f^*(r).$$

x	-1		1		r
$y' = 2(x - r)$		$-$		$-$	
$y = (x - r)^2$	$(1 + r)^2$	$\searrow$	$(1 - r)^2$	$\searrow$	0

TABLE 4.7: Variations of $y = (x - r)^2$ when $r \in (1, +\infty)$

x	r		-1		1
$y' = 2(x - r)$		$+$		$+$	
$y = (x - r)^2$	0	$\nearrow$	$(1 + r)^2$	$\nearrow$	$(1 - r)^2$

TABLE 4.8: Variations of $y = (x - r)^2$ when $r \in (-\infty, -1)$

** $r \in (1, +\infty)$: Using Table 4.7, we obtain

$$\max_{x \in [-1,1]} (x - r)^2 = (1 + r)^2 = f^*(r).$$

** $r \in (-\infty, -1)$: Table 4.8 shows that

$$\max_{x \in [-1,1]} (x - r)^2 = (1 - r)^2 = f^*(r).$$

Conclusion: Note that $(1 + r)^2 - (1 - r)^2 = 4r$, then

$$f^*(r) = \begin{cases} (1 - r)^2 & \text{if} & r < 0 \\ 1 & \text{if} & r = 0 \\ (1 + r)^2 & \text{if} & r > 0 \end{cases}$$

For $r \neq 0$, f^* is differentiable since it is a polynomial. For $r = 0$, we have

$$\frac{f^*(r) - f^*(0)}{r - 0} = \begin{cases} \dfrac{(1 - r)^2 - 1}{r} = -(2 - r) & \text{if} & r < 0 \\ \dfrac{(1 + r)^2 - 1}{r} = 2 + r & \text{if} & r > 0 \end{cases}$$

Hence

$$\lim_{r \to 0^-} \frac{f^*(r) - f^*(0)}{r - 0} = -2 \qquad \lim_{r \to 0^+} \frac{f^*(r) - f^*(0)}{r - 0} = 2$$

and f^* is not differentiable at 0.

This doesn't contradicts the theorem since the regularity of f^* was proved when x^* is an interior point for f, which is not the case here with $x^* = \pm 1$. Indeed, we have $f(x,0) = x^2$ and $f^*(0) = f(\pm 1, 0) = 1$.

ii) We have $\quad f(x) = f(x,0) = x^2$,

$$\min_{x \in [-1,1]} x^2 = 0 = f(0) = g^*(0), \quad 0 \in (-1,1), \quad f''(x) = 2 > 0.$$

So f attains its minimal value at the interior point 0, where the second derivatives test is satisfied. Moreover, f is convex on $[-1,1]$, which let's 0 be the global minimum point. Therefore, for r close to 0, that is $\exists \eta > 0$ such that $g^* \in C^1(-\eta, \eta)$. In fact, from i), we have exactly $g^*(r) = 0$ for $r \in (-1,1)$, which is a regular function.

2. – Find an approximate value of

$$\max_{\mathbb{R}^2} \ (1.05)^2 x + 5y \sin(0.01) - 2x^2 - 3y^2$$

Solution: Since, we are looking for an estimate of the maximal value, we will proceed using the linear approximation for a suitable function. First, we remark that $1.05 \approx 1$, $0.01 \approx 0$ and $\sin(0.01) \approx 0$. So, if we introduce the function

$$f(x, y, r, s) = r^2 x + 5y \sin(s) - 2x^2 - 3y^2$$

where r and s are parameters, then the problem seems like a perturbation of the simpler problem $\max f(x, y, 1, 0) = x - 2x^2 - 3y^2$ to the given problem $\max f(x, y, 1.05, 0.01)$.

Solving $\quad \max\limits_{\mathbb{R}^2} \ x - 2x^2 - 3y^2.$

Since $\mathbb{R}^2$ is an open set, a global extreme point of $f(x, y) = f(x, y, 1, 0)$ is also a local extreme point. Therefore, it is a stationary point of $f(x, y) = x - 2x^2 - 3y^2$ (f is a polynomial, it is C^∞). We have

$$\nabla f(x, y) = \langle 1 - 4x, -6y \rangle = \langle 0, 0 \rangle \quad \Longleftrightarrow \quad (x, y) = (\frac{1}{4}, 0).$$

The only stationary point is $(\frac{1}{4}, 0)$. The Hessian matrix is

$$H_f(x, y) = \begin{bmatrix} -4 & 0 \\ 0 & -6 \end{bmatrix}$$

The leading principal minors are $D_1(x,y) = -4 < 0$ and $D_2(x,y) = 24 > 0$. Hence, f is strictly concave on $\mathbb{R}^2$ and we conclude that $(x^*, y^*) = (\frac{1}{4}, 0)$ is a global maximum point, and the only one.

Linear approximation. We have

1. $H_{f(.,r,s)}(x,y) = \begin{bmatrix} -4 & 0 \\ 0 & -6 \end{bmatrix}$ $\implies$ $f(.,r,s)$ is concave on $\mathbb{R}^2$ for any $(r,s) \in \mathbb{R}^2$.

2. $f(\frac{1}{4}, 0) = \max_{\mathbb{R}^2} f(x,y,1,0) = \frac{1}{8}$.

3. The second order condition for strict maximality is satisfied when $(r,s) = (1,0)$ at the point $(x,y) = (1/4, 0)$.

As a consequence,

-   $\exists \eta > 0$ such that the function $f^*(r,s) = \max_{(x,y)\in\mathbb{R}^2} f(x,y,r,s)$

    is defined for any $(r,s) \in B_\eta(1,0)$

-   f^* is $C^1(B_\eta(1,0))$ and

-

$$\frac{\partial f^*}{\partial r}(1,0) = \frac{\partial f}{\partial r}\Big|_{(x,y)=(1/4,0),\ (r,s)=(1,0)} = 2rx\Big|_{(x,y)=(1/4,0),\ (r,s)=(1,0)} = \frac{1}{2}$$

$$\frac{\partial f^*}{\partial s}(1,0) = \frac{\partial f}{\partial s}\Big|_{(x,y)=(1/4,0),\ (r,s)=(1,0)} = 5y\cos(s)\Big|_{(x,y)=(1/4,0),\ (r,s)=(1,0)} = 0.$$

We can write the following approximation, for (r,s) close to $(1,0)$,

$$f^*(r,s) \approx f^*(1,0) + \frac{\partial f^*}{\partial r}(1,0)(r-1) + \frac{\partial f^*}{\partial s}(1,0)(s-0) = \frac{1}{8} + \frac{1}{2}(r-1).$$

In particular, for $(r,s) = (1.05, 0.01)$, the objective function f^* takes the following approximate value:

$$f^*(1.05, 0.01) \approx 0.125 + \frac{1}{2}(1.05 - 1) = 0.125 + 0.025 = 0.15$$

and the approximate change in the maximum value of the maximum profit function is

$$f^*(1.05, 0.01) - f^*(1,0) \approx \frac{1}{2}(1.05 - 1) = 0.025.$$

3. – Consider the problem

$$\min(\max) f(x, y, z) = e^x + y + z \qquad \text{s.t} \qquad \begin{cases} g_1 = x + y + z = 1 \\[2mm] g_2 = x^2 + y^2 + z^2 = 1. \end{cases}$$

i) Apply Lagrange's theorem to the problem to show that there are four points satisfying the necessary conditions.

ii) Show that each point is a regular point.

iii) What can you conclude about the global minimal and maximal values of f subject to $g_1 = g_2 = 1$? Justify your answer.

iv) Replace the constraints by $x + y + z = a$ and $x^2 + y^2 + z^2 = b$ with (a, b) close to $(1, 1)$ $(a > 0, \ b > 0)$.

    - What is the approximate change in the optimal value function

$$f^*(a, b) = \min_{g_1 = a, \, g_2 = b} f(x, y, z)?$$

    - What is the approximate change in the optimal value function

$$F^*(a, b) = \max_{g_1 = a, \, g_2 = b} f(x, y, z)?$$

Solution: i) Note that f, g_1 and g_2 are C^∞ in $\mathbb{R}^3$. Consider the Lagrangian

$$\mathcal{L}(x, y, z, \lambda_1, \lambda_2) = e^x + y + z - \lambda_1(x + y + z - 1) - \lambda_2(x^2 + y^2 + z^2 - 1)$$

and look for its stationary points solution of the system

$$\nabla \mathcal{L}(x, y, z, \lambda_1, \lambda_2) = 0_{\mathbb{R}^5} \quad \Longleftrightarrow \quad \begin{cases} (1) \quad \mathcal{L}_x = e^x - \lambda_1 - 2x\lambda_2 = 0 \\[2mm] (2) \quad \mathcal{L}_y = 1 - \lambda_1 - 2y\lambda_2 = 0 \\[2mm] (3) \quad \mathcal{L}_z = 1 - \lambda_1 - 2z\lambda_2 = 0 \\[2mm] (4) \quad \mathcal{L}_{\lambda_1} = -(x + y + z - 1) = 0 \\[2mm] (5) \quad \mathcal{L}_{\lambda_2} = -(x^2 + y^2 + z^2 - 1) = 0. \end{cases}$$

From equations (2) and (3), we deduce that

$$(z - y)\lambda_2 = 0 \qquad \Longrightarrow \qquad z = y \qquad \text{or} \qquad \lambda_2 = 0.$$

* If $\lambda_2 = 0$, we deduce from equation (2) that $\lambda_1 = 1$ and then from equation (1), that $x = 0$. Hence, equations (4) and (5) give

$$2y^2 - 2y = 0 \qquad \Longrightarrow \qquad y = 0 \qquad \text{or} \qquad y = 1.$$

Therefore, we have the two points

$$(0,1,0), \qquad (0,0,1) \qquad \text{with} \qquad (\lambda_1, \lambda_2) = (1,0).$$

** If $z = y$, then equations (4) and (5) give

$$x = 1 - 2y \quad \text{and} \quad 6y^2 - 4y = 0 \quad \Longrightarrow \quad y = 0 \ \text{or} \ y = \frac{2}{3}.$$

Therefore, we have the two points

$$(1,0,0) \qquad \text{with} \qquad \lambda_1 = 1 \quad \text{and} \quad \lambda_2 = \frac{1}{2}(e-1)$$

$$(-\frac{1}{3}, \frac{2}{3}, \frac{2}{3}) \qquad \text{with} \qquad \lambda_1 = \frac{1}{3}(1 + 2e^{-\frac{1}{3}}) \quad \text{and} \quad \lambda_2 = \frac{1}{2}(1 - e^{-\frac{1}{3}}).$$

ii) Consider the matrix

$$g'(x,y,z) = \begin{bmatrix} \frac{\partial g_1}{\partial x} & \frac{\partial g_1}{\partial y} & \frac{\partial g_1}{\partial z} \\ \frac{\partial g_2}{\partial x} & \frac{\partial g_2}{\partial y} & \frac{\partial g_2}{\partial z} \end{bmatrix} = \begin{bmatrix} 1 & 1 & 1 \\ 2x & 2y & 2z \end{bmatrix}$$

$$g'(0,1,0) = \begin{bmatrix} 1 & 1 & 1 \\ 0 & 2 & 0 \end{bmatrix} \qquad g'(0,0,1) = \begin{bmatrix} 1 & 1 & 1 \\ 0 & 0 & 2 \end{bmatrix}$$

$$g'(1,0,0) = \begin{bmatrix} 1 & 1 & 1 \\ 2 & 0 & 0 \end{bmatrix} \qquad g'(-\frac{1}{3}, \frac{2}{3}, \frac{2}{3}) = \begin{bmatrix} 1 & 1 & 1 \\ -\frac{2}{3} & \frac{4}{3} & \frac{4}{3} \end{bmatrix}.$$

Each critical point is regular, and we remark that the first two column vectors in the matrices $g'(-\frac{1}{3}, \frac{2}{3}, \frac{2}{3}))$, $g'(0,1,0)$ and $g'(1,0,0)$ are linearly independent, while they are linearly dependent in $g'(0,0,1)$. Therefore, we can keep the matrices without renumbering the variables when applying the second derivatives test in the first three matrices and change the variables in the last one.

iii) Now, f is continuous on the constraint set which is a closed and bounded curve of $\mathbb{R}^3$ as the intersection of the unit sphere $x^2 + y^2 + z^2 = 1$ and the plane $x + y + z - 1 = 0$. So f attains its optimal values by the extreme value theorem on points that are also critical points of the Lagrangian. Comparing the values of f on these points, we obtain

$$2 < f(-\frac{1}{3}, \frac{2}{3}, \frac{2}{3}) = e^{-\frac{1}{3}} + \frac{4}{3} \approx 2.0498 < e.$$

$$\min_{g_1=1,\,g_2=1} f(x,y,z) = f(0,1,0) = f(0,0,1) = 2$$

and

$$\max_{g_1=1,\,g_2=1} f(x,y,z) = f(1,0,0) = e.$$

iv) $*$ If we denote

$$f^*(a,b) = \min_{g_1=a,\,g_2=b} f(x,y,z)$$

then f^* is regular for (a,b) close to $(1,1)$ because we have:

1. for (a,b) close to $(1,1)$, there exists a solution to the constrained minimization problem by the extreme value theorem (because f is continuous on the closed bounded set $x+y+z=a$ and $x^2+2y^2+z^2=b$).

2. $(0,1,0)$ and $(0,0,1)$ are solutions to the constrained minimization problem when $(a,b)=(1,1)$ and are regular points.

3. the second order condition for minimality is satisfied when $(a,b)=(1,1)$ at $(0,1,0)$ and $(0,0,1)$. Indeed, $n=3$ and $m=2$, then we have to consider the sign of the following bordered Hessian determinant:

$$\mathbb{B}_3(x,y,z) = \begin{vmatrix} 0 & 0 & \frac{\partial g_1}{\partial x} & \frac{\partial g_1}{\partial y} & \frac{\partial g_1}{\partial z} \\[4pt] 0 & 0 & \frac{\partial g_2}{\partial x} & \frac{\partial g_2}{\partial y} & \frac{\partial g_2}{\partial z} \\[4pt] \frac{\partial g_1}{\partial x} & \frac{\partial g_2}{\partial x} & \mathcal{L}_{xx} & \mathcal{L}_{xy} & \mathcal{L}_{xz} \\[4pt] \frac{\partial g_1}{\partial y} & \frac{\partial g_2}{\partial y} & \mathcal{L}_{yx} & \mathcal{L}_{yy} & \mathcal{L}_{yz} \\[4pt] \frac{\partial g_1}{\partial z} & \frac{\partial g_2}{\partial z} & \mathcal{L}_{zx} & \mathcal{L}_{zy} & \mathcal{L}_{zz} \end{vmatrix} = \begin{vmatrix} 0 & 0 & 1 & 1 & 1 \\ 0 & 0 & 2x & 2y & 2z \\ 1 & 2x & e^x - 2\lambda_2 & 0 & 0 \\ 1 & 2y & 0 & -2\lambda_2 & 0 \\ 1 & 2z & 0 & 0 & -2\lambda_2 \end{vmatrix} = 4.$$

$$\mathbb{B}_3(0,1,0) = \begin{vmatrix} 0 & 0 & 1 & 1 & 1 \\ 0 & 0 & 0 & 2 & 0 \\ 1 & 0 & 1 & 0 & 0 \\ 1 & 2 & 0 & 0 & 0 \\ 1 & 0 & 0 & 0 & 0 \end{vmatrix} = 4 \qquad \Longrightarrow \qquad (-1)^2 \mathbb{B}_3(0,1,0) = 4 > 0.$$

We change the variables in the order (x,z,y) to compute $\mathbb{B}_3(0,0,1)$ and obtain

$$\mathbb{B}_3(0,0,1) = \begin{vmatrix} 0 & 0 & 1 & 1 & 1 \\ 0 & 0 & 0 & 2 & 0 \\ 1 & 0 & 1 & 0 & 0 \\ 1 & 2 & 0 & 0 & 0 \\ 1 & 0 & 0 & 0 & 0 \end{vmatrix} = 4 \qquad \Longrightarrow \qquad (-1)^2 \mathbb{B}_3(0,0,1) = 4 > 0.$$

Consequently, with the new Lagrangian,

$$\mathcal{L}_{a,b}(x, y, z, \lambda_1, \lambda_2) = e^x + y + z - \lambda_1(x + y + z - a) - \lambda_2(x^2 + y^2 + z^2 - b),$$

we have

$$f^*(1, 1) = f(0, 1, 0) = f(0, 0, 1) = 2 \qquad \text{with}$$

$$\lambda_1(1, 1) = 1 \quad \text{and} \quad \lambda_2(1, 1) = 0$$

$$\frac{\partial f^*}{\partial a}(1, 1) = \frac{\partial \mathcal{L}_{a,b}}{\partial a}\Bigg]_{(x,y,z,\lambda_1,\lambda_2)=(0,1,0,\lambda_1(1,1),\lambda_2(1,1))} = \lambda_1(1, 1) = 1$$

$$\frac{\partial f^*}{\partial b}(1, 1) = \frac{\partial \mathcal{L}_{a,b}}{\partial b}\Bigg]_{(x,y,z,\lambda_1,\lambda_2)=(0,1,0,\lambda_1(1,1),\lambda_2(1,1))} = \lambda_2(1, 1) = 0$$

$$f^*(a, b) \approx f^*(1, 1) + \frac{\partial f^*}{\partial a}(1, 1)(a - 1) + \frac{\partial f^*}{\partial b}(1, 1)(b - 1)$$

$$= 2 + (a - 1) + (0)(b - 1) = a + 1.$$

** If we denote
$$F^*(a, b) = \max_{g_1=a,\, g_2=b} f(x, y, z)$$

then F^* is regular for (a, b) close to $(1, 1)$ because we have:

1. for (a, b) close to $(1, 1)$, there exists a solution to the constrained maximization problem by the extreme value theorem (because f is continuous on the closed bounded set $x + y + z = a$ and $x^2 + 2y^2 + z^2 = b$).

2. $(1, 0, 0)$ is the solution to the constrained maximization problem when $(a, b) = (1, 1)$ and it is a regular point.

3. the second order condition for maximality is satisfied when $(a, b) = (1, 1)$ at $(1, 0, 0)$. Indeed, $n = 3$ and $m = 2$, then we have to consider the sign of the following bordered Hessian determinant:

$$\mathbb{B}_3(1, 0, 0) = \begin{vmatrix} 0 & 0 & 1 & 1 & 1 \\ 0 & 0 & 2 & 0 & 0 \\ 1 & 2 & 1 & 0 & 0 \\ 1 & 0 & 0 & 1-e & 0 \\ 1 & 0 & 0 & 0 & 1-e \end{vmatrix} = 8(1 - e) < 0 \qquad (-1)^3 \mathbb{B}_3 = 8(e - 1) > 0.$$

Consequently, we have

$$F^*(1, 1) = f(1, 0, 0) = e \quad \text{with}$$

$$\lambda_1(1, 1) = 1 \quad \text{and} \quad \lambda_2(1, 1) = \frac{1}{2}(e - 1)$$

$$\frac{\partial F^*}{\partial a}(1,1) = \left.\frac{\partial \mathcal{L}}{\partial a}\right]_{(x,y,z,\lambda_1,\lambda_2)=(1,0,0,\lambda_1(1,1),\lambda_2(1,1))} = \lambda_1(1,1) = 1$$

$$\frac{\partial F^*}{\partial b}(1,1) = \left.\frac{\partial \mathcal{L}}{\partial b}\right]_{(x,y,z,\lambda_1,\lambda_2)=(1,0,0,\lambda_1(1,1),\lambda_2(1,1))} = \lambda_2(1,1) = \frac{1}{2}(e-1)$$

$$F^*(a,b) \approx F^*(1,1) + \frac{\partial F^*}{\partial a}(1,1)(a-1) + \frac{\partial F^*}{\partial b}(1,1)(b-1)$$

$$= e + (a-1) + \frac{1}{2}(e-1)(b-1).$$

4. – Consider the problem

$$\min(\max)\ f(x,y) = 1 - (x-2)^2 - y^2 \quad \text{s.t} \quad \begin{cases} x^2 + y^2 \leqslant 8 \\ x - y \leqslant 0. \end{cases}$$

i) Sketch the feasible set and write down the necessary KKT conditions.

ii) Find the solutions candidates of the necessary KKT conditions.

iii) Use the second derivatives test to classify the points.

iv) Explore the concavity and convexity of the associated Lagrangian in (x,y).

v) What can you conclude about the solution of the maximization problem?

vi) Determine the approximate values of each problem.

$$\min(\max)\ 1 - (0.98)^3(x-2)^2 - e^{-0.01}y^2 \quad \text{s.t} \quad \begin{cases} x^2 + \sqrt{1.04}y^2 \leqslant 8 \\ (1.04)^2 x - y \leqslant 0. \end{cases}$$

Solution: i) Figure 4.25 describes the constraint set and locate the extreme points, approximately, following the variation of the objective function along the level curves.

Consider the Lagrangian

$$\mathcal{L}(x,y,\lambda,\beta) = 1 - (x-2)^2 - y^2 - \lambda(x^2+y^2-8) - \beta(x-y)$$

We look simultaneously for the possible minima and maxima candidates. Thus, the Karush-Kuhn-Tucker conditions are

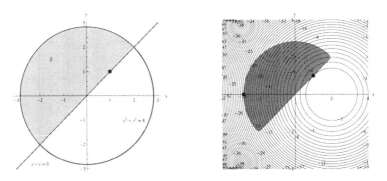

FIGURE 4.25: Level curve of highest profit

$$
\begin{cases}
(1) & \mathcal{L}_x = -2(x-2) - 2\lambda x - \beta = 0 \\[2mm]
(2) & \mathcal{L}_y = -2y - 2\lambda y + \beta = 0 \\[2mm]
(3) & \lambda = 0 \quad \text{if} \quad x^2 + y^2 < 8 \\[2mm]
(4) & \beta = 0 \quad \text{if} \quad x - y < 0
\end{cases}
$$

with

$$
\begin{aligned}
&\| \ (\lambda, \beta) \geqslant (0,0) \quad \text{for a maximum point} \\
&\| \ (\lambda, \beta) \leqslant (0,0) \quad \text{for a minimum point}
\end{aligned}
$$

ii) Solving the system.

∗ If $x^2 + y^2 < 8$ then $\lambda = 0$ and

$$
\begin{cases}
-2(x-2) - \beta = 0 \\[2mm]
-2y + \beta = 0
\end{cases}
\qquad \Longrightarrow \qquad y = -(x-2)
$$

then inserted in (4) we discuss

    – Suppose $x - y = x + (x+2) = 2(x+1) < 0$, then $x < -1$. From (2), we have $\beta = 2y = -2(x-2) > 0$. Thus, by (4), we get $x - y = 0$ which contradicts $x - y < 0$.

    – Suppose $x - y = 0$. We have then $y = x$ and $y = -x + 2$. Thus, we have a candidate point for optimality:

$$
(x, y) = (1, 1) \qquad \text{with} \qquad (\lambda, \beta) = (0, 2).
$$

∗ If $x^2 + y^2 = 8$ then

     – Suppose $x - y < 0$ then $\beta = 0$ and

$$\begin{cases} x - 2 + \lambda x = 0 \\ -2y(1 + \lambda) = 0 \end{cases} \iff y = 0 \quad \text{or} \quad \lambda = -1.$$

$\lambda = -1$ is not possible by $x - 2 + \lambda x = 0$. Thus $y = 0$.

With $x^2 + y^2 = 8$, we deduce that $x = \sqrt{8}$, which contradicts $x < y$, or $x = -\sqrt{8}$. Inserting the value $x = -\sqrt{8}$ into $x - 2 + \lambda x = 0$ gives $\lambda = -1 - \frac{1}{\sqrt{2}}$. So, we have another candidate

$$(x, y) = (-\sqrt{8}, 0) \qquad \text{with} \qquad (\lambda, \beta) = (-1 - \frac{1}{\sqrt{2}}, 0).$$

     – Suppose $x - y = 0$. With $x^2 + y^2 = 8$, we deduce that $x = 2$ or $x = -2$. Then, inserting in (1) and (2), we obtain

$$(x, y) = (2, 2) \quad \Longrightarrow \quad \begin{cases} -4\lambda - \beta = 0 \\ -4\lambda + \beta = 4 \end{cases} \quad \iff \quad (\lambda, \beta) = (-\frac{1}{2}, 2)$$

contradicting the common sign of λ and β.

$$(x, y) = (-2, -2) \quad \Longrightarrow \quad \begin{cases} 4\lambda - \beta = -8 \\ 4\lambda + \beta = -4 \end{cases} \quad \iff \quad (\lambda, \beta) = (-\frac{3}{2}, 2)$$

contradicting the common sign of λ and β.

Regularity of the candidate point $(1, 1)$. Note that the constraints $g_1(x, y) = x^2 + y^2$ and $g_2(x, y) = x - y$ are C^1 in $\mathbb{R}^2$ and that only the constraint g_2 is active at $(1, 1)$. We have

$$g_2'(x, y) = \begin{bmatrix} 1 & -1 \end{bmatrix} \qquad\qquad rank(g_2'(1, 1)) = 1$$

Thus the point $(1, 1)$ is a regular point.

Regularity of the candidate point $(-\sqrt{8}, 0)$. Only the constraint g_1 is active at $(-\sqrt{8}, 0)$. We have

$$g_1'(x, y) = \begin{bmatrix} 2x & 2y \end{bmatrix} \qquad\qquad rank(g_2'(-\sqrt{8}, 0)) = 1$$

Thus the point $(-\sqrt{8}, 0)$ is a regular point.

iii) Second derivatives test at $(1,1)$. With $p = 1$ (the number of the constraints), $n = 2$ (the dimension of the space), then $r = p + 1, n = 2, 2 \Longrightarrow$ $r = 2$ and we will consider the following determinant

$$
\mathbb{B}_2(x,y) = \begin{vmatrix} 0 & \frac{\partial g_2}{\partial x} & \frac{\partial g_2}{\partial y} \\ \frac{\partial g_2}{\partial x} & \mathcal{L}_{xx} & \mathcal{L}_{xy} \\ \frac{\partial g_2}{\partial y} & \mathcal{L}_{yx} & \mathcal{L}_{yy} \end{vmatrix} = \begin{vmatrix} 0 & 1 & -1 \\ 1 & -2 - 2\lambda & 0 \\ -1 & 0 & -2 - 2\lambda \end{vmatrix}
$$

∗ At $(1,1)$, we have $\lambda = 0$, then

$$
\mathbb{B}_2(1,1) = \begin{vmatrix} 0 & 1 & -1 \\ 1 & -2 & 0 \\ -1 & 0 & -2 \end{vmatrix} = 4 \quad \Longrightarrow \quad (-1)^2 \mathbb{B}_2(1,1) > 0
$$

and $(1,1)$ is a local maximum.

Second derivatives test at $(-\sqrt{8}, 0)$. We consider the following determinant

$$
\mathbb{B}_2(x,y) = \begin{vmatrix} 0 & \frac{\partial g_1}{\partial x} & \frac{\partial g_1}{\partial y} \\ \frac{\partial g_1}{\partial x} & \mathcal{L}_{xx} & \mathcal{L}_{xy} \\ \frac{\partial g_1}{\partial y} & \mathcal{L}_{yx} & \mathcal{L}_{yy} \end{vmatrix} = \begin{vmatrix} 0 & 2x & 2y \\ 2x & -2 - 2\lambda & 0 \\ 2y & 0 & -2 - 2\lambda \end{vmatrix}
$$

∗ At $(-\sqrt{8}, 0)$, we have $\lambda = -3/2$, then

$$
\mathbb{B}_2(1,1) = \begin{vmatrix} 0 & -2\sqrt{8} & 0 \\ -2\sqrt{8} & 1 & 0 \\ 0 & 0 & 1 \end{vmatrix} = -32 \quad \Longrightarrow \quad (-1)^1 \mathbb{B}_2(-\sqrt{8}, 0) > 0
$$

and $(-\sqrt{8}, 0)$ is a local minimum.

iv) and v) Let us explore the concavity and convexity of $\mathcal{L}$ with respect to (x, y) where the Hessian matrix of $\mathcal{L}$ in (x, y) is

$$
\mathcal{H}_{\mathcal{L}} = \begin{bmatrix} \mathcal{L}_{xx} & \mathcal{L}_{xy} \\ \mathcal{L}_{yx} & \mathcal{L}_{yy} \end{bmatrix} = \begin{bmatrix} -2 - 2\lambda & 0 \\ 0 & -2 - 2\lambda \end{bmatrix}
$$

• When $\lambda = 0$, the principal minors are $\Delta_1^1 = \mathcal{L}_{yy} = -2 < 0$, $\Delta_1^2 = \mathcal{L}_{xx} = -2 < 0$ and $\Delta_2 = 4 > 0$. So $(-1)^k \Delta_k \geqslant 0$ for $k = 1, 2$. Therefore, $\mathcal{L}$ is concave in (x, y) and then $(1, 1)$ is a global maximum for the constrained maximization problem.

• When $\lambda = -3/2$, the principal minors are $\Delta_1^1 = \mathcal{L}_{yy} = 1 > 0$, $\Delta_1^2 = \mathcal{L}_{xx} = 1 > 0$ and $\Delta_2 = 1 > 0$. So $(-1)^k \Delta_k \geqslant 0$ for $k = 1, 2$. Therefore, $\mathcal{L}$ is convex in (x, y) and then $(-\sqrt{8}, 0)$ is a global minimum for the constrained minimization problem.

vi) Note that $0.98 \approx 1$, $\quad 1.04 \approx 1$, $\quad 0.01 \approx 0$ and $e^{-0.01} \approx 1$. Thus, the new problems seem like a perturbation from the original problem. Therefore, we will use linear approximation to solve the problem when $r = 0.98$, $s = 1.04$ and $t = -0.01$. So, introduce the Lagrangian associated with the new constrained optimization problem

$$\mathcal{L}(x, y, \lambda, \beta, r, s, t) = 1 - r^3(x-2)^2 - e^t y^2 - \lambda(x^2 + \sqrt{s}y^2 - 8) - \beta(s^2 x - y)$$

• Set $f(x, y, r, s, t) = 1 - r^3(x-2)^2 - e^t y^2$, and the value function

$$f^*(r, s, t) = \min f(x, y, r, s, t) \quad \text{s.t} \quad \begin{cases} x^2 + \sqrt{s}y^2 \leqslant 8 \\ s^2 x - y \leqslant 0 \end{cases}$$

Then f^* is well defined and differentiable when (r, s, t) is close to $(1, 1, 0)$. Indeed, the following is satisfied

1. There is a unique solution $(x, y) = (-\sqrt{8}, 0)$ to the constrained minimization problem when $(r, s, t) = (1, 1, 0)$ and $(-\sqrt{8}, 0)$ is a regular point.

2. For (r, s, t) close to $(1, 1, 0)$, there exists a solution to the constrained minimization problem by the extreme value theorem since the set of constraints is a closed bounded set and the function is continuous.

3. the second order condition for minimality is satisfied at $(-\sqrt{8}, 0)$ when $(r, s, t) = (1, 1, 0)$.

As a consequence,

$$\frac{\partial f^*}{\partial r}(1, 1, 0) = \frac{\partial \mathcal{L}}{\partial r}\Big|_{(x,y,\lambda,\beta)=(-\sqrt{8},0,-1-1/\sqrt{2},0),\ (r,s,t)=(1,1,0)}$$

$$= -3r^2(x-2)^2\Big|_{(x,y,\lambda,\beta)=(-\sqrt{8},0,-1-1/\sqrt{2},0),\ (r,s,t)=(1,1,0)} = -3(\sqrt{8}+2)^2$$

$$\frac{\partial f^*}{\partial s}(1, 1, 0) = \frac{\partial \mathcal{L}}{\partial s}\Big|_{(x,y,\lambda,\beta)=(-\sqrt{8},0,-1-1/\sqrt{2},0,\ (r,s,t)=(1,1,0)}$$

$$= -\lambda\frac{1}{2\sqrt{s}}y^2 - 2\beta sx\Big|_{(x,y,\lambda,\mu)=(-\sqrt{8},0,-1-1/\sqrt{2},0),\ (r,s,t)=(1,1,0)} = 0$$

$$\frac{\partial f^*}{\partial t}(1, 1, 0) = \frac{\partial \mathcal{L}}{\partial t}\Big|_{(x,y,\lambda,\beta)=(-\sqrt{8},0,-1-1/\sqrt{2},0),\ (r,s,t)=(1,1,0)}$$

$$= -e^t y^2\Big|_{(x,y,z,\lambda,\beta)=(-\sqrt{8},0,-1-1/\sqrt{2},0),\ (r,s,t)=(1,1,0)} = 0.$$

Hence, for (r, s, t) close to $(1, 1, 0)$

$$f^*(r, s, t) \approx f^*(1, 1, 0) + \frac{\partial f^*}{\partial r}(1, 1, 0)(r - 1)$$

$$+ \frac{\partial f^*}{\partial s}(1, 1, 0)(s - 1) + \frac{\partial f^*}{\partial t}(1, 1, 0)(t - 0)$$

$$f^*(r, s, t) \approx 2 - 3(\sqrt{8} + 2)^2(r - 1)$$

$$f^*(0.98, 1.04, -0.01) \approx 2 - 3(\sqrt{8} + 2)^2(0.04).$$

- Set the value function

$$F^*(r, s, t) = \max f(x, y, r, s, t) \quad \text{s.t} \quad \begin{cases} x^2 + \sqrt{s}y^2 \leqslant 8 \\ s^2 x - y \leqslant 0 \end{cases}$$

Then F^* is well defined and differentiable when (r, s, t) is close to $(1, 1, 0)$. Indeed, the following is satisfied

1. There is a unique solution $(x, y) = (1, 1)$ to the constrained maximization problem when $(r, s, t) = (1, 1, 0)$ and $(1, 1)$ is a regular point.

2. For (r, s, t) close to $(1, 1, 0)$, there exists a solution to the constrained maximization problem by the extreme value theorem since the set of constraints is a closed bounded set and the function is continuous.

3. the second order condition for maximality is satisfied at $(1, 1)$ when $(r, s, t) = (1, 1, 0)$.

As a consequence,

$$\frac{\partial F^*}{\partial r}(1, 1, 0) = \frac{\partial \mathcal{L}}{\partial r}\Big|_{(x,y,\lambda,\beta)=(1,1,0,2),\ (r,s,t)=(1,1,0)}$$

$$= -3r^2(x - 2)^2\Big|_{(x,y,\lambda,\beta)=(1,1,0,2),\ (r,s,t)=(1,1,0)} = -3$$

$$\frac{\partial F^*}{\partial s}(1, 1, 0) = \frac{\partial \mathcal{L}}{\partial s}\Big|_{(x,y,\lambda,\beta)=(1,1,0,2),\ (r,s,t)=(1,1,0)}$$

$$= -\lambda\frac{1}{2\sqrt{s}}y^2 - 2\beta sx\Big|_{(x,y,\lambda,\mu)=(1,1,0,2),\ (r,s,t)=(1,1,0)} = -4$$

$$\frac{\partial F^*}{\partial t}(1, 1, 0) = \frac{\partial \mathcal{L}}{\partial t}\Big|_{(x,y,\lambda,\beta)=(1,1,0,2),\ (r,s,t)=(1,1,0)}$$

$$= -e^t y^2\Big|_{(x,y,\lambda,\beta)=(1,1,0,2),\ (r,s,t)=(1,1,0)} = -1.$$

Hence, for (r, s, t) close to $(1, 1, 0)$

$$F^*(r, s, t) \approx F^*(1, 1, 0) + \frac{\partial F^*}{\partial r}(1, 1, 0)(r - 1)$$

$$+ \frac{\partial F^*}{\partial s}(1, 1, 0)(s - 1) + \frac{\partial F^*}{\partial t}(1, 1, 0)(t - 0)$$

$$F^*(r, s, t) \approx -1 - 3(r - 1) - 4(s - 1) - (t - 0)$$

$$F^*(0.98, 1.04, -0.01) \approx -1 - 3(-0.02) - (0.04) - (-0.01) = 0.02.$$

Remark. The set of feasible solutions $S = \{(x, y) : \quad x^2 + y^2 \leqslant 8, \quad x - y \leqslant 0\}$ is a closed bounded set of $\mathbb{R}^2$ and f is continuous on S. Therefore, the extreme points are attained on this set by the extreme value theorem. Moreover, such points must occur either at points satisfying the KKT conditions or at points where the constraint qualification fails. Since, $(1, 1)$ and $(-\sqrt{8}, 0)$ are the only two points solution and they are regular, then they solve the problem.

For more practice, we refer the reader to [11], [27], [28], [26], [25], [24], [4].

Bibliography

[1] H. Anton, I. Bivens, and S. Davis. *Calculus. Early Transcendentals.* John Wiley & Sons, Inc. New York, NY, USA, 2005.

[2] R. G. Bartle and D. R. Sherbert. *Introduction to Real Analysis.* John Wiley & Sons, Inc, 2011.

[3] W. Briggs, L. Cochran, and B. Gillett. *Calculus. Early Transcendentals.* Addison-Wesley. Pearson, 2011.

[4] E. K. P. Chong and S. H. Żak. *An Introduction to Optimization.* Wiley, 2013.

[5] P. G. Ciarlet. *Introduction à l'analyse numérique matricielle et l'optimisation.* Masson, 1985.

[6] B. Dacorogna. *Introduction au calcul des variations.* Presses polytechniques et universitaires romandes. Lausanne, 1992.

[7] Jr. Ernest.F. Haeussler, S. P. Richard, and J. W. Richard. *Introductory Mathematical Analysis for Business, Economics, and the Life and Social Sciences.* Pearson, Prentice Hall, 2008.

[8] P. E. Fishback. *Linear and Nonlinear Programming with Maple™. An Interactive, Applications-Based Approach.* CRC Press, Taylor and Francis Group, 2010.

[9] A.S. Gupta. *Calculus of Variations with Applications.* Prentice-Hall of India, 2006.

[10] W. Keith Nicholson. *Linear Algebra with Applications.* McGraw-Hill Ryerson, 2014.

[11] D. Koo. *Elements of Optimisation with Applications in Economics and Business.* Springer-Verlag, 1977.

[12] R. J. Larsen and M. L. Marx. *An Introduction to Mathematical Statistics and its Applications.* Prentice Hall, 2001.

[13] S. Lipschutz. *Topologie, cours et problèmes.* McGraw-Hill, 1983.

[14] D.G. Luenberger. *Introduction to Linear and Nonlinear Programming.* Addison Wesley, 1973.

[15] J. E. Marsden. *Elementary Classical Analysis.* W. H. Freeman and Company, 1974.

[16] M. Mesterton-Gibbons. *A primer on the calculus of variations and optimal control theory.* Student Mathematical Library vol 50. American Mathematical Society, 2009.

[17] M. Minoux. *Mathematical Programming: Theory and Algorithms.* John Wiley and Sons, 1986.

[18] J. R. Munkres. *Topology of First Course.* Prentice Hall, 1975.

[19] J. Nocedal and S. J. Wright. *Numerical Optimization.* Springer, 1999.

[20] M.H. Protter and C.B. Morrey. *A First Course in Real Analysis.* Springer, 2000.

[21] S.L. Salas, E. Hille, and G.J. Etgen. *Calculus. One and Several Variables.* Tenth Edition. John Wiley & Sons, INC, 2007.

[22] J. A. Snyman. *Practical Mathematical Optimization: An Introduction to Basic Optimization Theory and Classical and New Gradient-Based Algorithms.* Springer, 2005.

[23] J. Stewart. *Essential Calculus.* Brooks/Cole, 2013.

[24] K. Sydsæter and P. Hammond. *Mathematics for Economic Analysis.* FT Prentice Hall, 1995.

[25] K. Sydsæter, P. Hammond, A. Seierstad, and A. Strøm. *Further Mathematics for Economic Analysis.* FT Prentice Hall, 2008.

[26] K. Sydsæter, P. Hammond, A. Seierstad, and A. Strøm. *Instructor's Manual: Further Mathematics for Economic Analysis.* Pearson, 2008. 2nd Edition.

[27] K. Sydsæter, A. Strøm, and P. Hammond. *Instructor's Manual: Essential Mathematics for Economic Analysis.* Pearson, 2008. 3rd Edition.

[28] K. Sydsæter, A. Strøm, and P. Hammond. *Instructor's Manual: Essential Mathematics for Economic Analysis.* Pearson, 2014. 4th Edition.

[29] W. L. Winston. *Operations Research: Applications and Algorithms.* Brooks/Cole, 2004.

Index